CHARGED AND REACTIVE POLYMERS

VOLUME 3

CHARGED GELS AND MEMBRANES – PART I

CHARGED AND REACTIVE POLYMERS

A SERIES EDITED BY ERIC SÉLÉGNY

VOLUME 3

CHARGED GELS AND MEMBRANES

PART I

Edited by

ERIC SÉLÉGNY

Université de Rouen, France

Co-edited by

GEORGE BOYD   and   HARRY P. GREGOR

University of Georgia, U.S.A.   *Columbia University, U.S.A.*

D. REIDEL PUBLISHING COMPANY

DORDRECHT-HOLLAND / BOSTON-U.S.A.

Library of Congress Cataloging in Publication Data

Nato Advanced Study Institute on Charged and Reactive
    Polymers, 2d, Forges-les-Eaux, France, 1973.
    Charged Gels and Membranes.

    (Charged and reactive polymers; v. 3–4).
    Includes bibliographical references and indexes.
    1.   Polymers and polymerization—Electric properties—
Congresses.   2.   Membranes (Technology)—Congresses.
3.   Membranes (Biology)—Congresses.   I.   Sélégny, Eric.   II.
Title.   III.   Series.
QD381.9.E38N2   1973   668   76–6086
ISBN–13:978–94–010–1466–3   e–ISBN–13:978–94–010–1464–9
DOI: 10.1007/978–94–010–1464–9

Published by D. Reidel Publishing Company,
P.O. Box 17, Dordrecht, Holland

Sold and distributed in the U.S.A., Canada, and Mexico
by D. Reidel Publishing Company, Inc.
Lincoln Building, 160 Old Derby Street, Hingham,
Mass. 02043, U.S.A.

To the 70th anniversary celebration of
Professors Karl Sollner and Torsten Teorell,
whose long constant and knowledgeable activity
has decidedly marked the science, history,
and development of Membranology

TABLE OF CONTENTS

PROFESSOR KARL SOLLNER

PROFESSOR TORSTEN TEORELL

INTRODUCTION

The series on 'Charged and Reactive Polymers' was set forth in two volumes concerning the fundamentals and applications of polyelectrolytes. A follow-up on 'Charged Gels and Membranes' would therefore seem appropriate, necessitating, however, some explanation for non-specialists.

Theories of the most dilute gels originate in that of concentrated polyelectrolytes: the methods and problems are similar in structural, spectroscopic or thermodynamic properties. The borderline can be situated in dialysis conducted with a 'bag' impermeable to polyelectrolytes but not to small ions, solutes and water.

One may recall Donnan's use of such a system to experiment and discover his famous law of unequal distribution of ions of different charge inside and out. Remarkably so, it is the difference in scale which characterizes the difference between polyelectrolyte solutions and gels and membranes: the colloidal solution of macromolecules is heterogeneous only on the microscopic level, whereas the gel-solution system is a macroscopically heterogeneous one.

A gel is formed when weak or strong cohesive forces counterbalance the dispersing ones (usually by crosslinking) without inhibiting the penetration of solvent and of small solutes into the polymeric network. The solvophile macromolecules cannot invade the total volume of liquid. As a result of phase-segregation excess solution and gel coexist and interact. The macroscopic swelling depends on gel cross-linking as well as on ionic concentration and type and ion-selectivities are observed.

During transport phenomena, the gel network is macroscopically immobile in contrast to the sedimentation or the electrophoretic movements of soluble polyelectrolytes. However, this distinction disappears if smaller scale and relative movements are considered, as small ions and molecules move along and in between the immobilized chains and fixed charged groups.

It becomes clear that two groups of theories and investigations can coexist.

One of them considers the chains with their fixed charges, and after considering them lengthwise or distributing them following some pattern, tries to estimate chain extensions and interactions with mobile species and consequently their microscopic distribution. The result permits deduction of the type of solution corresponding to the equilibrium.

This approach already has been considered with respect to different models in Volume I of the series, as it concerns polyelectrolytes and membranes as well.

A second group of methods to be analyzed further is more specific to gel and membrane specialists. It proceeds to examine the interdependence of the two components of the heterogeneous membrane-solution system, on the macroscopic level. The *phenomenologic homogeneity* of the 'gel phase' or 'membrane phase' is assumed, in a first approximation at least. Statistical mean concentration of ionizable fixed groups (exchange capacity) and water and ion content are averaged throughout the volume of the gel or its interstitial liquid. These quantities are directly accessible to experimental measurements and usable in formulations.

In the absence of externally applied forces the thermodynamic equilibrium properties of gels do not depend on their physical forms and shapes; thus we find *no significant difference in formulation and phenomena of gels and membranes of identical composition*. Each situation and experiment is analyzed in terms of nature, structure and composition of the macromolecular network on one hand and, of distribution between gel and solution of mobile components ions, and non-electrolytes, on the other hand.

A synthesis of knowledge concerning the differences in gels or solutions is achieved by comparing the chemical effects on distributions and selectivities in each major group of strong or weak polyacid or polybase gels.

The aim is ultimate explanation and prediction of the results for all compositions and gel structures, on the basis of a limited number of experiments.

Anomalies of distribution or swelling reveal specific interactions or complex formation. Nevertheless, these indirect deductions from thermodynamics must be correlated with *spectroscopic* investigations if more detailed information is to be gained.

From application of such *forces* as electric potential, chemical potential or pressure gradients, modifications of equilibria and *fluxes* of ions, non-electrolytes (volume) and electric current result. They are generally if not exclusively studied with membranes as opposed to gels. One of the major advantages is that formulations can be reduced to a single dimension by means of experiments using fluxes and forces in the plan of the membrane or transversal to it.

Until recently, these fluxes were mainly analyzed in electrochemical terms.

But the methods of *thermodynamics of irreversible processes* are particularly well adapted to formulate the couplings of these phenomena. The yet recent theoretical explicitations for membranes and numerical applications based on classical measurements have led lately to more direct experimental evaluations; the whole field is in full development.

The assumption of *phenomenologic homogeneity* can be accepted even in those cases where this is an evident approximation.

It suffices that the membrane does not change its character within the limits of the experimental precision of measurements; which is to say the type of mathematical interrelations of fluxes and forces remains the same. In other words perfect equivalent models can describe the behavior of a less perfect membrane, the *degree of correlation* being included in the macroscopic formulation.

Fundamental physical chemists must experiment with symmetrical membranes of great structural homogeneity in order to verify the model and with various structures to enable the establishment and the standardization of the correlation factors. Only the tested formulations can be universally adapted for practical use of quantitative predictions of phenomena. This explains and justifies some of the recent investigations.

It appears that in the stationary state the linear global correlation can switch to a non-linear one in two main instances. *In the first case* because of a large increase of force gradients, a new continuum (e.g. polarization layers), a new membrane structure (dissipative structure: e.g. pores closing or opening under pressure), or, a new gradient of functional properties of states (dissipative functional structures: e.g. modification of ionization of fixed charges) appears and modifies the degree of phenomenological homogeneity. *In the second case* none of the pre-existing continuous structures controls the vectorial properties of a nonhomogeneous membrane, namely with respect to *fluxes* related to vectorial properties in series* (relaxation times, selectivities...) or to forces related to vectorial properties in parallel** (potentials, mobilities...) or both.

At present, independent approaches indicate that relations of non-equilibrium thermodynamics *remain surprisingly linear* far from equilibrium. This is true even for non-homogeneous membranes when both of the above considerations are correctly taken into account.

Non-linearity inherent to time-dependent evolutions can be reduced and explained similarly in terms of a lack of homogeneity.

Once all the correlations for all principal types of membranes (porous, non-porous, charged, uncharged, containing mobile transporters or reactions etc....) are known, testing and description of their combinations will become possible. From recognition and computation of typical deviations one can gain powerful means of analytical identification of unknown membranes, or quality evaluations for applications or synthesis formulation.

Efforts to increase precision or complexity are not unwarranted considering the necessities of membrane *applications* in purification, separation (electrodialysis, reverse osmosis, ultrafiltration, gas separations etc....), or analysis (membrane electrodes).

A similar conclusion may be drawn with respect to the exciting field of *biological membranes*.

Everyone today is aware of the importance of membranes in the molecular life sciences, e.g. biology and biophysics but one must remember that this has not always been so. Continuous confrontations assured a constant progress of understanding: inspired analysis of mass and current flows during nerve, muscle and cell-wall phenomena on one hand, and first intentionally naive, but later more composite and imaginatively varied modelling with synthetic and artificial membranes and systems, on the other hand.

To create bioanalogous phenomena, studies on positively or negatively charged, amphoteric, sandwich, mosaic, lipidic, bilayer and later reactive or fluctuating membranes and their combinations became more and more justifiable.

This is a field in which applications for human well-being and deeper understanding of natural phenomena equally justify scientific activity, while providing an added appeal.

---

* Reverse osmosis properties of thick cellulos-acetate membranes are due to their thin active layer.
** A pinhole in an ion exchange membrane is sufficient to destroy the membrane potential.

Scientific interest in all membrane problems would increase were entry into the field made from the fundamental synthetic or biophysical point of view. So complementary are these two aspects that competence gained in one favors progress in the other, such that nowadays scientists of the same laboratory, department, or institute frequently investigate both.

It is not uncommon that this interdisciplinary aspect and very satisfactory and efficient use of competence should as yet surprise the funders of classical research. This leads to a careful consideration of the proposition of Professor Teorell in Forges that the science of Membranology may be declared open.

The second Advanced Study Institute in Forges les Eaux (France) on Charged and Reactive Polymers dealing with Charged Gels and Membranes was held in September 1973 under the direction of Professor E. Sélégny.

Both co-directors professors G. E. Boyd and K. S. Spiegler have lectured at the previous (1972) meeting. Their personal initiatives and knowledgeable contributions to the organization were determinant in the constitution of the list of speakers, and indicative of a high-level programme continuous with the previous Institute.

The Scientific Committee organised the sessions devoted to Equilibria, Spectroscopy, Non-equilibrium Phenomena, Fluctuations, Biological Membranes, or Membranes of biophysical interest or of practical application.

During the formal opening session Professor Champetier, member of the French Academy of Science, presented a welcoming address to the participants.

The scientific programme opened with two introductory lectures. Professor Sollner authoritatively reviewed the history and the evolution of the discoveries in early membrane research. Professor Teorell described the birth of Membrane Biophysics as the union of experimental physiology with physical chemistry, and successfully designated, as well as interconnected most of the present-day problems in logical order.

All those interested in membranes are familiar with the names and the work of these two giants in the field. During the whole meeting speakers referred to the *fixed charge theory* of Teorell (and Siever and Mayers) or to one or another variety of collodion membranes or particular phenomena studied by Sollner and his coworkers. They participated in the discussions with the enthusiasm of young scientists or complimented younger collegues on their new or remarkable progress, comparing more recent research approaches to those of the past treating the same problem.

The presence of these seniors and also of Professor Staverman communicated a strong feeling of continuity to all participants. They not only actively lived a major episode in the history of membrane research but also learned about most of those who initiated modern physical chemistry at the beginning of the century.

The progress of knowledge became more evident and a sort of a fervor was generated, stimulating intense and high level discussions. All lectures or contributors presented the most out standing elements of their science.

A high point was reached at the closing dinner when Professor Teorell, in a sort of a 'Saga of a Northman', recalled how he had been so greatly impressed as a young man by the personality of 'Old Donnan'. In so doing, Professor Teorell transmitted to today's young scientists his exhortation to contribute to the development of Membranology and Biophysics for the betterment of man.

All participants were receptive to this message. They parted richer not only in knowledge, but in friendships as well. The spirit of Forges had been at work and the Institute had attained its objective.

The publication of papers issuing from the oral contributions made at the meeting had been decided in advance. It became rapidly evident that the material was too abundant for a single volume, necessitating division into two parts. This division could be neither strict nor perfect. The one finally adopted encompasses many diverse opinions plus the contraints of extent, pagination as well as unavoidable delays, modifications and absence of manuscripts; these latter having now been replaced.

Althoug a few written contributions were available in Forges, most of them were prepared during the following year. A few of those found in volume II principally, were joined at a later date and are specified in the Table of Contents.

The finished work includes not only the proceedings of the Institute of Forges, but also inspirations ensuing from it.

In addition to presenting recent results our primary ambition in the books has been to collect a selection of works, covering an extensive time-period, which represent a sampling of most of the major directions of activity and means of research on gels and membranes. These have been assembled into a good quality edition in order to justify the reader's interest over a sufficient period of years. Both volumes together, constitute the result of this effort and should therefore not be separated in most libraries.

The first volume begins with the written version of the introductory lectures of Professors K. Sollner and T. Teorell. The remainder of the volume treats the fundamental physical chemistry of gels and membranes containing one type of charged group only, membranes for electrodialysis, ultrafiltration or reverse osmosis. This volume is synthetic, isothermal-thermodynamic, non-equilibrium thermodynamic, and application oriented.

The second volume begins with articles on the spectroscopy of complexes of natural or synthetic molecules, followed by papers on more complex or more directly biophysically oriented membranes: lipid membranes with carriers or associated with charges; liquid carrier facilitated transport; mosaic or redox membranes and their preparation; non-isothermal studies on model or bio-membranes; fluctuations, nerve membranes or pressure driven heart-model oscillators. This subject, initiated with enzyme membrane driven active-transport at the end of volume I, is extended here, providing the junction with biophysical chemistry.

The contributions are regrouped around physico-chemical aims and techniques, independent of the synthetic or biophysical nature of the material studied. This conforms with the line and presentation of the series and constitutes one of its particularities.

The editors hope that the presentation and the content of these two volumes will receive the attention of physical chemists, biophysicists, biologists and engineers.

It is a pleasure to acknowledge the help of all those who have contributed to the realization of the Institute and of the books and more specifically:

Mrs Sélégny organized, guided and lead the Social Programme of the Institute. She also participated in the editing and correspondence.

Miss Albane Sélégny, Drs Demarty and Ripoll and Mr Labbé were responsible for the index, or the technical organization of the meeting and Mrs Varin for typing.

Professor H. P. Gregor spent a part of his sabatical year in Rouen to help in editing at a particularly difficult moment.

The generous financial help of the Centre National de la Recherche Scientifique, Scientific Division of NATO, National Science Foundation (U.S.A.), Université de Rouen, Faculté des Sciences and Institut Scientifique de Haute-Normandie made the Institute and books possible.

E. SÉLÉGNY

SCIENTIFIC AND ORGANIZING COMMITTEE

G. E. Boyd (Oak Ridge), Co-director of the Institute
H. P. Gregor (Columbia)
J. Neel (Nancy)
E. Sélégny (Rouen), Director of the Institute
K. Sollner (N.I.H.)
K. S. Spiegler (Berkeley), Co-director of the Institute
T. Teorell (Uppsala)

LIST OF PARTICIPANTS

Belgium

Mr L. Baeten, Universiteit te Leuven, Laboratorium voor Macromoleculaire en Organische Scheikunde, 3030 Weverlee, Celestijnenlaan 200 F, Leuven

Mr A. Jenard, Université Libre de Bruxelles, Faculté des Sciences, Avenue F. Roosevelt, 50, 1050 Bruxelles

Dr A. Sanfeld, Faculté des Sciences, Université Libre de Bruxelles

Canada

Professor* and Mrs J. W. Lorimer, Department of Chemistry, University of Western Ontario, London 72, Ontario

Denmark

Dr H. Waldmann–Meyer**, Fysisk-Kemisk Institut, DTH 206, DK 2800, Lyngby

France

Mr P. Canova, CNRS Cermav, Domaine Universitaire, Cédex 53, 38 Grenoble

Professeur and Mrs G. Champetier, ESPCI, 10, rue Vauquelin, Paris 5°

Dr P. Gramain, CNRS, Centre de Recherches sur les Macromolécules, 6, rue Boussingault, 67083 Strasbourg Cédex

Mr H. Kranck, Laboratoire de Biophysique, Université de Nice, Parc Valrose, 06034 Nice Cédex

Dr M. Milas, CNRS Cermav, Domaine Universitaire, Cédex 53, 38, Grenoble

Professor J. Neel, Laboratoire de Chimie-Physique Macromoléculaire, ENSIC, Université de Nancy, 1, rue Grandville, 54, Nancy

Professor M. Rinaudo*, CNRS, Cermav, Domaine Universitaire, Cédex 53, 38, Grenoble

Mrs M. A. Rix, Laboratoire de Biophysique, Université de Nice, Parc Valrose, 06034, Nice Cédex

Professor* and Mrs E. Sélégny, Laboratoire de Chimie Macromoléculaire, Faculté des Sciences et des Techniques, Université de Rouen 76130 Mont Saint Aignan

Professor D. Vasilescu, Laboratoire de Biophysique, Université de Nice, Parc Valrose, 06034 Nice Cédex

Dr. R. Varoqui[†], CNRS, CRM, 6, rue Boussingault, 67033 Strasbourg Cédex

Germany

Mr U. Demisch, Max-Planck-Institut für Biophysik, Kennedyallee 70, 6 Frankfurt am Main 70

Professor N. Heckmann, Universität Regensburg, Fachbereich Chemie, Universitätstrasse 31, Postfach 8400, Regensburg

Dr K. P. Hofmann, Institut für Biophysik und Strahlenbiologie der Universität
Freiburg, Albertstrasse 23, 78 Freiburg i.Br.

Mrs C. Lüschow, Universität Regensburg, Fachbereich Chemie, Universitätstrasse 31,
Postfach 8400, Regensburg

Professor and Mrs G. Manecke, Institut für Organische Chemie der freien Universität
Berlin, Thielallee 63–67, 1 Berlin 33, Dahlem, West Berlin

Dr W. Pusch*, Max-Planck-Institut für Biophysik, Kennedyallee 70, 6 Frankfurt am
Main 70

Dr F. Sauer‡**, Max-Planck-Institut für Biophysik, Kennedyallee 70, 6 Frankfurt am
Main 70

Dr F. Staude, Kalle Aktiengesellschaft, Rheingaustrasse 190–196, D6202 Wiesbaden-
Biebrich, Postfach 9165, Wiesbaden

Professor G. Zundel*, Institute for Physical Chemistry, University of Munich 2,
Sophienstrasse II, Munich

Holland

Dr P. H. Bijsterbosch**, Agricultural University, Laboratory for Physical and
Colloid Chemistry, De Dreijen 6, Wageningen

Dr J. L. Cohen, Akzo Research Laboratories, Velperweg 76, Postbus 60, Arnhem

Dr J. C. Eysermans, Gorlaeus Laboratoria der Rijksuniversiteit, Wassenaarseweg,
Postbus 75, Leiden

Dr J. de Goede**, Chemie-Complex der Rijksuniversiteit, Wassenaarseweg, Postbus 75, Leiden

Professor Leyte, Chemie-Complex der Rijksuniversiteit, Fys. Chem. III, Wassenaarseweg, Postbus 75, Leiden

Professor M. Mandel, Chemie-Complex der Rijksuniversiteit, Fys. Chem. III,
Wassenaarseweg, Postbus 75, Leiden

Dr J. A. M. Smit**, Gorlaeus Laboratoria der Rijksuniversiteit, Wassenaarseweg,
Postbus 75, Leiden

Professor A. J. Staverman, Gorlaeus Laboratoria der Rijksuniversiteit, Wassenaarseweg, Postbus 75, Leiden

Dr J. van der Touw, Gorlaeus Laboratoria der Rijksuniversiteit, Fys. Chem. III,
Wassenaarseweg, Postbus 75, Leiden

Irak

Dr M. Rasoul Hamoud, College of Sciences, Chemistry Department, Adamiyha,
Baghdad

Israel

Mr M. S. Dariel, Atomic Energy Commission, Nuclear Research Centre-Negev,
P.O. Box, 9001, Beer-Sheva 84190

Dr C. Forgacs**, Negev Research Institute, P.O. Box 1025 Beer-Sheva

Professor and Mr Heitner-Wirguin, Dept. of Inorganic and Analytical Chemistry, The Hebrew University, Jerusalem

Professor O. Kedem*, Weizmann Institute of Science, Rehovot

Italy

Professor G. Barone, Instituto Chimico, Via Mezzocannone, 4, Università di Napoli, 80134 Napoli

Professor J. Celentano, Università di Milano, Instituto di Fisiologia Generale, Via Mangiagalli 32, 20133 Milano

Mrs L. Constantino, Instituto Chimico, Università di Napoli, Via Mezzocannone, 4, 80135 Napoli

Professor V. Crescenzi, Università degli Studi di Trieste, Instituto di Chimica, Chimica delle Macromolecole, Trieste

Professor F. Drioli, Università di Napoli, Instituto di Principi di Ingegneria Chimica, Piazzale Tecchio, I. 80125 Napoli

Dr V. Elia, Università di Napoli, Instituto Chimico, Via Mezacannone, 4, Napoli

Dr di Francesco, Università di Milano, Instituto di Fisiologia Generale, 20133 Milano

Mr F. Gambale, Consiglio Nazionale delle Ricerche, Laboratorio di Cibernetica en Biofisica, Via Mazzini 72, I. 16032 Camogli

Dr** and Mrs E. Patrone**, Instituto di Chimica Industriale, Università Genova, Via A. Pastore 3, Genova

Dr R. Rolandi, Università di Genova, Instituto di Scienze Fisiche, Viale Benedetto XV, 5 Genova

Dr R. Sartorio, Instituto Chimico, Università di Napoli, Via Mezacannone, 4, Napoli 80134

Mr G. Tiravanti, Consiglio Nazionale delle Ricerche, Instituto di Richerche sulle Acque, Via Francesco di Blasio, 5, Zona Industriale, Bari

Dr V. Vitagliano, Instituto Chimico, Università di Napoli, Via Mezacannone, 4 30134 Napoli

Japan

Professor F. Oosawa*, Institute of Molecular Biology, Faculty of Science, Nagoya University, Chikusa-Ku, Nagoya

Jugoslavia

Professor D. Dolar, Dept. of Chemistry, University of Ljubljana, Murnikova 6-POB 537, 61001 Ljubljana

Norway

Professor T. S. Brun*, Chemical Institute, University of Bergen, Bergen

Mr H. Hiland, Chemical Institute, University of Bergen, Bergen

Mr E. Lindberg, Institute of Physical Chemistry, University of Trondheim, N 7034 Trondheim

Sweden

Professor* and Mrs T. Teorell, Biomedicum, Box 572, S-751 23 Uppsala

U.K.

Mr I. S. Burke, Chemistry Dept., The University, Glasgow, W 2

Dr J. B. Craig, University of Aberdeen, Dept. of Chemistry, Meston Walk, Old Aberdeen, ABO 2 UE

Mr S. Cutler, University of Aberdeen, Dept. of Chemistry, Meston Walk, Old Aberdeen, ABO 2 UE

Dr T. Foley, Imperial College, Dept. of Chemistry, South Kensington, London S.W. 7

Dr J. Klinowski, Imperial College, Dept. of Chemistry, South Kensington, London S.W. 7

Professor and Mrs P. Meares, Dept. of Chemistry, University of Aberdeen, Meston Walk, Old Aberdeen ABO 2 UE

Dr R. Paterson, Dept. of Chemistry, Glasgow University, Glasgow W. 2

U.S.A.

Dr E. P. Allen, Medical College of Virginia, School of Dentistry, Richmond, Virginia 23219

Mr and Mrs J. A. Baker, College of Engineering, Dept of Chemical Engineering, The University of Iowa, Iowa City, Iowa 52240

Dr G. E. Boyd*, Oak Ridge National Laboratory, Post Office Box X, Oak Ridge, Tennessee 37830

Dr D. Cahen, Northwestern University, Dept. of Chemistry, Evanston, Illinois 60602

Professor S. R. Caplan*, Biophysical Laboratory Harvard University, Boston, Massachusetts 02115

Professor G. Eisenman*, Dept. of Physiology, University of California, Los Angeles, California 90024

Professor H. P. Gregor*, Dept. of Chemical Engineering, Columbia University, 356 Terrace Building, New York NY 10027

Mr J. Leibovitz, Sea Water Conversion Laboratory, 1301 South 46th Street, Richmond, California 94 804

Mr P. W. Rauf**, Dept. of Chemical Engineering, Columbia University, 356 Terrace Building, New York NY 10027

Dr A. Rembaum, Polymer Research Section, Jet Propulsion Laboratory, California Institute of Technology, 4800 Oak Grove Drive, Pasadena, California 91103

Professor* and Mrs K. Sollner, National Institute of Health, Building 2, Room 112, Bethesda, Maryland 20014

Professor A. A. Sonin*, Dept. of Mechanical Engineering, M.I.T., Cambridge, Massachusetts 02139

Professor* and Mrs K. S. Spiegler, Sea Water Conversion Laboratory, University of California, 1301 South 46th Street, Richmond, California 94804

Dr G. E. Stoner, University of Virginia, School of Engineering and Applied Science, Thornton Hall, Charlottesville, Virginia 22901

Dr A. Thorhaug*, Dept. of Microbiology, School of Medicine, University of Miami, Miami, Fla., U.S.A.

Professor I. Tasaki*, Dept. of Health, Education and Welfare, NIH – Room 36/ID-02 9000 Rockeville Pike, Bethesda, Maryland 20014

Professor and Mrs G. T. Schelling, College of Agriculture, Dept of Animal Sciences, University of Kentucky, Lexington, Kentucky 40506

Université de Rouen, Faculté des Sciences, Laboratoire de Chimie Macromoleculaire, ERA 417, 76130 Mont Saint Aignan

Dr Beaumais, Dr Bourdillon*, Dr Braud, Dr Demarty**, Dr Fenyo, Mr Folliard, Mr Labbé, Mr Langevin, Dr Mme Meffroy-Biget, Dr Merle**, Dr Metayer**, Dr Muller**, Dr Ripoll**, Dr Vert, Mr Vincent.

 * Read a lecture at the Institute and Contributor to the Books.
** Oral lecture or communication.
 ‡ Written contribution after the closing of the Institute.
 † Participant.
†† Not participant in Forges.

OPENING LECTURES

THE EARLY DEVELOPMENTS OF THE ELECTROCHEMISTRY
OF POLYMER MEMBRANES

KARL SOLLNER

National Institute of Arthritis, Metabolism and Digestive Diseases, National Institutes of Health,
Bethesda, Md. 20014, U.S.A.

Abstract. After some personal recollections of some of the great masters in the field of membranes, this paper reviews the pre-1930 history of the electrochemistry of polymer membranes which virtually coincides with the physical chemistry of porous membranes in electrolytic systems.

Since the eighteen twenties, several investigators, foremost Thomas Graham, studied the rates of the cross-membrane diffusion of electrolytes and the accompanying osmosis and 'negative' osmosis. The results were most confusing. After the development of the classical theory of elctrolytic solutions in the late eighteen eighties, a new phase started with Ostwald's speculations of 1890 concerning the electromotive properties of semipermeable membranes. They were followed by Nernst's investigations on the electromotive forces and the polarization effects which arise at the interfaces of two ionically conducting phases (1896, 1900), Bernstein's studies on the electromotive properties of living membranes (1902), and Donnan's work on the membrane equilibrium (1911). In 1914, Bethe and Toropoff, utilizing information from several historically disparate fields, created the still universally accepted concept of the electrochemical nature of charged porous membranes. The systematic electrochemistry of porous membranes was developed by Michaelis in the nineteen twenties in his classical investigations on dried, 'ion sieve', collodion membranes which became directly or indirectly the basis of all further work in this field, such as the author's studies, around 1930, on the mechanism of anomalous osmosis and on mosaic membranes and, a few years later, of the Teorell, and Meyer and Sievers fixed charge theory of ionic membranes.

1. Introduction: Recollections of Some Classical Masters of Membranology

At the very beginning of my talk I should like to state the main reasons for the choice of my topic. One is essentially subjective. After a lifetime of research in a particular field, one has acquired a feeling for the slow and circuitous interweaving of the originally rather disparate ideas and lines of work which have furnished the basis of today's knowledge; a desire arises to gain a broad and balanced historical overview of the growth of the concepts which today are dominant. In recent years I have therefore spent much time and effort on compiling a comprehensive review of the early history of the physical chemistry of membranes, and gladly take the opportunity to give here a somewhat condensed presentation of a specific topic, 'The Early Developments of the Electrochemistry of Polymer Membranes'.

A second reason for the choice of this topic was that some of my most respected scientific friends, mainly Torsten Teorell, Karl Kammermeyer, and the late Aharon Katchalsky encouraged me repeatedly to write about the historical development of the physical chemistry of membranes. Dr. Katchalsky emphasized time and again that some familiarity with the history of our field should be very useful to some of our younger and youngest colleagues. In addition, some of my friends were at times unkind enough to indicate that I am a kind of still living 'missing link' between an already half forgotten past and the present. This, in a way, is true. I started work on the electrochemistry of membranes more than 45 years ago and in the early years had personal contacts with many of the classical great investigators of physicochemical membranology.

I still heard Wilhelm Ostwald (Figure 1), one of the founding fathers of physical chemistry who in 1909 had received the Nobel Prize for his work on catalysis. Ostwald, in 1890, had been the first to apply, though only in a qualitative and speculative manner, the then new 'classical' theory of electrolytic solutions to membrane problems [1].

Walther Nernst (Figure 2), who had received the 1920 Nobel Prize for formulating

Fig. 1.　Wilhelm Ostwald, 1853–1932.　　　　Fig. 2.　Walther Nernst, 1864–1941.

the third law of thermodynamics, I saw often in seminars. His striking intellectual superiority was most conspicuous; there was no part of physical chemistry in the broadest sense, where he could not make immediately worthwhile suggestions. In earlier years, around 1900, Nernst invented, as one may say, the electrochemistry of liquid membranes. In connection with the problem of nerve excitation he first had become interested in polarization effects at the phase boundaries between two ionically conducting phases [2] and later investigated with Riesenfeld [3, 4] the electromotive behavior of solvent membranes in concentration cells. These studies were the first examples of a quantitative application of the classical theory of electrolytic solution to membrane problems.

Another prominent man in the membrane field was Fritz Haber (Figure 3), Nobel Laureate in 1918 for the synthesis of ammonia from the elements. I knew Prof. Haber rather well since I worked from 1927 to 1933 at his Kaiser Wilhelm (now Max Planck) Institute for physical and electro-chemistry at Berlin. Haber, like Wilhelm Ostwald, a man of very wide and diversified interests, had published in 1908 classical studies on the electromotive behavior of solid salt phases [5], and in 1909 his well-known paper with Klemensiewicz [6] on the glass electrode and its theory. This paper also described the electromotive behavior of certain liquid interphases which in concentration cells respond in a nearly Nernstian manner to concentration changes [6]. This

Fig. 3.   Fritz Haber, 1868–1934.   Fig. 4.   Frederick G. Donnan, 1870–1956.

work on liquid membranes was later continued and broadened by his former student R. Beutner [7] whom I met repeatedly.

Donnan (Figure 4), famous for his theory of the membrane equilibrium and the membrane potential [8–10], I knew fairly well; I worked from 1933 to 1937 in his laboratory at University College, London, however not engaged in membrane work.

I met Leonor Michaelis (Figure 5) in 1932. Later he became a much admired fatherly friend. His classical work on the systematic electrochemistry of membranes [11–15] particularly on dried 'molecular sieve' collodion membranes, laid the broad groundwork for the renaissance of the electrochemistry of porous membranes, which was initiated in 1935–36 by Teorell [16] and by Meyer and Sievers [17, 18]. This development, however, is already outside the period covered by this review which ends in the early nineteen thirties.

Freundlich's (Figure 6) monumental 'Capillary Chemistry' [19] was my own introduction to the physical chemistry of membranes and the origin of my interest in this field. I worked from 1927 to 1933 in Prof. Freundlich's laboratory at Haber's institute and remained associated with him for four more years while both of us were guest workers at Donnan's Laboratory in London. Though Prof. Freundlich never worked extensively on membranes, he had a deep interest in this field; I owe much to his encouragement and our long discussions on membranes.

Fig. 5.   Leonor Michaelis, 1875–1949.   Fig. 6.   Herbert Freundlich, 1880–1941.

Rudolf Höber (Figure 7), the dean of the biological membrane research, became in the late nineteen thirties a much beloved friend in spite of the great difference in age. Höber and his collaborators also had made significant contributions to the basic physical chemistry of membranes. His book *The Physical Chemistry of Cells and Tissues* [20], first published in 1902, was for several decades 'The Bible' in this field. It was always a delight to discuss with Prof. Höber membrane problems and the involved history of this field.

The personal contacts with these great men have given me not only some most interesting insights into the workings of their minds, but also some special glimpses into the developments of the conceptual framework of the physical chemistry of membranes.

Fig. 7.   Rudolf Höber, 1873–1953.

Before turning to the history of the electrochemistry of polymer membranes itself, I must first make a few remarks concerning the nature of such membranes and their place in the systematic classification of membranes.

2. Charged Polymer Membranes, Their Nature and Place in the Systematic Physical Chemistry of Membranes

It is a commonly overlooked fact that virtually all natural and artificial porous membranes consist, mostly *in toto*, of polymers, and that all the porous membranes investigated in the past carry inherently at least some ionizable groups and therefore are charged, except in the rarest instances [21, 22]. Accordingly, the physical chemistry of charged polymer membranes in electrolytic systems may for all practical purposes be equated with the electrochemistry of porous membranes.

Porous membranes in electrolytic systems have been investigated for well over 100 years. Thus it is evident that the study of charged polymer membranes antedates by many decades the creation of the concept 'polymer' and of the still newer term 'polyelectrolyte', or the art of making membranes by the methods used in preparing ion exchange resins.

The typical functional properties of membranes, their permeability and their electromotive actions, can be studied only in systems in which a membrane functions as a barrier [23–24]. There are four conceptually sharply distinguished classes of systems in which a membrane separates two solutions in the same solvent, the only systems to be considered here:

(1) Solvent, ordinarily liquid, membranes in non-electrolytic systems;

(2) Solvent, ordinarily liquid, membranes in electrolytic systems;

(3) Porous membranes in non-electrolytic systems;

and

(4) Porous membranes in electrolytic systems.

This grouping of membrane systems into four basic classes, which seems obvious today, was not readily visible to the early investigators; it was one of the fruits of their labors. It took many decades before it was generally realized how much more complicated and varied the effects are which arise in electrolytic than those which arise in non-electrolytic membrane systems. Likewise, many early investigators paid little attention, or were not sure, whether their membranes were essentially porous in nature or more like interphases consisting of a different solvent. In this context it is interesting to note that, as late as 1914, Bethe and Toropoff [25, 26] in one of the most fundamental studies on the basic electrochemistry of porous membranes, to which we shall return below in detail, still found it necessary to adduce in detail the various arguments for the porous nature of such membranes as pigs bladders, gelatin and collodion membranes, parchment paper, clay diaphragms, etc. The reader will note that the organic and inorganic substances of which these membranes consist are polymeric materials.

3. The Period before the Creation of the Classical Theory of Solutions in the Late Nineteen Eighties

Space does not permit me to review here the vast literature on bioelectrical effects which had been studied since Galvani discovered in 1786 the excitability of living tissues. It is noteworthy that for a long time it was not realized that virtually all bioelectric effects are due to the presence of membranes. It is quite characteristic that Biedermann in his monumental *Electrophysiology* of 1895 [27] does not mention the term 'membrane' in the subject index; there is nowhere an indication that he saw any connection between membranes and bioelectric phenomena. Höber [28] in the first edition of his later famous *Physical Chemistry of Cells and Tissues* of 1902 was similarly mute on these points, and Verworn [29], in his days probably the most prominent general physiologist, stated in 1903 specifically that..." the idea that the cell membrane is a universal component of cells has been dropped completely'. (Author's translation) Bioelectricity remained an essentially descriptive science till Bernstein [30–32] in 1902 enounced the 'membrane theory' of bioelectric potentials; we shall return later to his work, mainly its physicochemical aspects.

The systematic study of *electrolytic membrane systems* can be said to have started in the late eighteen twenties when Dutrochet [33] published a series of papers on 'Endosmosis' and 'Exosmosis', that is the diffusion of electrolytes across membranes and the concomitant osmotic movement of water. Dutrochet's papers were followed by similar studies by Jolly [34], Graham [35, 36], Eckhard [37] and many others quoted in a comprehensive review by Waitz [38]. These investigators tried to correlate the rates of the cross-membrane diffusion of electrolytes and of osmosis. Contrary to the situation observed with non-electrolytes the experimental results were confusing, nearly erratic. The efforts to gain meaningful insights remained essentially futile.

Generally, the mentioned authors believed that their membranes were porous structures though some doubt remained in the case of very dense membranes of animal origin, such as pig's bladders. Graham (Figure 8), whose papers are outstanding in

Fig. 8.   Thomas Graham, 1805–1869.

this group and by far of the greatest importance for the further development of membranology, paid considerable attention to this problem [35]:

The force of liquid diffusibility will still act if we interpose between the two liquids a porous sheet of animal membrane or unglazed earthenware; for the pores of such a septum are occupied by water, and we continue to have an uninterrupted liquid communication between the water on one side of the septum and the saline solution on the other side.

In the same paper [35] Graham emphasizes the erratic character of the osmotic effects which arise in systems with electrolytic solutes, and describes the occurrence of *negative osmosis*:

With animal septa, frequent examples of outward flow of liquid from the osmometer present themselves, causing the liquid column to fall instead of rise in the tube.

Graham suspected that the erratic osmotic effects arising with electrolyte solutions may somehow be connected with the "electrical endosmosis, as described recently by Wiedemann' [39]. Parenthetically we may add that this effect had actually been reported first by Reuss [40] in 1809.

One must admire Graham's intuitive insight, and ability to see a connection between seemingly disparate effects. More than sixty years elapsed before Bethe and Toropoff [25, 26] demonstrated definitely that the charge of membranes, as determined in electrokinetic experiments, is the basic physical factor which governs their behavior in electrolytic systems.

In the intervening years many outstanding physicists furthered the study of the electrokinetic phenomena without particular reference to other membranes effects. Quincke in 1860 and 1861 reported extensive studies on streaming potentials [41] and on electroosmosis [42]. He also proposed the idea of an electrical double layer as the basis of all those effects which today are referred to as 'electrokinetic'. Helmholtz [43] in 1879 formulated this concept quantitatively. It was improved in detail by others, foremost by v. Smoluchowski [44], and Perrin [45] after the classical theory of electrolytic solutions had been developed.

According to this concept an electrical double layer exists at all phase boundaries. At a solid-liquid phase boundary, charges of one sign are affixed immovably, by some hardly discussed mechanism, to the solid phase, on it or just inside it. An equal number of charges of the opposite sign are located in the liquid, according to the Quincke and Helmholtz conception at an invariant exceedingly small but finite distance from the phase boundary. This classical picture of a condensor-like electrical double layer is sketched in Figure 9 representing a longitudinal section of a (very hypothetical) cylindrical pore. This extremely useful classical concept was refined in 1910 by Gouy [46] who introduced the concept of the 'diffuse' electrical double layer. The older literature on electrokinetic effects has been reviewed repeatedly [19, 44, 47].

It would serve no useful purpose to report in greater detail on the largely futile literature of the eighteen fifties to eighteen eighties on electrolytic membrane systems. General ideas or meaningful wider concepts are missing. The cause for this state of affairs is obvious; it was the lack of a reasonably adequate theory of electrolytic solutions.

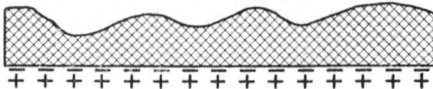

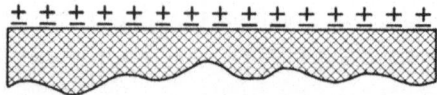

Fig. 9.   The classical, Helmholtz, conception of the electrical double layer as the basis of all electro-kinetic effects, e.g., electroosmosis.

4. The Classical Theory of Electrolytic Solutions and Its Application to some General Electrochemical Membrane Problems

A meaningful attempt at a rational electrochemistry of membranes had to be preceded by the development of the physical chemistry of solutions in general, and that of electrolytic solutions in particular.

In the context of this review it is of some interest to note that van't Hoff's [48] classical paper of 1887 which outlined the modern theory of solutions [48, 49, 50] is based on the measurements of the osmotic pressure of sucrose solutions which

Fig. 10.   Moritz Traube, 1826–1894.

Wilhelm Pfeffer [51] had obtained by the use of copperferrocyanide membranes. These membranes had been first described and thoroughly investigated in the eighteen sixties by Moritz Traube (Figure 10) [52], one of the fathers of modern membrane studies.

The development of the theory of electrolytic solutes is associated mainly with the names of Kohlrausch [53, 54], Rauolt [50], Arrhenius [55], Ostwald [56, 57] and Nernst [58, 59]. The classical theory of electrolytic solutions found its capstone in 1889 in Nernst's (Figure 2) [59] paper on 'The Electromotive Action of the Ions' in solution. In this paper Nernst developed the theory of the 'liquid junction' or 'diffusion potential' which arises when an electrolyte in solution diffuses from a higher to a lower concentration. For a uni-univalent electrolyte Nernst's equation for the liquid junction potential, E_{junc}, is given by the expression:

$$E_{junc} = \frac{RT}{F} \frac{u-v}{u+v} \ln \frac{c_1}{c_2} \qquad (1)$$

in which u and v are the ionic mobilities of the cation and the anion, and c_1 and c_2 the concentrations of the two solutions. The sign of the charge of the more dilute solution is that of the faster moving ion.

In the same paper Nernst developed also the kinetic theory and the well known equation for the change in electrode potential, ΔE, when the concentration of the potential determining ion is changed from c_1 to c_2:

$$\Delta E = \frac{RT}{nF} \ln \frac{c_1}{c_2}, \qquad (2)$$

where n is the valency of the potential determining ion.

It is today not generally remembered that Nernst was fully aware of the fact that the concentration ratios in Equations (1) and (2) are not an accurate measure of the electromotive activity of ions in solution. The ratios of the osmotic pressures of the two solutions, or the ratios of their conductivities, according to Nernst, are preferable, because at higher concentration the fraction of undissociated, electromotively inactive molecules increases at the expense of the dissociated ones. In this context Nernst also draws attention to Helmholtz's [60] now often overlooked equations of 1877 in which the lowering of the vapor pressure of the solvent is used instead of the concentration of the solute.

Nernst's basic conception of the liquid junction potential was soon expanded by Planck [61] and by Planck's collaborator Henderson [62] to the general case of the interdiffusion of different electrolytes. The general equation for the liquid junction potentials arising under these conditions, the so-called Planck-Henderson equation, together with Nernst's equations turned out in the long run to be cornerstones of the theoretical electrochemistry of membranes.

Now the stage was set for the development of a rational electrochemistry of membranes. The first move in this direction was taken by Wilhelm Ostwald [1], (Figure 1) who used the newly developed ideas concerning the nature of electrolytic solutions as the basis for some ingenious speculations concerning the electromotive properties of semipermeable membranes. Ostwald's ideas were of such a general nature that he did not have to make any specific assumptions concerning the molecular mechanism of the restraining action of the membrane; his considerations apply equally to porous and to liquid membranes.

Ostwald started with the statement that a membrane, in order to be impermeable to an electrolyte in solution, needs not necessarily be impermeable both to its anions *and* to its cations. It is sufficient if the membrane is impermeable either to the cations or to the anions, because cations and anions cannot be separated from each other in quantities detectable by chemical methods. Macroscopic electroneutrality must be maintained.

According to Ostwald, a minute quantity of the permeable ions penetrates however across a 'semipermeable' membrane of the type described by Traube [52] and used by Pfeffer [51] when such a membrane is interposed between two solutions of the same electrolyte at different concentrations. Thereby a static electrical double layer is established at the membrane, the permeable ions impressing their charge on the more dilute solution. This membrane potential can be considered as the limiting case of a liquid junction potential in which the mobility of one species of ions, the nonpermeable ions, is zero in the membranes. Ostwald, however, strange as it might seem, did not formulate his ideas quantitatively.

With respect to the bioelectric significance of his ideas, Ostwald surmised that:

not only the currents in muscles and nerves but also the puzzling actions of the electric fishes will find their explanation in the... outlined properties of semipermeable membranes (Author's translation).

In addition, Ostwald pointed out a very important fact that, if one of his semipermeable membranes, say one intrinsically permeable to anions and impermeable to cations, separates the solutions of two electrolytes having different anions, the two species of anions may cross the membrane and exchange in quantity [1]:

For each negative ion which enters the cell from the outside, one of the inside negative ions will leave it, and an undisturbed interdiffusion process begins; it is restricted only by the law that the same number of ions must penetrate across the membrane in the one direction as in the other. (Author's translation).

The neglect of this basic fact has caused considerable confusion in many later investigations.

The just outlined considerations of Ostwald, presented in 1890, are the logical and historical starting point of the rational electrochemistry of membranes.

The recognition of the fundamental importance of Ostwald's paper, however, was slow in coming, probably delayed by the mode of its presentation. Intermingled with some most original and in the long run immensely fruitful ideas, Ostwald also presented a rather labored and diffuse *ad hoc* explanation of the so-called 'electrocapillary Becquerel effect' [19, 63]. This puzzling phenomenon, as was shown later [64], can be readily explained on the basis of conventional electrochemistry. Thus it is of no wider significance or general interest. For the author, however, the Becquerel effect was of major importance, since an investigation on its mechanism [64] was his first step towards a lifetime of work in the electrochemistry of membranes.

Before turning to the post 1890 developments which are specific for the electrochemistry of porous membranes, we shall consider first two other, highly important investigations that by their very nature are equally valid with porous as with liquid membranes, as were Ostwald's [1] just outlined considerations: Nernst's [65] considerations on the polarization effect at the phase boundaries between two ionically conducting phases, and the Donnan membrane equilibrium [8, 9].

The detailed experimental exploration of the excitation of cells and tissues by impulses of current and by alternating current had been for decades the main topic of electrobiology, when Nernst (Figure 2) [65] in 1899 proposed the first meaningful molecular theory of these effects which was based on the classical theory of electrolytic solutions. Nernst pointed out (a) that all component parts of tissues are electrolytic conductors, without making any specific assumptions as to their nature and (b) that the contributions of the various species of ions to the transportation of current in different electrolytic conductors, that is their transport numbers in the different phases, are ordinarily different. Accordingly, when a current is passed, changes in electrolyte concentration, i.e., concentration polarization, arise at the two sides of the phase boundaries. This effect is analogous to the polarization in a liquid phase at a metallic electrode when current is passed. Nernst showed that the degree of polarization is a function of the intensity and frequency of the current and of the counteracting influence of diffusion. He theorized that the concentration changes due to polarization at the phase boundaries were the cause of electrical excitation by alternating currents, an idea which he later confirmed experimentally [66].

Nernst's work on concentration polarization exerted a considerable influence on the later development of the systematic electrochemistry of porous membranes. His ideas on this topic were utilized in 1914 by Bethe and Toropoff [25] in forming for the first time a clear and coherent concept of the electrochemical structure of porous membranes and their mode of action in electrolytic systems. This most important work of Bethe and Toropoff will be discussed below in detail.

The other remarkable piece of work of great generality to be discussed here, Donnan's (Figure 4) theory of the membrane equilibrium ('Donnan equilibrium') and membrane potential ('Donnan potential') was published in 1911 [8]. It is a rigorous quantitative expansion of Ostwald's [1] before outlined qualitative ideas.

Donnan considered a completely dissociated binary electrolyte, NaR, whose anion R^- can not penetrate across the membrane. This 'non-diffusible', 'colloidal' ion R^- might be visualized as a large dye-stuff ion which is screened out mechanically by a porous membrane. In the simplest case, a compartment of invariant size filled with solution of NaR is separated from an equal volume of NaCl solution by a membrane which is permeable to all small ions and impermeable to R^-. Under these conditions as Donnan pointed out, Na^+ and Cl^- move across the membrane until a distribution equilibrium is established which conforms to the law of macroscopic electroneutrality and gives rise to a potential difference between the two solutions, the Donnan membrane potential.

Donnan's quantitative derivation [8–10] of his well known equations for the equilibrium ion distributions and the concomitant membrane potentials which arise in such simple and in related more complicated membrane cells, must not be repeated here. As a reminder, we present here only Donnan's equation for the equilibrium distribution of ions, including the ions of water, between solution 1 and solution 2

(Equation (3)) and for the membrane potential (Equation (4)):

$$\frac{[Na^+]_1}{[Na^+]_2} = \frac{[Cl^-]_2}{[Cl^-]_1} = \frac{[H^+]_1}{[H^+]_2} = \frac{[OH^-]_2}{[OH^-]_1} \tag{3}$$

$$E = \frac{RT}{F}\ln\frac{[Na^+]_1}{[Na^+]_2} = \frac{RT}{F}\ln\frac{[Cl^-]_2}{[Cl^-]_1} = \frac{RT}{F}\ln\frac{[H^+]_1}{[H^+]_2} = \frac{RT}{F}\ln\frac{[OH^-]_2}{[OH^-]_1}. \tag{4}$$

Today Donnan's equations are ordinarily written in terms of activities and activity ratios instead of the concentrations and concentration ratios used by Donnan. The reader will note that Donnan's equation for the membrane potential (Equation (4)) is identical with Nernst's equation (Equation (2)) for the electrode potential.

Much less well known than Donnan's theory, however, is the fact that Donnan and collaborators [9, 67, 68] soon proved experimentally the essential correctness of this theory. Other investigators confirmed these findings and also dealt with the application of Donnan's concept to various physicochemical and biological systems [69–72]. In the present context, it is of interest to note that in all these studies, with one exception [68], porous membranes were used in conjunction with large dyestuff or protein nondiffusible ions.

The influence of Donnan's work on the development of the electrochemistry of membranes remained rather limited for many years. One might surmise that this was due to two factors: Donnan's papers did not provide any lead to systematic broad-scale experimental work on the electrochemistry of membranes, and his theory was seemingly too far ahead of its time and too abstract for the graps of many contemporary investigators.

5. Permeability Studies: Dialysis and Electrodialysis

Before turning to the behavior of porous membranes in electrolytic systems we must first touch upon the permeability of porous, particularly dense porous membranes for nonelectrolytes.

The question whether some membranes of low water permeability and low water content should be considered as porous structures or as solventlike entities had puzzled many early investigators. Graham [35, 36] preferred to think of the common membranes as porous structures and Moritz Traube (Figure 10) [52] in his classical investigations on the nonelectrolyte permeability of copperferrocyanide precipitation membranes offered strong, in his opinion decisive evidence for the porous nature of these membranes.

Over the years there developed a general consensus that virtually all commonly used membranes are porous, though this question was not definitely settled until Bethe and Toropoff [25] in 1914 demonstrated their porous character beyond any reasonable doubt. It was also generally assumed that aside from membrane porosity, the molecular size of the permeant is the main factor in determining the rate of penetration of nonelectrolytic solutes across dense porous membranes.

Definitive investigations on the permeability of such membranes for nonelectrolytes and very weak electrolytes, which permeate essentially in the undissociated state, were published by Runar Collander (Figure 11) in 1924 to 1926. Collander used in his studies first copperferrocyanide membranes [73] and later collodion membranes of three grades of porosities [74], thoroughly air dried membranes, and dried membranes which had been swelled in 68% and 80% ethyl alcohol before reimmersion in

Fig. 11.   Runar Collander, 1894–1973.

water which stabilized their new, slightly more open structures. Collander's papers also contain condensed, very instructive reviews of the literature.

Collander's data clearly demonstrated that the rates of crossmembrane diffusion of the various solutes depended primarily on their molecular sizes. With increasing molecular size of the solutes, the rates of their transmembrane diffusion decrease much more steeply than their diffusion velocities in free solution. This effect is the more pronounced the denser the membrane. The rates of transmembrane diffusion of solutes above a certain molecular size, characteristic for each degree of membrane porosity, was virtually zero. Some apparent discrepancies in Collander's data, as already suspected by him [74], were later shown to be due to the interaction of collodion with certain solutes which cause it to swell [75]. A limited number of data on the nonelectrolyte permeability of dense collodion membranes published at the same time by Fujita [76] are in excellent agreement with Collander's results.

Collander's experimental results on the nonelectrolyte permeability of copperfer-rocyanide membranes [74] paralleled those obtained with his collodion membranes [75], and thereby implicitly also provided overwhelming evidence that these mem-branes too, as already stated by Traube, act essentially as 'molecular sieves'.

By the mid-nineteen twenties, mainly due to Collander's work, the nonelectrolyte permeability of inert porous membranes could be considered to be an essentially solved problem. If the permeant is strongly adsorbed on the membrane or otherwise interacts with it, the situation is more involved and each system must be considered separately [75, 77].

The information on the *electrolyte permeability of porous membranes* available in the early nineteen twenties was meager and very confusing. The literature provided not more than an ever-increasing amount of qualitative observations and semiquan-titative data. No comprehensive review had ever been attempted and no systematic experimental investigations had ever been carried out which by any stretch of imagina-tion could be compared to Collander's [73, 74] work on nonelectrolytes.

The first pertinent investigations on the electrolyte permeability of membranes of low porosity which were undertaken after the creation of the classical physical chemis-try of electrolytic solutions were started in Ostwald's laboratory by Tammann [78] and, on a much larger scale, by Walden [79]. They studied the electrolyte permeability of various ferrocyanide precipitation membranes which, as we know today, are porous in nature, though Tammann and, to some extent, also Walden preferred to consider them as solvent-like.

The explicitly stated purpose of Walden [79] was sixfold: (1) to repeat Traube's experiments on the electrolyte permeability copper ferrocyanide membranes; (2) to expand them to 'as many substances as possible'; (3) to find new membranes; (4) to test Traube's opinion that the membranes are 'atom (molecular) sieves'; (5) to test Ostwald's [1] idea that the crossmembrane diffusion of electrolytes depends mainly on their ions; and (6) to investigate whether any correlation could be detected between the rates of permeation of various solutes across a given membrane and the other properties of the latter. In carrying out this program, Walden tested the permeation of numerous substances, mainly weak and strong acids and some salts of these acids, across a great variety of precipitation membranes. He used not only the copperfer-rocyanide and tannic acid-glue (gelatine) membranes studied by Traube, but also some new ones consisting of cobalt silicate, nickel silicate, ferrocyanezinc, and cobaltic-cyanecadmium, etc. Most of Walden's data were only qualitative. Rather amazing was his report that acids of relatively high molecular weight, such as citric acid (M.V. = = 192) and mellitic acid (M.W. = 342), permeate across all his membranes very much faster than nonelectrolytes of similar molecular weights. This surely was a most amaz-ing result and totally incompatible with the notion that the various precipitation membranes are molecular sieves. It is probably for this reason that Walden considered his results as a proof that the various precipitation membranes are not of a porous nature but act as solvent membranes. However, he does not discuss, much less explain, the peculiar solvent properties of these hypothetical solvents. Except where he summarizes experimental data, most of Walden's discussion was, and still is, difficult to follow. His conclusions were confused and arbitrary as was his polemic against Ostwald's concept of electrolyte impermeability.

The somewhat startling, if not mysterious results of Walden's papers, their origin in Ostwald's laboratory, and Walden's rapidly increasing reputation as an electrochemist [80] were probably the main reasons why for a prolonged period no further systematic physico-chemical study of the electrolyte permeability of precipitation or other porous membranes was attempted.

More than 30 years elapsed before Collander [81] showed in 1925 that Walden's alleged high rates of transmembrane diffusion of high molecular weight acids were based on the erroneous interpretation of experimental data obtained by an inappropriate experimental technique.

In Walden's experiments, as Collander pointed out, the membrane was interposed between a rather dilute $K_4Fe(CN)_6$ solution which contained an indicator, and a solution which contained both some $CuSO_4$ and the acid under investigation. A change in the color of the indicator was taken by Walden as proof that the acid had penetrated across the membrane. Collander, however, demonstrated that only an exchange of H^+ for the readily dialyzing K^+ occurs. If the $K_4Fe(CN)_6$ solution in Collander's experiments contained the high molecular weight acid, it took a long time before the $CuSO_4$ solution became significantly acid; if it contained a low molecular weight acid, this process was much faster. Today all this seems obvious; it certainly was not trivial nearly 50 years ago.

The most pertinent results of Collander's investigation on the electrolyte permeability of the copperferrocyanide precipitation membrane were summarized by him with the statement that this membrane "behaves with respect to organic acids in solution exactly as with dissolved nonelectrolytes."... "The permeability of strong mineral acids depends in all probability upon the volume of the anion." (Author's translation)

In this manner Collander's study provided at long last a high degree of clarity and orderliness as far as the permeability of copperferrocyanide and similar membranes for weak acid organic acids is concerned. For the case of the strong mineral acids, it provided suggestive evidence but it did not yield much new information of potentially fundamental electrochemical significance; a search in this direction was outside the framework of Collander's study.

The question of permeability of porous membranes for strong electrolytes was taken up in 1924 by A. Bethe, H. Bethe, and Y. Terada [82], who studied the rates of diffusion of strong electrolytes across porous membranes systematically from a basic physicochemical point of view. They compared the regularities in the diffusion of nonelectrolytes and of electrolytes across fairly dense parchment paper membranes, being particularly interested in testing the validity of Fick's law.

Though limited in scope, this competently conceived, executed, and presented investigation clearly revealed the basic facts and stated pertinent theoretical considerations. In the case of the transmembrane diffusion of nonelectrolytes well defined permeability constants, k, could be determined; Fick's law holds true. From this Bethe, Bethe, and Terada concluded that in the dialysis of nonelectrolytes one is dealing merely with hydrodiffusion in the pores of the parchment paper membrane.

With electrolytic solutes the 'permeability constants' computed from the experimental data obtained after different periods of dialysis varied and showed different trends characteristic for the various electrolytes; in other words, meaningful permeability constants could not be obtained. Possible changes in volume of the experimental solutions by osmosis or anomalous osmosis, incidentally, were not considered by Bethe, Bethe, and Terada.

Bethe, Bethe, and Terada came to the conclusion that in the case of electrolytic permeants "aside from diffusion, also membrane forces are active (capillary electric processes and with large molecules an increased influence of friction?)" (Author's translation). Their very brief remarks on these various factors are rather hazy and end with the factual statement that the temperature coefficients of capillary electric processes generally seem to be the same as that of hydrodiffusion. Over all, their discussion does not present any novel theoretical ideas or suggestions for further experimental work. It also clearly revealed the rather rudimentary state of the electrochemistry of porous membranes as it existed fifty years ago.

Historically, and from the point of view of the psychology of research, the paper of Bethe, Bethe, and Terada is of unusual interest. These investigators in 1924 were somehow oblivious of some of the most significant and constructive thoughts on the molecular basis of the characteristic electrochemical properties of porous membranes which had been enounced in 1914 by A. Bethe and Toropoff in the repeatedly mentioned paper on membrane polarization [25] which is reviewed below. Consequently, the paper by Bethe, Bethe, and Terada shows a clear regression from the level of insights outlined, however briefly, ten years earlier by A. Bethe and Toropoff.

The literature on *dialysis* contained hardly any systematic experimental data of basic electrochemical interest since dialysis had been studied virtually exclusively for its use as a preparative laboratory or industrial method [36, 83–88]. Nevertheless, several regularities were apparent: the rates of cross-membrane permeation of uni-univalent inorganic electrolytes were always higher than those of uni-bivalent or bi-bivalent salts. The differences between various electrolytes were decidely more pronounced with membranes of low porosities than with membranes of high porosities or diaphragms. Brown [84], for instance, in one of the most instructive papers of the early period, reported that NaCl can be dialyzed out from mixed solutions of NaCl and Na_2SO_4 by the use of collodion membranes of suitable porosity.

In the mid-nineteen twenties it was generally recognized that the exhaustive removal of electrolytes from colloids by dialysis is at best an excessively slow procedure and with unstable substances, such as proteins, a virtually impossible task. Much superior for this purpose was *electrodialysis*, a procedure that has been used for some time [89–94], particularly by W. Pauli and collaborators [89, 92, 94]. It represented a great improvement over dialysis, even at the rather low current densities initially used. High current densities, however, which result in correspondingly higher rates of electrolyte removal caused excessive local heating at the membranes and considerable, most undesirable pH changes near the membrane-solution interfaces due to membrane polarization, an effect which will be discussed below in detail.

The literature on electrodialysis contained some information of basic electrochemical importance. It showed that the dominant parameter in determining the behavior of a membrane in electrodialysis, aside from its porosity, is the sign of its electrical charge. Also, it was evident that univalent ions electromigrate across membranes much more readily than bivalent and polyvalent ions. It furthermore was apparent that to make electrodialysis efficient, the membrane through which cations move from the middle compartment of an electrodialysis cell toward the cathode should be electronegative, and electropositive in the case of the anions migrating toward the anode. The reason for this, though stated in the literature, was not clearly perceived by some experimenters.

The strangely confused state of the information on electrodialysis available in 1931, when a great deal more was known then a few years earlier, is evident from an exhaustive survey by Prausnitz and Reitstötter [95] which is still a very useful source of references and guide to the rather extensive patent literature.

6. The Electromotive, Electroosmotic and Polarization Properties of Charged Porous Membranes, and the Origin of Their Charge

The typically electrochemical characteristics of porous membranes, which as we know today depend on their charge, are their electromotive actions when interposed between electrolyte solutions of different composition, and the electroosmotic and polarization effects which arise at them on the passage of current. Insight into the molecular mechanisms of these effects was slow in coming. The reasons for this, in retrospect, appear fairly obvious. The intrinsically complex and then rather enigmatic structure of the microheterogeneous porous membranes prevented any straightforward use of conventional solution chemistry and of thermodynamics in explaining the functional properties of such membranes. It was impossible to rely on reasoning from first principles and sweeping generalizations in a manner analogous to that used so very successfully by Nernst and collaborators [2–4] in the case of the electromotive properties of simple solvent membranes and of the polarization effects at the phase boundaries of two electrolytically conducting phases in general [65, 66], or by Haber and collaborators [5, 6] in the case of the electromotive properties of solid electrolyte interphases and of certain liquid membranes which in concentration cells reacted electromotively similar to solid salt phases.

Physicists and physical chemists by and large avoided electrochemical studies with porous membranes except in the case of electroosmosis and streaming potentials [39–47]. Ostwald's [1] above reported ideas on the semipermeability and electromotive properties of membranes, published in 1890, were a rare exception. However, for many years Ostwald's conceptions were neither used nor developed further either by physical chemists or biologists who were not aware of the bioelectrical significance of membranes [27–29].

The picture changed abruptly in 1902 with the publication of the pioneering paper by the physiologist Julius Bernstein (Figure 12) [30] in which he took up Ostwald's ideas and presented the 'membrane theory' of bioelectric potentials. Thereafter, electrochemical membrane studies were taken up by a rapidly increasing number of

Fig. 12.   Julius Bernstein, 1839–1917.

mainly biological investigators. Most of this still essentially descriptive work can be disregarded here. We shall focus our attention primarily on the work of the three out-standing men whose intuitive insight, knowledgeable use of physicochemical back-ground information, and skillful experimentation laid the foundations of the electro-chemistry of porous membranes. It is of interest to note that all three were by training medical-biological investigators: the just named Julius Bernstein [30–32], the phy-siologist Albrecht Bethe (Figure 13) [25, 26], and Leonor Michaelis (Figure 5) [11–15] who after starting as a medical investigator [96] had later turned to biochemical problems and physico-chemical investigations of biophysical interest. In addition, we shall mention the work on anomalous osmosis by the physical chemists F. E. Bartell [97, 98] and Herbert Freundlich [99], and the extensive experimental studies mainly on the same topic by the biologist Jacques Loeb (Figure 14) [21, 100, 101] famous for his discovery of chemically induced parthenogenesis [102].

Fig. 13.   Albrecht Bethe, 1872–1954.   Fig. 14.   Jacques Loeb, 1859–1924.

6.1. JULIUS BERNSTEIN

Julius Bernstein had been an active worker in electrophysiology for forty years, when he published in 1902 his classical paper 'Investigations Concerning the Thermodynamics of Bioelectric Currents' (Author's translation) [30]. The main purpose of this and two later publications [31, 32] was to clarify the two most fundamental problems of electrophysiology, namely the mechanisms and the locale of the processes which give rise to bioelectrical effects. In the first of his contributions [30], Bernstein had to define these then all but clearly conceived problems which had been largely neglected by the essentially descriptive nineteenth century electrobiology.

Bernstein was the first biologist to apply to bioelectric problems the, then still rather new, theory of electrolytic solutions. He was thoroughly familiar with the kinetic theory of solutions in general, Nernst's [59] concept of the origin of the liquid junction potential, and Ostwald's [1] ideas concerning the electromotive properties of semipermeable membranes which were outlined above. The physicochemical part of his work consists largely of a quantitative interpretation of Ostwald's concept and its extension to membranes which are not strictly impermeable either to cations or to anions. In dealing with this problem, Bernstein made also a major original contribution to the basic electrochemistry of porous membranes.

That part of Bernstein's work which is of greatest importance from the physicochemical point of view is his explanation of the origin of the electromotive forces which arise at porous membranes which separate electrolyte solutions of different composition. Bernstein based his considerations on the Nernst-Planck [59, 61] concept of the liquid junction potential as expressed in Equation (1) for a concentration cell with solutions of a uni-univalent electrolyte at the concentrations c_1 and c_2. According to Equation (1) the magnitude of the liquid junction potential at a given temperature is determined not by the absolute values of the ionic mobilities of the ions in solution but by a function involving the ratio of these ionic mobilities.

If a membrane is inserted between the two solutions the mobilities of the cations and the anions within the membrane, u' and v', will ordinarily differ from the ionic mobilities in free solution u and v, and the potential which arises at the membrane, E_{mem}, will be given by the expression:

$$E_{mem} = RT \frac{u' - v'}{u' + v'} \ln \frac{c_1}{c_2}. \tag{5}$$

It is obvious that Equation (5), as Equation (1), can not be used to calculate from potential data the magnitude of the mobilities, u' and v' but only the ratio of these mobilities, u'/v'. The empirical data showed that the ratios of the mobilities (more correctly of the diffusion velocities) of cations and anions in membranes differ from those in free solution. Bernstein observed that E_{mem} is ordinarily higher than the corresponding liquid junction potential, E_{junc}. From this he drew the conclusion that the ratio of the ionic mobilities within the membrane is ordinarily higher than in free solution. If the mobility of the one or the other ion within the membrane approaches zero, the case discussed by Ostwald [1], E_{mem} is given by Nernst's equation for the electrode potential, Equation (2).

Bernstein also considered the potentials which arise when a membrane separates solutions of different uni-univalent electrolytes. Lack of space, however, does not permit a discussion of this topic.

Though Bernstein's approach was entirely formalistic, his papers show an amazing degree of novel insight into the electromotive action of porous membranes. It was unfortunate that he hardly considered, and certainly did not explain, the physical cause of the differences in the ratios of the transference numbers in free solution and in the membranes. This was probably the main reason why Bernstein's highly original and constructive work made only a limited impact on the further development of the electrochemistry of porous membranes.

The strictly bioelectric achievements of Bernstein can be mentioned here only in passing. He showed that the magnitude of bioelectric potentials, as that of concentration cells, is proportional to the absolute temperature, and thereby proved that these electromotive forces in living systems are of an osmotic not a biochemical nature. He further showed that bioelectric potentials in many instances vary with the potassium concentration in a nearly Nernstian manner, that is approximately according to Equation (2). Bernstein further demonstrated the overwhelming probability that the bioelectric potentials originate at the uninjured not at the injured surfaces of cells and tissues. On this basis he proposed the ever since accepted 'membrane theory' of the 'pre-existing' bioelectric potentials, and definitively clarified his second main problem, the locus at which these potentials arise. Bernstein also tried to apply his membrane theory to the problems of excretion and secretion and suggested that the membrane potential drives water by electroosmosis across membranes, a concept which he called the 'elektroosmotische Membrantheorie'. His ideas on the role of electroosmosis in the transport of water across living membranes were the origin of many later bio-electrical investigations.

6.2. ALBRECHT BETHE

The work of Bernstein [30–32] encouraged a slowly increasing number of biologists to undertake a variety of investigations with the aim to elucidate physicochemical membrane effects likely to cast light on the molecular mechanisms of the bioelectric phenomena. The gradual progress of this work was reviewed repeatedly by Höber (Figure 6) [20, 103] who especially in later years, made with his collaborators also significant contributions to the physical chemistry of membranes.

A big move forward was made in 1914 and 1915 by Bethe (Figure 13) and Toropoff in two papers 'On Electrolytic Processes at Diaphragms' [25, 26]. The proximate object of these investigations was to gain a clear insight into the mechanism of the polarization effects which arise at porous membranes on the passage of current. Their ultimate aim, like that of Nernst's [65, 66] earlier work on interfacial polarization in general, was to provide a sound understanding of the physicochemical basis of nerve excitation, a phenomenon of the greatest bioelectric interest.

The papers of Bethe and Toropoff contain a great deal more than their title promises. We shall emphasize here the most basic aspects of their work which are of paramount significance in the electrochemistry of porous membranes. In reviewing these ideas we shall closely follow the trend of thought of Bethe and Toropoff. This will give the reader some glimpses both at the rather rudimentary state of the electrochemistry of porous membranes 60 years ago, and at the birth of the concepts which form the basis of all later developments in this field. In the author's opinion the papers of Bethe and Toropoff are one of the most outstanding intellectual contributions to the development of the fundamental electrochemistry of membranes.

Bethe and Toropoff describe the observational basis from which they started in these words (Author's translation):

It has been observed repeatedly [104] that at the boundary electrolytic solution-diaphragm changes in the concentration of the solution may occur when an outside E.M.F. is applied.*
   A few years ago, one of us [106] was able to add a few new observations to the known effects.
   1. In neutral solutions on the one (plus) side [of the diaphragm] alkali is formed, on the other (minus) side acid.
   2. The nature of the electrolyte in solution exerts a substantial influence on the speed with which this disturbance of neutrality arises.
   3. If in neutral or alkaline solution the one ion, say the anion of a salt shows an increase in concentration at the cathodic side of the diaphragm in acid solution, this increase may switch over to the anodic side. Not only the ions of the dissolved electrolyte take part in these concentration changes

---

* This effect, unbeknown to Bethe and Toropoff, had been described already much earlier by Du Bois-Reymond [105].

but to a prominent extent also the ions of the water. Also, the H·- and OH′-ions of the original solution exert a decisive influence on the degree and on the direction of the concentration changes of the other ions.

4. The water movement (electroendosmosis) occurs in the same direction as the concentration changes. When the increase in concentration occurs on the minus side of the diaphragm, the stream of water moves in the direction of the positive current; when the increase in electrolyte concentration occurs at the plus side, the water moves in the direction of the negative current.

Bethe and Toropoff pointed out that there existed no clear concept of the molecular mechanism of the polarization effects at porous membranes and diaphragms. They were thoroughly familiar with the before mentioned studies of Nernst [65, 66] and Nernst and Riesenfeld [2–4] on polarization effects at the boundaries between two electrolytically conducting phases in general which were explained by these investigators as the result of the differences in the ratios of the mobilities of the anions and the cations in the two liquid media. Bethe and Toropoff also refer to the electromotive action of glass described by Cremer [107] and to the work of Haber and collaborators [5, 6] on the electromotive action of solid salt phases in general, the glass electrode and liquid membranes of similar electromotive behavior.

Bethe and Toropoff stressed that any explanation of the polarization effect at porous membranes which would be based on the conception of the membrane as a second liquid phase or as a solid salt phase would make it necessary to assume an additional independent mechanism for the explanation of the concomitant electroosmosis.

It is interesting to note that 60 years ago Bethe and Toropoff still found it necessary to prove the porous nature of their collodion, parchment paper, and gelatin membranes, and of pigs bladders and similar membranes of animal origin. They emphasized that it seems impossible to consider these membranes as solvent membranes because the 'apparent' ionic mobilities of ions in these membranes are not constant and independent of the other species of ions present, as they are in solution, but that these 'apparent' mobilities are interdependent and dependent on the absolute concentration of the solutions used. Bethe and Toropoff also pointed out that polarization effects identical with those observed with their membranes are likewise observed with manifestly porous materials, such as plates of porous clay or carbon. They further referred to the studies of Perrin [108] on electroosmosis and the observations by Girard [109] and others [110] on the changes of the E.M.F. of membrane concentration cells in response to changes in hydrogen ion concentration.

Having established the microheterogeneous, porous structure of their membranes, Bethe and Toropoff concluded that the explanation of the polarization effects must be looked for in the field of 'capillary electricity', specifically in the Quincke-Helmholtz concept of the electrical double layer that was reviewed above (see also Figure 9). Referring to this concept, Bethe and Toropoff say:

...the transportation of current through the diaphragm can be effected only by those ions which are freely movable in the interior of the pore. ... Consequently the within the diaphragm freely movable species of ions have relatively enhanced mobilities, those affixed to the pore walls have no mobility or greatly reduced relative mobilities; and this in turn must lead to the concentration changes in different directions at the two surfaces of the diaphragm. (Free translation by the author)

At this point Bethe and Toropoff discussed all too briefly the involved role of the transportation of water of hydration in electroosmosis.

Thereafter, they took up a heretofore rather neglected topic of paramount importance in the electrochemistry of membranes, namely, the origin of the immovable wall charges in the pores and the mechanism of their fixation to the pore walls. Concerning these points there existed various more or less *ad hoc* suggestions for specific cases but nobody seemingly had looked at these questions in a comprehensive manner as did Bethe and Toropoff.

Referring to the work on colloids by Freundlich [111] and Pauli's [112] studies on proteins, Bethe and Toropoff state:

The charge of the immobilized layer of liquid at the pore walls may be visualized either to be due to adsorption or in suitable cases, for instance with protein, gelatin, etc., as due to dissociation of the material of which the pore walls consists. For the latter possibility the nature of the proteins as amphoteric electrolytes yields a particularly clear concept, whereas with other materials, for instance collodion, one will have to rely on the still rather unclear idea of adsorption. In principle the endresult in both instances is the same.... In both instances a prominent role in the process of charging the wall must be attributed to the H·- and OH′-ions. If the liquid layer wetting the pore walls is positively charged one must assume a preponderance of cations in this immobile or less mobile layer of liquid, and in the interior of the pores a preponderance of anions. The converse holds true in the case of a negatively charged wall layer. (Author's translation)

In a footnote to this statement Bethe and Toropoff make the following statement:

A corresponding view, in our opinion will also hold true with respect to the changes in the potentials of concentration cells with interposed diaphragms.

In this manner, and through the use of some ancillary arguments, Bethe and Toropoff organized in a most admirable manner information from several apparently disparate fields into a coherent concept of the molecular mechanism of the movement of ions through the water-filled pores of structural membranes. These ideas have stood the test of time exceedingly well; they are the ultimate basis of all current work in this field. For the further development of the fundamental electrochemistry of porous membranes it was, however, most unfortunate that Bethe and Toropoff presented their ideas only in a sketchy form, as a six page introduction to their 97 page report on membrane polarization [25, 26], and that they never returned to these most basic aspects of their work.

Contemporary and many later investigators, including the author failed fully to understand and to utilize systematically the integrating conceptual scheme of Bethe and Toropoff. One might conclude that some of the ideas of Bethe and Toropoff, without further elaboration, were too far ahead of the times to be fully appreciated. As indicated before, even Bethe, *et al.* [82] ten years later made virtually no attempt to correlate their findings on the electrolyte permeability of porous membranes and the ideas of Bethe and Toropoff. The few reviewers [19, 20] who presented the work of Bethe and Toropoff paid no attention to the outlined most basic part of their work, but presented in detail the formal algebraic schemes which these investigators developed to describe the various polarization effects, mainly the 'disturbance of neutrality'.

Space alone would not permit a detailed presentation of the various concentration changes and concomitant electroosmotic effects which are expounded in the Bethe-Toropoff papers. Moreover such a presentation seems today also unnecessary, since the 'Bethe-Toropoff effect' is now widely known and has been treated extensively in numerous readily available reviews and books [113–116]. Thus, it will suffice here to present mainly the basic argument as formulated by Bethe and Toropoff (in a free translation by the author).

An electrolyte with the anion A and the cation Q is dissolved in water and this solution is in a state of distribution equilibrium with the diaphragm. Assume that one of these ions or an ion of water or several species of ions are immobilized in the liquid layer which wets the pore wall. In this manner the relative mobilities of the ions remaining in the interior of the pore are increased. We designate the product of the concentration of each ion and its relative mobility divided by the sum of the products for all ions in the same cross section of the free liquid by the letters m, n, a, and b, and in the diaphragm, by m_1, n_1, a_1, and b_1. If we send one faraday of electricity through the liquid and the diaphragm, the four species of ions will participate in the transportation of electricity according to the scheme:

Free Liquid	Diaphragm
$mQ^{\cdot}$	$m_1Q^{\cdot}$
$nH^{\cdot}$	$n_1H^{\cdot}$
aA'	a_1A'
bOH'	b_1OH'

Since the same quantity of electricity must pass through each cross section the following condition obtains:

$$m+n+a+b=m_1+n_1+a_1+b_1=1.\tag{5}$$

Bethe and Toropoff showed that at the boundary liquid-diaphragm at the side of the positive pole of their cell the following movements of ions occur:

	Free liquid	Diaphragm	
+ Pole	$+mQ^{\cdot}\rightarrow$ $+nH^{\cdot}\rightarrow$ $-aA'\leftarrow$ $-bOH'\leftarrow$	$\rightarrow-m_1Q^{\cdot}$ $\rightarrow-n_1H^{\cdot}$ $\leftarrow+a_1A'$ $\leftarrow+b_1OH'$	− Pole

By presenting this scheme in an algebraic form Bethe and Toropoff arrive at a set of equations which prove, what is intuitively obvious: if the expression $(n-n_1)-(b_1-b)$ is positive an increase of the $H^{\cdot}$ ion concentration occurs at the plus side of the diaphragm; if this expression is negative, a decrease is observed. A concentration change in the opposite direction occurs at the minus side of the diaphragm.

The quantitative proof of these ideas is experimentally difficult since the ratios of the transference numbers of the various ions at the two membrane solution boundaries change as the experiment progresses. This matter, however, as many other aspects of the Bethe and Toropoff effect, for the lack of space, can not be discussed further.

From a broader point of view one of the most important results of the remarkable papers of Bethe and Toropoff was the demonstration that the electrochemistry of porous membranes was ready for a systematic exploration on the basis of a set of clear and well-founded concepts.

6.3. F. E. BARTELL, HERBERT FREUNDLICH, AND JACQUES LOEB

As first described by Thomas Graham [53] in 1854 the osmotic flow across a membrane that is caused by an electrolyte solution is in some instances much greater than that due to nonelectrolyte solutions of the same concentration, in other instances 'negative osmosis' occurs, the volume of the more concentrated solution decreases. These effects attracted the attention of an increasing number of investigators. Biologists looked at anomalous, especially negative osmosis as a potential clue to the mechanism of secretory and excretory processes of cells and tissues. Physical chemists tried to understand the mechanism of the anomalous osmotic solvent movements which were contrary to any common sense expectation based on the kinetic theory of solutions. The literature on anomalous osmosis was reviewed by Freundlich [19], by Höber [20] and more extensively by the author [117].

We shall refer here only to the work of three investigators who published between 1914 and 1923, Bartell [97, 98], Freundlich [99], and Loeb [21, 70, 100, 101]. Their investigations are important not so much as constructive contributions to the basic electrochemistry of membranes, but because they were instrumental in drawing the attention of a wider audience to electrochemical membrane problems.

F. E. Bartell [97, 98] was the first to study anomalous osmosis in a systematic manner from a physicochemical point of view. In a long series of careful investigations he established the empirical rules which describe whether the anomalous osmosis is positive or negative. If the electrokinetic charge of the membrane is negative, and the water in the pores therefore positive, the anomalous osmosis will be positive when the electrolyte diffusing across the membrane impresses a positive charge on the more dilute solution. When the diffusing electrolyte impresses a negative charge on the more dilute solution, negative anomalous osmosis occurs. The converse holds true if the charge of the membrane is positive. Bartell's rules were confirmed by all later investigators.

Bartell also studied the influence of the porosity of the membranes on the membrane potential and thereby, with certain electrolytes, on the direction of the anomalous osmotic water flow; but space does not permit to present this matter at length nor how the concept of charged membranes evolved in his writings. In a review paper of 1923 [98] he presented some thoughts on the ionic mechanism which causes anomalous osmosis which were criticized by the author [117] and today are of little interest. From a broader point of view Bartell's papers are important as another pioneering effort in the physical chemistry of porous membranes and an irrefutable proof of the charged membrane concept.

Freundlich [99] in 1916 pointed out that a membrane potential by itself can not cause an electroosmotic water movement, as assumed by Bernstein and essentially also by Bartell, but that any electroosmotic flow must be caused by a current. He took the

Bethe-Toropoff concept of a charged membrane for granted and speculated that a current caused by the membrane potential flows through the pore in the one direction and in the other direction through the porewalls, the material of which the membrane consists. This conception of the mechanism of anomalous osmosis too, as pointed out later by the author [117], was flawed by some erroneous assumptions, foremost the notion that the wall materials, e.g. clay or collodion, have a significant conductivity, whereas in fact they are insulators. However this may be, Freundlich's paper and his ensueing treatment of membrane problems in his book [19] were instrumental in drawing the attention of physical chemists to the electrochemistry of porous membranes.

Most widely known, however, were probably the studies by the biologist Jacques Loeb (Figure 12) [21, 70, 100, 101] which he undertook in the course of his classical investigations on proteins and the colloidal state [101]. Most of Loeb's experimental membrane studies were concerned with the 'electrification of water' in the pores of collodion and 'proteinized' collodion membranes and a detailed study of anomalous osmosis.

Loeb, following the example of others, prepared the proteinized membranes by the adsorption of various proteins, such as gelatin, egg albumin, casein and oxyhemoglobin, etc., from aqueous solutions on collodion membranes of high porosity. As he had anticipated, the isoelectric points of the proteinized membranes, determined by the absence of electroosmotic water movement, coincided closely with those of the various proteins used. The membranes were electronegative in neutral solutions (except the oxyhemoglobin membranes which are nearly uncharged at pH 7), and electropositive only in solution on the acid side of the respective isoelectric points, in agreement with the ideas of Bethe and Toropoff on the origin of the charge of protein membranes.

Loeb confirmed Bartell's [97, 98] earlier qualitative observations on the close correlation of the direction of the anomalous osmotic water flow, and the sign of the charge of the membrane and the dynamic membrane potential which arises across it in a given system. In Loeb's experiments, as he stressed, the rate of anomalous osmotic water movement was proportional to the product of the electrokinetic charge of a membrane and the potential which arises across it in any given system. Loeb, however, did not present any theoretical concepts concerning the molecular mechanism of anomalous osmosis or any systematic views concerning the basic aspects of the physical chemistry of porous membranes. Nevertheless, his widely reported [19, 20] work on anomalous osmosis must be credited with promoting the general acceptance of the notion of a close interrelationship of the electrokinetic and electromotive properties of porous membranes.

It is interesting to note that Bernstein, Bethe, Bartell, Freundlich, and Loeb, as well as various other, mostly biological investigators were not primarily interested in an understanding of the electrochemistry of porous membranes from a fundamental point of view, but focused their efforts more narrowly on those properties and functions of such membranes which appeared promising in solving specific more limited problems.

6.4. LEONOR MICHAELIS

The long overdue systematic study of the electrochemistry of porous membranes was undertaken by Leonor Michaelis and collaborators in the mid nineteen twenties. Michaelis was interested primarily in the most conspicuous functional property of these membranes, their electromotive action in various cells, and the ionic mechanisms which cause these effects. Michaelis was well prepared for this task. In the second edition of his classical book on hydrogen ion concentration [118] published in 1922 he had critically reviewed, in a still most noteworthy presentation, such topics as phase boundary potentials and the electromotive properties of liquid membranes, and membrane potentials in general.

Michaelis' approach to the electrochemistry of porous membranes can hardly be better presented than by the subsequent verbatim quotations from a review-like paper [11] which he published in 1925 when many of his detailed studies were still unpublished.

The starting point of Michaelis' investigation were the observations of Loeb and Beutner [119] on the electromotive properties of the apple skin:

The potential difference of the apple skin against an electrolyte solution depends on the nature and the concentration of the dissolved electrolytes. The anions have no influence at all, the cations exert an effect which depends on their valency and on their chemical nature. When varying the concentrations of a given cation, the P. D. changes by 57 millivolts for each power of ten of the concentration if the cation is univalent.

In their attempts to gain some insight into underlying mechanism,

Michaelis and Fujita showed by direct diffusion experiments that the apple skin is absolutely impermeable for the electrolytes of the apple juice when the diffusion takes place against pure water. On the other hand an exchange of the cations occurs when the diffusion takes place against an electrolyte solution, e.g. increasing amounts of $K^.$ can be detected, from day to day, in a NaCl solution in which the apple is dipped. Anions are never exchanged.

In searching for an artificial membrane which has the same functional properties as the apple skin, Michaelis and Fujita found that parchment paper is only qualitatively similar to the apple skin.

When two solutions of any electrolyte in different concentrations are separated from each other by a membrane of parchment paper, the P. D. of the solutions is different from that obtained in free contact of the solutions, and that always in such a direction as though the mobility of the anion were relatively diminished in comparison to the mobility of the cation, under the influence of the membrane.

A further search led to the 'dried' collodion membrane, which is an excellent model for the apple skin and soon became the object with which Michaelis and collaborators carried out the bulk of their investigations [11–15]. Michaelis [11] summarized the main functional characteristics of these now famous 'dried' collodion membranes as follows:

1. When two electrolyte solutions are separated from each other by a membrane of collodion which has previously been thoroughly dried, the P. D. of the solutions differs from that obtained without a membrane in a much higher degree than in the case of parchment paper. The P.D. depends only on the nature and the concentration of the cations. The anions are of no importance at all, and this is approximately true even for OH' ions.
    2. Two solutions of one electrolyte in different concentrations being given, the potential difference

for a proportion of the concentrations of 1:10 is approximately 57 millivolts for any univalent cation, that is the thermodynamically possible maximum value for a chain reversible for cations. This maximum effect is reached practically completely for the HCl concentration chain, not quite so completely for chains with neutral salts. For polyvalent cations the P.D. is smaller and inconstant, especially for bivalent ions, while trivalent ions give a more reproducible P.D. being about 1/3 of the value for univalent cations.

3. In a chain consisting of two solutions of different electrolytes in equal concentration with a common cation but with different anions, the P.D. is always practically zero, even if the anions are of different valency.

4. Two solutions of different electrolytes in equal concentrations with any anion and with two different cations being given, the potential difference depends not only on the valency but also, in high degree, on the chemical nature of the cation.

5. An HCl solution or a solution of any neutral salt does not diffuse across such a collodion membrane against pure water at all.

6. When the solutions of two different electrolytes of equal concentration are separated by a collodion membrane, only the cations are exchanged by diffusion, the anions are not.

7. In opposition to that, NaOH diffuses easily against pure water, but it does not diffuse against solutions of neutral salts...

Having thus summarized the main functional properties of his dried collodion membranes, Michaelis proceeds to outline his ideas concerning the mechanism which causes the characteristic behavior of these membranes and of porous membranes in general in electrolytic systems.

1. As an attempt to interpret all of these facts the hypothesis may be made that the mobility of the anions in the capillary canals within the membrane is much more inhibited than that of the cations. In parchment paper the mobility of the anions is diminished in comparison with that of the cations; in the apple skin and in collodion it is practically annulled.

This, in connection with some auxiliary hypotheses, is the foundation of the following theoretical discussion.

2. Two circumstances may be recognized to be effective for the alteration of the ionic mobility in capillary spaces. The first is the hydration of the ions. Any ion drags a water envelope when moving through the water. The attraction of the ions for the water molecules may be interpreted as an electrostatic attraction of the ion to the electric dipole represented by the water molecule. The dipoles are oriented in a wide extent round the ion as to the direction of their axis and constrained in their local distribution; this effect decreases with the distance from the ion. When an ion moves in a capillary canal, the frictional resistance is increased, if the sphere of attraction to the water molecules extends to those water molecules which by adhesion are fixed to the wall of the canal.

The second matter is the fact that the walls of the canal are generally covered with an electric double layer. The retardation of any anion in parchment paper and collodion may be supposed to be the consequence of the negative charge of these membranes in aqueous solutions. Amphoteric membranes the charge of which can be changed by varied hydrogen ion concentration may show a quite different behavior. In all events this effect of the capillaries, the conditions being given, is always directed either only to the anions or only to the cations.

The mechanism of the retarding effect of the electric charge of the wall may be represented by a scheme already suggested by Bethe and Toropoff [25, 26] in order to explain the deviations from the neutral reaction on either side of a membrane in an originally neutral solution when an electric current flows across the membrane. If the wall is negative this charge can only be brought about by adsorbing negative ions. These ions appertain to the electrolyte solution or are the OH′ ions of the water. Therefore, there must be a lack of anions in the movable liquid within the canal. The mobility of the anions observed in the experiments is an average value of the different mobilities of the fixed and the free anions. Even if the mobility of the *free* anions were wholly unchanged, yet the average value of the anions would be diminished. Furthermore, if the canal is narrow enough, it may happen that all of the anions are fixed by the wall and only cations can move. Let us admit this to be true for the membrane of previously dried collodion, which is practically impermeable for any anion. This membrane is an excellent model for the apple skin...

With respect to the origin of the charge of the pore walls, Michaelis expressed then and later [15] a widely held erroneous belief that the charge of collodion must be due to preferential adsorption of anions. This assumption obviously was based on the uncritical acceptance of the textbook formulas of cellulose and cellulose derivatives which presented these materials as free of ionizable groups inspite of the fact that these substances carry inherently a small number of carboxyl groups, as was well known in the technological literature [120, 121]. This error, however, was fortunately not a significant impediment in Michaelis' main theoretical considerations. It also did not prevent him from remarking very astutely:

The material of the membrane being given, the relative retardation of the anion, is a scale for the size of the pores. The smaller the pores the greater is the retardation of the anions. For the retardation of the anions is an average value, and the smaller the size of the pores the greater is the surface of the walls of the pores and the amount of anions fixed by the wall. Membranes of different material cannot be compared by this method, as the amount of the fixed anions depends on the specific nature of the membrane itself.

Michaelis stressed the 'diffuse' structure of the electrical double layer and emphasized that one must distinguish sharply between the charge density at interfaces, e.g. the interface, between a solid and a liquid, and the magnitude of the electrokinetic potential of the solid as determined in electrokinetic, for instance, in electroosmotic experiments. For the electromotive properties of porous membranes, as Michaelis stated, the charge density at the pore walls is the determinant factor.

Michaelis introduced in the following words his basic concept of the ion permeability of porous membranes:

Let the mobility of the cation and of the anion of an electrolyte in a free aqueous solution be U and V and let these mobilities be within the membrane u and v on the average. The further development of the theory is different for 'permeable' membranes where u and v have finite values and for 'semipermeable' membranes where one of these values vanishes. A transition is represented by cases where the permeability for one kind of ion is so strongly decreased that the membrane appears as a semipermeable one in experiments of short duration, but as a permeable one in experiments of a long duration. ... In practice no difficulty has occurred hitherto. The membranes of our experiments were either permeable for any ion in a measurable degree or they were practically impermeable for any anion. But transition cases may also be met.

Michaelis' experimentally supported conclusion that in semipermeable membranes the mobility of the one or the other ion is vanishing, agrees with the speculative conception concerning the origin of the electromotive actions of semipermeable membranes which Ostwald [1] had enounced 35 years earlier. Michaelis, however, as we have seen, had now a well defined molecular picture in mind. He stated specifically that the movement of ions across the membrane occurs throughout in an aqueous medium without the interference of transverse phase boundaries and pointed out that with 'permeable' membranes the influence of unstirred diffusion layers must be considered.

Figures 15 and 16 are schematic representations of Michaelis' conception of charged pores. Though drawn by the author, these figures don't contain anything that was not clearly stated by Michaelis in 1925 [11].

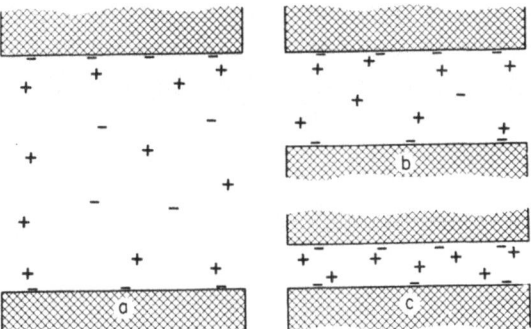

Fig. 15.   Michaelis' conception of the distribution of ions in three pores of different diameter at the same concentration of the outside solution.

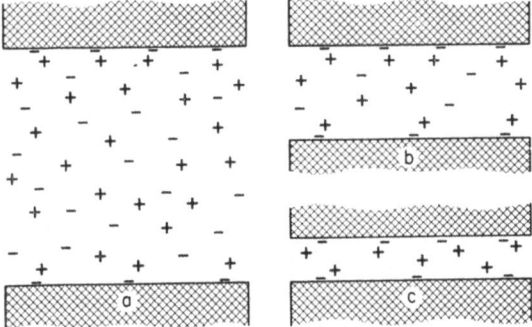

Fig. 16.   Michaelis' conception of the distribution of ions in the same three pores as shown in Figure 15, but at a higher outside concentration.

In applying these ideas in formal electrochemical terms to the electromotive force arising in concentration cells, Michaelis comes to the same conclusion as did Bernstein [30–32], and before him Nernst and Riesenfeld [2, 3, 4] in the case of solvent membranes, namely, that the cross-membrane diffusion potential, E_{mem}, is the liquid junction potential, E_{junc}, arising within the membrane.

There remained, however, in Michaelis' presentation a point of unclarity, more conceptual than merely semantic, of the *average mobilities*, u and v, of the cations and anions in the membrane and their correlation to membrane potential.

According to Nernst, the cause of the liquid junction potential in free solution is the difference in the *diffusion velocities* of the anions and the cations; specifically, as is shown in Nernst's Equation (1), E_{junc} is determined by a function in which u and v appear in both in the enumerator and the divisor. Nernst [59], in his paper on the electromotive action of the ions in solution had pointed out the numerical proportionality of the diffusion velocity of an ion and its electrolytic mobility (with bi- and higher valent ions their valencies must be taken into account). In developing Equation (1), he therefore used the then already well-known ionic mobilities – a custom which has been followed ever since in textbooks. Nernst plucked the known absolute ionic mobilities into Equation (1) to obtain the calculated values of the liquid junction potentials which he could then compare with experimental data.

The absolute ionic mobilities of ions in solution may be used to calculate liquid junction potentials. Liquid junction potentials, however, as stated before, cannot be used to calculate absolute ionic mobilities or diffusion velocities. Some meaningful information, however, may be obtained from E_{junc} data if we rewrite Equation (1) in terms of the transference number, t^+ and t^-, of the two (univalent) ions in solution. Substituting in Equation (1), t^+ for $u/(u+v)$ and t^- for $v/(u+v)$ we obtain:

$$t^- = 0.5 - \frac{E_{junc}}{\frac{2RT}{F} \ln \frac{c_1}{c_2}} \tag{6}$$

and

$$t^+ = 0.5 + \frac{E_{junc}}{\frac{2RT}{F} \ln \frac{c_1}{c_2}}, \tag{7}$$

where, to use Michaelis' terminology c_1 and c_2 are the 'adjusted thermo-dynamic concentrations' (activities) of the uni-univalent electrolyte in the two solutions. Thus in the case of solutions, the transformed Nernst diffusion equation permits by the use of the rearranged expression 6 and 7 to compute from E_{junc} data the transference numbers t^- and t^+ and thereby the *ratio* of the intrinsic individual *diffusion velocities* of the cations and anions in solution. This ratio, according to Nernst's [59] basic concept, is identical with the ratio of their ionic mobilities, u and v, which are readily determined by conventional electrochemical methods.

With porous membranes, contrary to the case of solutions, we have no way to determine ionic mobilities in an independent manner. Nevertheless Equations (6) and (7) when applied to membranes, tell us the ratio of the rates at which anions and cations, if not tied together by the law of macroscopic electroneutrality, would move independently across the membrane.

Michaelis, showed that these (virtual) transport numbers computed by Equation (6) or (7) from E_{cell} measurements with concentration cells, can be compared to transference numbers obtained in electrical transfer experiments under comparable conditions of concentration. For this purpose a direct current is passed through the membrane cell for some time, and the quantities of cations and anions which were moved by the current in opposite direction across the membrane are determined by chemical analytical methods. These quantities, expressed in equivalents when divided by their sum, are the experimental transference numbers of the cations and anions in the membrane.

We turn now to the experimental study of concentration cells with 'permeable' membranes. Membranes such as parchment paper, showed that the potentials obtained with each membrane were reproducible; however, different, nominally identical membranes, such as parchment paper extraction thimbles of the same batch, yielded substantially different results. Such 'permeable' membranes turned out not to be desirable objects for the type of fundamental study that Michaelis had in mind. Therefore he and his collaborators soon focused their attention on the very dense 'dried' collodion membranes which behave electromotively like the apple skin.

Dried membranes prepared from one brand of collodion, 'Celloidin' (Schering), gave consistently the highest concentration potentials, in many instances approaching the thermodynamically possible maximum described by Equations (2) and (4). Another preparation, gun cotton, 'generally has given the poorest potential differences'. Michaelis did not look into the cause of this difference between nominally identical products. The resistances of Michaelis' dried membranes were extremely high, even with thin membranes of the order of $10^5 \ \Omega \ cm^2$. The establishment of final stable potentials took ordinarily some days; thereafter they were constant over many months and reproducible within a millivolt.

Some illustrative data on concentration cells are given in Tables I and II which are reproduced here as published by Michaelis. Table I also contains the t^- values calculated from the potential data using Equation (6). Michaelis' potential data were seemingly not corrected for the asymmetry of the liquid junction potentials at the ends of the KCl bridges connecting the solutions to calomel electrodes.

As could be anticipated on the basis of Michaelis' general views, see also Figures

TABLE I

Concentration potentials ($c_1:c_2 = 2:1$) with solutions of several chlorides across collodion bag membranes and the corresponding transference numbers of the Cl$^-$ ion, t^-, at 20.5°C, after Michaelis [14, 15][a]

Concentr. of Solutions c_1/c_2 equiv/liter	LiCl mv.	t^-	NaCl mv.	t^-	KCl mv.	t^-	HCl mv.	t^-
0.02/0.01	15.1	.066	15.5	0.55	16.05	.026	17.2	.006
0.04/0.02	13.4	.115	14.8	.075	15.4	.058	17.1	.009
0.08/0.04	11.4	.173	13.5	.112	13.6	.109	15.65	.050
0.16/0.08	8.3	.262	10.8	.198	10.8	.190	15.5	.055
0.32/0.16	5.2	.351	6.2	.322	7.3	.290	–	–

[a] Theoretically possible maximum value of the concentration potential 17.4 mv.

TABLE II
After Michaelis [11]

Concentration of HCl.	Activity	P.D. Calculated	P.D. Observed
0.1:0.01	0.0814:0.00924	54,0	54,8
0.0:0.001	0.0814:0.000984	106	108

(15) and (16), the potentials in Table I are higher at lower solution concentrations. They approach the theoretical maximum at the lowest concentration fairly closely, and the corresponding values of t^- approach zero within a few percent. The data in Table II indicate t^- values of zero. The membranes in these latter cells act as 'semi-permeable membranes' or 'membranes of ideal ionic selectivity' as they are customarily called today. In this case the nonpermeating ion can be said to play the role of Donnan's nondiffusible 'colloidal' ion, and Equations (6) and (7) reduce to the expression for the Donnan potential, Equation (4) which is identical with Nernst's equation for the electrode potential, Equation (2).

Michaelis *et al.* compared computed transference numbers, like those shown in Table II, with transference data obtained in electrical transference experiments. The agreement between the two sets of data was reasonably good, as good as was to be expected in view of the inherent experimental difficulties, mainly polarization at the two membrane solution interfaces. Thus Michaelis' basic conception of the nature of porous membranes seemed validated by the experimental work of Michaelis and collaborators on concentration cells.

Aside from the concentration cells Michaelis and collaborators also studied 'Chemical Chains', cells in which a membrane of high ionic selectivity, like his best dried collodion membranes, separate two solutions of different electrolytes at the same concentration. In such cells potential differences arise which greatly differ from the liquid junction potentials observed in the absence of a membrane. The nature of the anions is of small importance, likewise, the concentration of the solutions, provided it is not too high ($\leq 0.1\ M$). The 'chemical potentials' arising in such cells vary widely with different membranes; as a rule they are higher with membranes giving the highest concentration potentials. They are very stable and reproducible with any given membrane. Column 3 of Table III shows the potentials obtained with one of Michaelis'

TABLE III

'Chemical potentials' in cells of the type 0.1 $\underline{M}$ A$^+$Cl$^-$ | dried collodion membrane | 0.1 B$^+$Cl$^-$
$T = 20.00\ °C$. After Michaelis [14, 15]

1 Soln. 1	2 Soln. 2	3 mv[a]	4 $\dfrac{u_2}{u_1}$	5 Relative rate of ionic 'mobility' of B$^+$, u_2, in free solution; $u_{K^+} = 1$
KCl	HCl	−93	42.5	4.9
KCl	RbCl	−8	2.8	1.04
KCl	NH$_4$Cl	−6	2.2	1.00
KCl	KCl	0	–	1.00
KCl	NaCl	+48	0.14	0.65
KCl	LiCl	+74	0.048	0.52
RbCl	HCl	−87 (85)	33	
NaCl	HCl	−140 (141)	288	
LiCl	HCl	−165 (167)	794	

[a] The sign refers to Solution 2

'best' membranes. With other membranes these 'chemical potentials' were lower often by 50% and more. In all instances "we can arrange the cations in a series in which each cation is positive with respect to any preceding cation but negative to any following cation", [11] as was done for the first six cells in Table III.

Michaelis interpreted these 'chemical potentials' as being due to differences in the mobilities of the different cations within the membrane. He applied the formula for the liquid junction potential arising between the solutions of equal concentration of two uni-univalent electrolytes having a common anion and different cations,

$$E = \frac{RT}{F} \ln \frac{u_1 + v}{u_2 + v}, \tag{8}$$

where u_1 and u_2 are the mobilities of the two cations and v the mobility of the anion in the membrane. Since the mobility of the anion in the membrane is very small, Michaelis neglected v as a rough approximation and computed the relative mobilities of the cations in the membrane as shown in Column 4 of Table III, using with the first six cells K^+ as the reference ions. Column 5 gives the ratios of the mobilities in free solution of the several cations in Solution 2 relative to the mobility of K^+. The 'chemical potentials' of the last three cells of Table III indicate that these potentials are additive, as is evident from the figures in parenthesis which were computed from the data in the upper part of this table.

Michaelis warned aginst overrating the validity of Equation (8), He stated that fewer channels are available to ions carrying large water shells, but omitted any systematic attempt to explain in detail the mechanism of the origin of his 'chemical potentials'.

Michaelis' concept of the mobilities u_1 and u_2 in the membranes of 'chemical' cells is rather unclear, still more unclear than the 'average mobilities' in the case of the concentration cells. In the case of chemical cells too, one is obviously dealing with magnitudes which might best be described as virtual transport numbers of the two cations in the membrane, τ_1^+, and τ_2^+. This approach was taken many years later by the author [122] who based his considerations concerning the 'chemical potential', renamed 'bi-ionic potential', on the Teorell [16], and Meyer and Sievers [17, 18] fixed charge theory of porous membranes. This matter, however, is outside the time-frame of this review and is mentioned here only to introduce the following recollection. In 1946 when the author's ideas on the origin of the bi-ionic potential had crystallized, he tried during a walk to outline them to Prof. Michaelis. After briefly sketching the Teorell and Meyer-Sievers conception of porous membranes as ion exchange bodies, the author started summarizing the known facts concerning the adsorption of ions by ion exchange bodies. Before he had ended, Prof. Michaelis stopped and said "Of course, you are right!" Immediately, without any hesitation, he started a most interesting and productive discussion, which demonstrated an incredible agility of mind and depth of knowledge that had been a hallmark of Michaelis' work in many different areas during a period of more than half a century. With this remark we must close this review of the most important aspects of Michaelis' work on membranes, many sides of which had to be omitted here.

7. The State of the Electrochemistry of Porous Membranes in the Later Nineteen Twenties and Early Thirties

The author, as mentioned before, had entered the field of membranes in 1928 from the side of conventional electrochemistry through work on the mechanism of the electrocapillary Becquerel effect [64]. Having to my surprise solved this problem within a few weeks, I felt inclined to clarify the mechanism of another of the several peculiar and for my feeling unexplained membrane effects which were described in Freundlich's book [19], namely electrostenolysis, that is the formation of metallic deposits at membranes and diaphragms on the passage of direct current. To prepare myself for this task I had to read widely on electrochemistry in general, to search the literature for unconventional electrochemical effects, and to learn as much as possible about the physical chemistry, especially the electrochemistry of membranes, primarily of porous membranes.

From the preceding sections it is clear that physicochemical membrane research in the late nineteen twenties was far from being the organized field of systematic studies which it is today. The electrochemistry of liquid membranes, that is the work of Nernst and collaborators [2–4], of Haber and collaborators [5, 6], and of Beutner [7] had been reviewed in 1922 in a masterly manner by Michaelis [118]. The electrochemistry of porous membranes, however, about which much less was known, had never been reviewed systematically. The nearest approach to this were Michaelis' reviews of his own work [11–14].

Freundlich in his book [19] did not try to present an integrated view on the electro-chemistry of porous membranes, but dealt systematically only with a few specific membrane topics, mainly the Donnan membrane equilibrium, the Haber glass electrode, and the Bethe and Toropoff [25, 26] disturbance of neutrality effect. In addition, in a chapter on electrokinetic phenomena, Freundlich discussed briefly several poorly understood effects arising with porous membranes: the electrocapillary Becquerel phenomenon, electrostenolysis, anomalous osmosis, and thermoosmosis which had attracted my interest from the moment I found them in Freundlich's book.

The very voluminous biological literature contained scattered pieces of basic information on the physical chemistry of membranes some of which Höber [20] had presented in his classical book on the *Physical Chemistry of Cells and Tissues*.

Overall, the book and review literature 45 years ago was only of limited help to a newcomer in membrane studies. I therefore started an extensive survey of the journal literature. After many months a fairly well organized picture arose in my mind, similar to that presented above.

My enthusiasm for Michaelis' work was not much dampened by the stepwise realization of some of its shortcomings: the lack of clarity of the concept of 'average' ionic mobilities and of the correlation of mobilities and diffusion velocities, and the limited and vacillating use of the concept of transference numbers. I also was much bothered by the assumption that whatever electromotive forces may arise at the two macroscopic membrane-solution interfaces are identical in magnitude. I realized that the measurable potentials ordinarily consist of three parts: two electromotive forces of opposite sign. at the two membrane-solution interfaces which ordinarily are not of equal magnitude and a liquid junction potential arising within the pores of the membrane. In 1930 [117], I presented this matter in a casual manner. The quantitative treatments of this problem which were published several years later by Teorell [16] and by Meyer and Sievers [17, 18] are already outside the scope of this paper.

Another point left painfully open by Michaelis was the question why the electromotive behavior of different collodion preparations should vary widely. The origin of the charge of collodion membranes remained for Michaelis a matter of doubt. Even in the last review of his work on dried collodion membranes, published in 1933, Michaelis [15] remarked: "It must be left to the further study by the experts in the manufacture of collodion to explore how this effect is influenced by the nature of the manufacturing process" (author's translation). Thus, Michaelis left entirely open the question how to prepare 'good', electromotively highly active membranes, a point of primary importance for the experimenter.

When I tried in 1929 to repeat Michaelis' work on dried collodion membranes and Loeb's experiments on anomalous osmosis with wide-pored membranes even my best dried membranes consistently yielded concentration potentials somewhat lower than those reported by Michaelis, and the anomalous osmotic effects obtained with wide-pored collodion membranes were less pronounced than those described by Loeb. My dried membranes had still higher resistances and much lower rates of exchange of cations across them than those of Michaelis. These results were rather bewildering and discouraged experimental work.

With respect to the preparation of preferentially anion permeable membranes Michaelis [15] remarked as late as 1933: "...strange as it may seem, there exists no substance suitable for the preparation of membranes which under all conditions carry in aqueous solutions a positive charge" (author's translation).

The very important problem of the preparation of selectively anion permeable membranes, as a matter of fact, had been taken up already in 1928 in Höber's laboratory by Mond and Hoffmann [123], with the intent to produce electropositive analogues of Michaelis' dried collodion membranes, Mond and Hoffmann clearly realized that membranes which are positive both in moderately basic as well as in acidic milieus should be obtainable by the incorporation of strongly absorbable basic materials into collodion membranes. They cast membranes from a collodion solution containing the basic dyestuff Rhodamine B. After thorough drying, these membranes were reasonably permeable to anions and fairly impermeable to cations, as was evident from the rates of exchange of anions and cations across them. The concentration potentials observed with these membranes in KNO_3 cells with the concentration ratios 1 $N \mid 0.1 \ N$, $0.1 \ N \mid 0.01 \ N$ and $0.01 \ N \mid 0.001 \ N$ were -40, -38, and -11 mV, the sign referring to the more dilute solution. The influence of the concentration on the concentration potentials, for unexplained reasons, was opposite to that observed by Michaelis.

Mond and Hoffmann [123] also reported the following 'chemical' (bi-ionic) poten-

tials obtained with 0.1 N solutions, the sign referring to the second solution: NaCl |
| NaSCN, $+60$ mV; NaCl | NaNO$_3$, $+51$ mV; NaCl | NaI, $+33$ mV; NaCl | NaBr,
$+20$ mV; NaCl | Na Acetate, -8 mV; NaCl | NaSO$_4$, -38 mV. Mond and Hoffmann pointed out that the order of the electromotive efficacies of the anions in these
cells is that of the Hofmeister series. They were, however, unable to explain why the
sequence of the 'mobilities' of the anions in the narrow pores of a membrane differs
from that in water.

When I prepared some dried Rhodamine B membranes, the concentration potentials were initially somewhat higher than those reported by Mond and Hoffmann,
and the effect of the absolute concentration on the magnitude of the concentration
potential was decidedly less pronounced. Stored in water, my membranes released
some of the dyestuff and the potentials became progressively lower over a period of a
few days. The resistance of these membranes was very high, and the rates of the cross-
membrane exchange of anions exceedingly low. Wide-pored Rhodamine B membranes
released the dyestuff readily and soon lost their positive charge.

Though Mond and Hoffmann had shown in principle a way to prepare highly
anion selective membranes, it was obvious that their method would require a vast
improvement before reasonably useful anion selective membranes could be produced.
After some experimentation I came to the conclusion that it would be extremely difficult to improve significantly the experimental work of Michaelis and collaborators,
or of Mond and Hoffmann and to expand it in a really productive manner.

Of the various other investigations on porous membranes none seemed to be of
particular importance for the electrochemistry of membranes. Even the systematic
large-scale investigation 'On Collodion Membranes', later renamed 'On Capillary
Systems', started in 1927 by Bjerrum and Manegold [124] and continued by Manegold with numerous collaborators [125], contained for several years nothing of basic
electrochemical interest.

Having acquired the outlined information on the electrochemistry of membranes
and a good knowledge of electrochemistry in general, I felt ready for an attempt to
elucidate the mechanism of 'electrostenolysis'. Prof. Freundlich warned me that this
may be too ambitious a plan, but I went ahead with my project and started a critical
review of the widely scattered literature [126].

The author's work on electrostenolysis as well as some of his subsequent membrane
studies will be reviewed here at a length that might appear disproportionate. However,
since the author around 1930 was one of the very few physical chemists who took an
active interest in membrane research, he feels that such a mode of presentation might
be the best way to transmit to the reader a meaningful picture of the general state of
innovative physicochemical membrane research which was possible at that time, of its
problems and its limitations.

Grotthuss [127] reported in 1819 that silver is deposited if a current is passed
through a cracked glass tube separating two silver nitrate solutions. In 1891 this effect
was designated by Braun [128] as 'Elektrostenolyse'. It occurs also on porous plates
and membranes. Similar 'electrostenolytic' depositions of heavy metals on various

heavy metal precipitation membranes and other membranes on the passage of current had been frequently described in the literature. These effects seemed somewhat outside the framework of conventional electrochemistry, and some biologists fancied that they may explain biological oxidation processes.

The mechanism of electrostenolysis had been discussed by Wilhelm Ostwald [1], the prominent electrophysicist F. Braun [128], A. Coehn [129], a well-known electrochemist, and others. Ostwald's ideas on this point were based on his concept of membrane semipermeability; and those of Coehn, which were favored by Freundlich [19], stemmed from electrokinetic considerations. When analyzed closely, neither of these ideas made any physicochemical sense as I showed in detail in my review of the literature [126]. The explanation suggested by Braun was on the right track, at least for some of the effects.

On repeating many of the experiments described in the literature, it became clear to me that the electrostenolytic effects fall into two distinct groups, those in which the deposition of metals appears at applied voltages below three volts and those which occur only at much higher voltages, 20 V or more, depending on the heavy metal in solution [126].

With the high voltage effects the sequence of events is as follows: under the influence of the high local potential gradients sparking, often visible sparking, occurs which leads to the formation of primary metallic nuclei which, once formed, grow readily on the further passage of current. At low applied voltages, the formation of metallic deposits is preceded by the primary formation of a precipitate in the form of membranes or at a membrane which is a metallic or semimetallic conductor. Such precipitates are formed on the interaction of solutions of many heavy metals with solutions of alkali sulfides, selenides, and, in some instances, hydroxides which slowly turn to oxides. Hydroxide precipitates may also be formed as the result of the Bethe-Toropoff disturbance of neutrality effect. On the passage of current, these metallically or semimetallically conducting precipitates act as middle electrodes for the deposition of metal [126].

The fact that the mechanisms of high voltage, true electrostenolysis, of the other low voltage electrostenolytic effects, as well as of the electrocapillary Becquerel phenomenon could be explained in a straightforward manner on the basis of well established electrochemical knowledge was of some general interest, mainly for two reasons. First, it demonstrated that the named effects, contrary to a diffuse but widespread belief, are not a group of somewhat mysterious special phenomena outside the framework of classical physical chemistry which could be explained only on the basis of some novel, fanciful *ad hoc* assumptions. Second, the clarification of the mechanisms of the named effects also destroyed the basis of some interesting but essentially baseless speculations concerning the role of membranes in biological oxidation and reduction processes.

With greatly strengthened self-confidence I was now ready to attack the next of the essentially unexplained membrane phenomena discussed in Freundlich's book [19], anomalous osmosis.

In the middle of September 1929 I told Prof. Freundlich of my plan to elucidate the mechanism of anomalous osmosis. He cautioned me that several competent investigators, such as Bartell [97, 98], Loeb [21, 100, 101], and he himself [99], had struggled rather recently with this problem in which he was still very interested. To his warning I replied that I had an intuitive feeling that the work of Michaelis, which postdated these prior investigations, must contain somehow the clue to the explanation of anomalous osmosis, and I delved full force into this problem which had now become a personal challenge.

The empirical rules concerning the direction and magnitude of anomalous osmosis stated by Bartell [97, 98] and by Loeb [27, 100, 101] were not in doubt and could be readily verified by simple experiments. The concentration dependence of the magnitude, and in certain instances of the sign, of the dynamic membrane potential arising at electrolyte permeable membranes on the diffusion of electrolytes across them, was readily understandable on the basis of the Bethe-Michaelis concept represented in Figures 15 and 16. The open question was the ionic mechanism which causes anomalous osmosis.

As stated before, Loeb did not go beyond stating his empirical results and did not propose any specific ionic mechanism of anomalous osmosis. Freundlich [99] had assumed local currents, flowing in one direction through the solution in the pores and in the opposite direction through the pore walls. This scheme seemed faulty if for no other reason than the fact that pore walls consisting of collodion, parchment paper, or glass are for all practical purposes insulators and therefore can not facilitate the passage of current. The mechanism suggested by Bartell [98] was based on the idea that the current is transported in one direction by the movable ions of the electrical double layer near the pore wall-solution phase boundary, in the other direction by the ions in the solution filling the core of the pore. This picture too seemed unrealistic; whatever potential differences exist would tend to equalize at any transverse cross section of the pore. Thus, neither of these suggested ionic mechanisms of anomalous osmosis made any electrochemical sense. However both ideas contained an element that seemed more or less self-evident to me, namely, that a current must flow when electroosmosis is observed.

Experimenting with collodion membranes while ruminating about this problem, I was struck with their conspicuous macroscopic inhomogeneity which obviously extends down into the microscopic range. In a sudden flash of insight I realized that different parts of the membrane and even adjacent pores, according to the Bethe-Michaelis concept visualized in Figures 15 and 16, must yield different pore potentials. Accordingly, adjacent pores or areas of a membrane which yield different electromotive forces must interact with each other through the two adjacent solutions. This means that local currents must flow, in one direction, through the narrower pores, in the other direction through the wider pores. This idea is expressed in Figure 17 for the simplest possible case, an electrolyte having univalent anions and cations of equal diffusion velocities. This figure, redrawn from my first paper on anomalous osmosis [117], shows a narrow pore yielding a high pore potential, E, interacting with a wider pore yielding a lower pore potential, e; the arrows indicate the direction of the flow of the (positive) current. According to the accepted electrokinetic theory, an electroosmotic flow must occur through the wider pore, from the side of the more dilute to the side of the more concentrated solution, in other words, anomalous positive osmosis arises in this case.

It was an essentially mechanical, relatively easy task to expand the indicated line of reasoning to systems with electrolytes whose ions have different diffusion velocities, to develop by deduction in which instances positive anomalous osmosis and in which

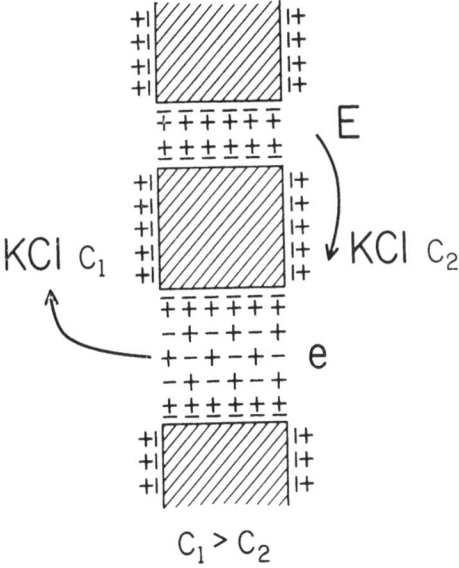

$$C_1 > C_2$$

Fig. 17. Schematic representation of the mechanism of anomalous positive osmosis in the system KCl c_1‖KCl c_2, $c_1 > c_2$. The narrower pore creates a higher pore potential, E, than the pore potential arising at the wider pore, e. The arrows indicate the direction of the flow of the (positive) current. The current drives water by electro-osmosis through the wider pore from the side of the more dilute to the side of the more concentrated solution.

negative osmosis occur, and to represent these various cases in figures analogous to Figure 17 [130]. These deductively developed rules fully agreed with all the available experimental data and gave the rational explanation of the empirical regularities deduced by Bartell [97, 98] and Loeb [21, 100, 101] from their experiments.

When I informed Prof. Freundlich about my conception of the ionic mechanism of anomalous osmosis, he accepted it readily and, to my great delight, called it a significant achievement. My long first paper on anomalous osmosis in which I developed the basic thesis [117] was received by the journal seven weeks after I had started work on this problem, the second one followed after five more weeks [130].

On the basis of my four electrochemical membrane papers I was awarded the 'Weston Fellowship in Electrochemistry' of the Electrochemical Society (U.S.) for the academic yaer 1930–31, to continue this work, specifically to test experimentally my theory of anomalous osmosis. This award was a very welcome indication that my electrochemical membrane work was beginning to be recognized also abroad.

The general plan on which the experimental test of my theory of anomalous osmosis had to be based, could be derived from simple electrochemical considerations and the available body of information on the electrochemical properties of porous membranes.

In any macro-model system in which two membranes yielding different membrane potentials would be arranged to interact electrically, the distances between the interacting units would be for the order of centimeters, whereas the distances between interacting parts or pores of a single membrane could be assumed to be of the order of micrometers. Since the driving forces in comparable macro- and micro-systems would be of similar magnitudes, the current flowing in the macro-system could be expected to be several orders of magnitude weaker than the sum of all the local micro-currents in the single membrane system, and the electroosmotic effect to be correspondingly smaller.

The smallness of the electroosmotic transport effects to be expected in macro-model systems even under the most favorable conditions indicated that the existence of such effects hardly could be proven in an unequivocal manner in models imitating anomalous positive osmosis. The electroosmotic water transport in such models would represent only a relatively small increment to the normal, unavoidably positive osmosis which ordinarily is quite substantial, particularly at higher solution concentrations. Therefore it appeared advisable to build a macro-model which would imitate negative osmosis. Such a model would be much better suited to prove the occurrence of an electroosmotic water transport due to the interaction of two membranes.

The uni-univalent electrolyte most suitable for this purpose was obviously LiCl which in the range of higher concentrations yields with suitable membranes pronounced negative osmosis. With not too dilute solutions of LiCl a fairly wide-pored electronegative membrane would yield a dynamic membrane potential approaching the liquid junction potential in free solution, -16 mV for a 10:0 concentration ratio, the more dilute solution being negative. This membrane would be the source of the driving E.M.F. The other, denser membrane would be chosen to yield only a minimal, ideally a zero membrane potential; current would be driven through it and cause an electroosmotic movement of the water through its pores.

The attempt to make collodion membranes suitable for the planned model experiments showed me that the basic concepts of the electrochemistry of porous membranes and my ability to manipulate these concepts were much farther developed than the art of preparing membranes of predetermined characteristics. After several months of unsuccessful work I was already rather discouraged when Dr Arthur Grollman (now Professor at the Southwestern Medical School at Dallas, Texas) on sabbatical leave from Johns Hopkins University, quite unexpectedly joined Prof. Freundlich's laboratory as a temporary guest worker. Since the time he could spend there was very limited, he was anxious to join some ongoing research. After hearing about Dr Grollman's scientific background, Prof. Freundlich introduced us to each other and I explained to Dr Grollman my theory of anomalous osmosis and my difficulties in making, or finding, suitable membranes. He in turn told me that he had several years of experience in experimental membrane investigations [130–132] starting with his Ph.D. thesis work on the measurement of osmotic pressure [131] carried out under Dr Frazer [132] at Johns Hopkins University. Dr Frazer was a member of a well-known group which for 30 years, first under the leadership of Morse [134], had used magnesium silicate and furnace baked magnesium oxide precipitation membranes deposited in porous clay cells for the determination of the osmotic pressure of solutions [134]. I learned that Dr Grollman frequently had observed rather strong anomalous osmotic effects which had been very disturbing in his own studies. He also had become convinced of the heteroporosity of membranes; however, he had not drawn any conclusions from this notion.

Dr Grollman readily agreed to join me in an attempt to test my theory of anomalous osmosis by the construction of two-membrane macro-model systems. Thus, by a stroke of good luck I gained the cooperation of an investigator with a rare, virtually ideal background for the planned work, and we started an all too brief period of most happy and fruitful collaboration.

Relying on the methods so familiar to Dr Grollman, we prepared a variety of precipitation membranes in porous clay cells, 100 mm high and 40 mm wide. The electroosmotic, osmotic, and hydraulic properties of these membranes varied over a considerable range. Some of them gave very strong anomalous osmotic effects and, most interesting for us, negative osmoses [135, 136] several times larger than those known to us from the literature.

After these preliminary experiments, we turned to the stepwise construction of our final model system. This model consisted of two precipitation membranes of different properties deposited in clay cells which were filled with 0.01 N LiCl solution and placed side by side into a beaker containing 0.1 N LiCl. As the denser membrane we selected one which when interposed between 0.01 and 0.1 N LiCl solutions, yielded neither a readily measurable membrane potential nor a detectable osmotic effect. The other, rather widepored membrane acted as source of the driving E.M.F. The cell with the denser membrane was closed with a rubber stopper that carried a narrow manometer tube and a saturated KCl – agar bridge, an inverted U tube, whose other end dipped into the 0.01 N KCl solution inside the other, open clay cell. This KCl-agar bridge established a direct outside electrical connection between the two inside solution compartments. Thus, it completed a closed circuit arrangement, in which the current flowed in the one direction through the two membranes and the interposed outside solution, in the other direction through the KCl-bridge. This current could be interrupted by placing a glass thimble over the accessible end of the KCl-agar bridge

This model reproducibly gave the predicted electroosmotic effect created by the electrical interaction of two membranes; it amounted to several mm^3 over a period of 20 min [136, 137].

Thus, with the indispensable collaboration of Dr Grollman, the basic idea of my theory of anomalous osmosis was verified, the occurrence of electroosmosis due to the electrical interaction of different parts of a membrane. That the rate of electroosmotic water transport in our model was rather small had been anticipated for the before out-lined reasons. Parenthetically it may be added that many years later, when high charge density wide-pored and permselective collodion membranes of low resistance had been developed [138], Carr and Sollner [139] built another structurally improved model which gave about ten times stronger effects.

When I presented the theoretical and experimental work on anomalous osmosis before several rather critical audiences, including Haber, Nernst, and Planck I received an unexpected amount of acclaim. It was recognized that my theory of the mechanism of anomalous osmosis was the first instance in which electrochemical membrane theory had been applied successfully not only in a physicochemically meaningful ex-planation of a long controversial rather involved membrane effect, but that it also had been used as the basis for the construction of macro-models. Rather suddenly I was considered an expert in the electrochemistry of membranes and received much en-couragement to continue work in this field, particularly by Prof. Freundlich.

Having clarified to my satisfaction three of the four puzzling electrical membrane effects discussed in Freundlich's book, I started to think about the fourth, thermo-osmosis, the transport of liquid across a membrane which separates two solutions of identical composition but different temperature. The well-known French physicist G. Lippmann [140] and M. Aubert [141] in Lippmann's laboratory had studied this effect experimentally with porous membranes, and found it to be much stronger with solutions of electrolytes than of nonelectrolytes with which thermoosmosis in many instances did not arise at all. The direction of the thermoosmotic movement of liquid was always from the warmer to the colder solution. The thermoosmotic flow of liquid was strongest at solution concentrations around 0.01 N, and proportional to the tem-perature difference across the membrane. Lippmann and Aubert were inclined to consider thermoosmosis as an electroosmotic effect, Freundlich [19], more specifically, linked it to negative osmosis. Thinking about this problem, I came to the conclusion that thermoosmosis in all probability is a special case of anomalous osmosis caused by differences in the thermoelectric electromotive forces arising with pores of different width. Useful background information on nonisothermal effects in solutions was virtually nonexistent and I did not see any way to attack the problem of thermoosmosis without first going into a large scale investigation of non-isothermal solution chemist-ry, an undertaking for which I lacked time, facilities, and interest. Moreover, there existed then among physical chemists a kind of tabu against the study of nonisother-mal phenomena. Thus I refrained at that time from a study of thermoosmosis. I might add that Dr Charles W. Carr and I returned to this topic right after World War II. We found a great similarity in the concentration dependence of the rates of thermo-osmosis and anomalous osmosis in comparable systems. This parallelism clearly con-firmed the Lippmann-Aubert conception of the electroosmotic nature of thermo-osmosis, and Freundlich's suggestion that it may be closely akin to anomalous os-mosis. In 1946, I presented our results orally but, in view of the lack of background information on non-isothermal electrochemistry, refrained from suggesting any specific ionic mechanism of thermoosmosis. We postponed publication till 1962 [142] by which time the study of non-isothermal phenomena had become 'respectable'. – The mechanism of thermoosmosis still awaits clarification.

Already in my first paper on anomalous osmosis [117] it was stated that all heterogeneities in membranes which give rise to locally different electromotive forces, not only heteroporosity, must of necessity produce local electrical circuits, a point which was reemphasized by Sollner and Grollman [136]. Having discarded the thought of trying to treat thermoosmosis from this point of view, I tried to find another problem to which the concept of membrane inhomogeneity could be applied profitably.

Among my notes on strange membranes and membranes effects I had a reference to a paper by Höber and Hoffmann [143] on the electrolyte permeability of mosaic membranes composed of highly anion selective and highly cation selective parts, the theoretical reasoning of which seemed to me basically erroneous. I therefore started to look into the problem of mosaic membranes of this general type.

Intuitively, it was immediately evident that by far the simplest charge mosaic membranes are those which are composed of ideally anion selective and ideally cation selective parts. Such idealized membranes were obviously also the most suitable for a theoretical investigation. The main questions were: are such mosaic membranes permeable to electrolytes, and if so, what is the correlation of their electrolyte permeability to the electrochemical properties of the component parts of which such membranes consist. This matter was developed by reference to the sequence of line drawings of Figure 18 which are redrawn for my original paper on mosaic membranes [144].

Figure 18 illustrates schematically a system in which a mosaic membrane separates a lower compartment (of invariable volume) from an upper compartment. The striated structure in the figure indicates the membrane; the electro-negative, cation-permeable (anion-impermeable) parts of the membrane are indicated by minus signs, and the

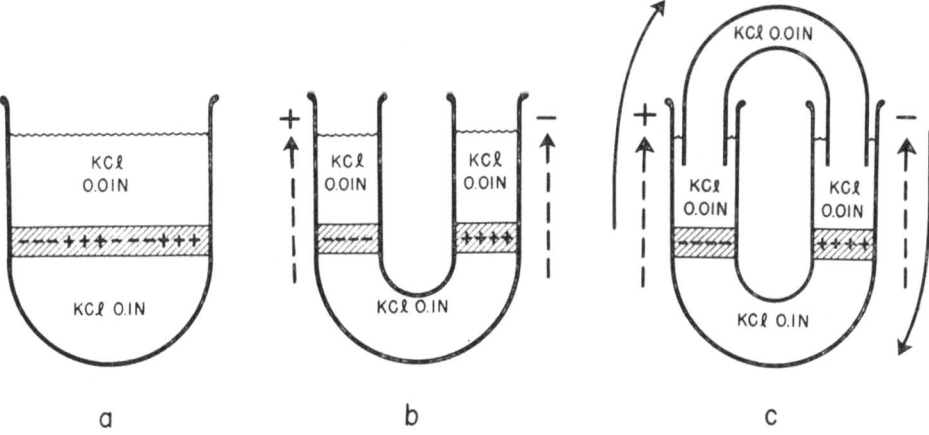

a   b   c

Fig. 18.   Pictorial development of the theory of electrolyte permeability of mosaic membranes that are composed of ideally anion-selective and ideally cation-selective parts. (a) Mosaic membrane with adjacent cation-selective and anion-selective parts. (b) Spatial and electrical separation of the cation-selective and anion-selective parts of the membranes. (c) The spatially separated cation-selective and anion-selective parts joined electrically by a bridge of solution in an all-electrolytic circuit.

electropositive, anion-permeable (cation-impermeable) parts by plus signs. The lower compartment is filled with 0.1 N and the upper one with 0.01 N KCl solution.

Inspection of Figure 18a leads to the conclusion that such mosaic membranes facilitate the permeation of electrolyte from the lower compartment to the upper one. Cations move through the electronegative parts of the membrane and anions through the electropositive parts, neutralizing each other electrically. Thus a continuous movement of the electrolyte must occur across the membrane until equilibrium between the two compartments is established. This answers the first of the before stated questions in the affirmative.

To answer the second question, namely the quantitative relationship of the electrolyte permeability of the mosaic membranes and the electrochemical properties of their component parts, required the consideration of systems in which the cation-permeable and anion-permeable parts are rearranged and separated from each other. Figure 18b shows a U-tube containing in its left arm an assembly of the electronegative, cation-permeable parts of the membrane of Figure 18a, and in its right arm the assembly of the electropositive, anion-permeable parts of the membrane of this figure, both types of membranes being assumed to be of ideal ionic selectivity. The lower part of the system, having an invariable volume, is filled with 0.1 N KCl chloride solution; the two compartments above the membrane contain 0.01 N KCl. The only processes that can occur are the establishment of the membrane concentration potentials across the two membranes, and the build-up of a hydrostatic pressure in the lower compartment; thereafter the system is at rest.

The sign and the magnitude of the two membrane potentials in the system of Figure 18b is given by Equation (4), $+55.1$ and -55.1 mv at 25 °C; their signs are shown by broken-line arrows pointing at a plus and a minus sign.

The essential feature of the situation represented in Figure 18a, in which the cation and the anion permeable membrane parts may interact with each other, can now be re-established by connecting the two compartments which contain 0.01 N solution by means of a liquid conduit filled with 0.01 N KCl solution. This is shown in Figure 18c. The system of Figure 18c is an all-electrolytic electrical circuit; a (positive) current flows in a clockwise direction through the system, as is indicated by the solid arrows. The current, I, is given by Ohm's law: $I = 2E/\varrho$ where ϱ is the total resistance of the system.

The current that flows in a clockwise direction in the system of Figure 18c is transported through the negative membrane in the left arm of the system exclusively by cations which move clockwise, in the direction of the brokenline arrow; through the positive membrane in the right arm the electricity is transported exclusively by an equivalent quantity of anions that move in a counterclockwise direction, as indicated by a broken-line arrow. Thus, the number of the equivalents of electrolyte that move in a given time in the mosaic system of Figure 18c from the concentrated to the dilute solution must be numerically identical with the number of faradays that flow in the system during the same period.

This theory of mosaic membranes was the first instance in which a complex membrane system had been shown to be amenable to a quantitative theoretical analysis. I considered it as a big step forward in spite of the fact that the theory could not be verified experimentally at that time. The most higly cation selective membranes then available had unduly high resistances and still leaked considerable quantities of anions, and suitable highly anion selective membranes did not exist. The theory had outrun the experimental possibilities.

The experimental verification of the outlined theory of mosaic membranes became possible many years later after 'permselective' membranes of extreme cationic and anionic selectivities had been developed in the author's laboratory [138]. Using these membranes, Neihof and Sollner [145] showed in model experiments that the quantities of electrolyte that permeated from the more concentrated into the more dilute solution across the two membranes agreed, according to Faraday's law, within $\pm 2\%$ with the quantities of electricity that had flowed in these cells during the experimental period. Later, the author used a mosaic membrane concept in the construction of a model system which, imitating the behavior of living cells, accumulates electrolytes, both anions and cations, against concentration gradients [146]. The theory of this model, incidentally, is based entirely on concepts which were clearly understood 40 years ago; it could have been developed quite as well at that time. Then, however, an experimental test would have been impossible for the lack of permselective membranes of extreme ionic selectivity and low resistance.

The rapidly expanding literature on charge mosaic membranes of the type shown in Figure 18 was recently reviewed by Weinstein et al. [147].

The general state of the physical chemistry of membranes in the very early nineteen thirties may also be illustrated by a few personal recollections.

After I had finished the experimental work on anomalous osmosis and written the paper on mosaic membranes, I discussed with Prof. Freundlich my further scientific plans. At first he suggested that I should continue working on membranes since I had been conspicuously successful in my membrane work. However, I had to confess that I did not know of any sharply defined problems except thermoosmosis which, for the before stated reasons, did not seem attractive to me. Reasonably clearly I saw only two obviously important wide open problems. The one was to find suitable acidic and basic organic substances for incorporation in liquid membranes to produce improved Haber-Klemensiewicz-Beutner type liquid membranes which in concentration cells would respond electromotively virtually as membranes of ideal ionic selectivity according to Nernst's equation for the electrode potential, Equation (2), and Donnan's equation for the membrane potential, Equation (4). Exploratory experiments, however, in which I used some of the rather limited number of then available substances had proven utterly negative. The other obvious, wide open problem was the preparation of anion selective positively charged membranes which would be stable and useful in systematic experimental work. However, in this instance too, preliminary test experiments, based on the use of thorium oxide diaphragms, polyvalent cation, and the incorporation of basic dyestuffs in collodion membranes had given me only discouraging results. Moreover, I also did not see any hope for finding the anwers to these questions by that kind of abstract reasoning which had been my main tool in dealing with membrane problems. I had the feeling that for the time being I might have accomplished in membrane research what I could accomplish speedily. For a long drawnout, more or less groping experimental approach I had neither interest nor time.

When I asked Prof. Freundlich whether he knew of any good membrane problems, he suggested after some deliberation work on the Donnan equilibrium. However, when asked what specifically could be done, he was not able to suggest any new approach, nor could I. It seemed that about everything that one would want to do on this topic had already been done, at least in principle, and I was not interested in merely expanding other people's work in a routine manner. It seemed that further significant progress would require either some novel fundamental insight or a vast improvement in the art of membrane preparation, for which neither Prof. Freundlich nor I could see any clue.

At the same time I became aware of the fact that specializing in the physical chemistry of membranes might not be a good way to foster my career plans aiming at a University chair in physical chemistry. Research on membranes was rather looked down upon by most physical chemists. This was brought clearly home to me in a discussion that I had about in 1930 with a slightly older friend, the late K. F. Bonhoeffer, who soon was to become Professor of Physical Chemistry in Frankfurt and finally, after World War II, Director of the Max Planck Institute for Physical Chemistry in Göttingen. One day when we talked about our work, he said "You know, Karl, 'membranes' is really not a respectable field of research for a physical chemist", and urged me to take up some other line of work; otherwise I would hurt my prospects in the academic field. Bonhoeffer around 1930 did not foresee that about twenty years later he would lead one of the most active and gifted groups of membrane investigators ever assembled in one laboratory.

Since also other interests had arisen, I decided to stop my electrochemical membrane work, at least for the time being. I summarized it in a paper published in 1933 in which I also discussed a variety of then new ideas of membrane transport [77] which I hoped to follow up some time in a somewhat nebulous future when I would have collaborators and larger facilities.

With respect to the Haber-Beutner type of liquid membranes I then anticipated that sooner or later, as the number of commercially available organic compounds increased, there would become available some highly carbophilic high molecular weight compounds suitable for this purpose. It was thirty years later, with the advent of the liquid ion exchangers, that such greatly improved liquid ion exchanger membranes became a reality [148, 149, 150].

With respect to the porous membranes the hoped for breakthrough came much sooner, in 1935 and 1936, when Teorell [16] and Meyer and Sievers [17, 18] recognized that the charge of porous membranes consisting of supposedly inert materials, such as cellulose or collodion, is due to the dissociable groups which are an integral part of these substances, and that charged membranes in general must be considered as ion exchange bodies. This insight became the basis of the now universally accepted Teorell [16], and Meyer and Sievers [17, 18] fixed charge theory of ionic membranes which indicated implicitly also the way to prepare membranes of pronounced electrochemical activity, namely, by increasing the charge density at the pore walls. This approach was taken right away by Meyer and Sievers [17]. All these developments, however, which have opened up a new era in the electrochemistry of charged porous membranes are already outside the time frame of this review.

Acknowledgments

The author thanks Dr Gerald M. Shean for reading the manuscript and for many helpful suggestions. He also thanks the New York Academy of Sciences and the Editors of the *Annals of the New York Academy of Sciences* and The Marcel Dekker, Inc. and the Editors of the *Journal of Macromolecular Science* [138] for the permission freely to use material from several of his previous publications. Likewise, the author is greatly indebted to the Akademische Verlagsgesellschaft, Frankfurt am Main, for permission verbatim to quote (in translation) from a paper by A. Bethe and T. Toropoff published in the *Zeitschrift für physikalische Chemie* [25]; and to the Rockefeller University Press and the Editors of the *Journal of General Physiology* for the permission for lengthy verbatim quotations from a paper by L. Michaelis [11].

The author thanks mrs Runar Collander for the photograph of Prof. Collander, and all those who have kindly provided pictures of the following investigators: Julius Bernstein: Mr Heindorf, Universitätsarchiv Halle, Martin Luther Universität Halle-Wittenberg (UA Halle, Rep 40 Nr. BI Nr. 18); Albrecht Bethe: Prof. Dr Erich Heinz, Goethe Universität, Frankfurt am Main; Thomas Graham: Mr Monroe H. Fabian, Associate Curator, National Portrait Gallery, Smithsonian Institution, Washington, D.C.; Walter Nernst: Dipl.-Phys. H. Klages, Physikalisch Technische Bundesanstalt, Braunschweig; Wilhelm Ostwald: Prof. Dr G. Geiseler, Karl Marx Universität, Leipzig.

References

1. Ostwald, W.: *Z. phys. Chem.* **6**, 71 (1890).
2. Nernst, W.: *Nachr. Akad. Wiss. Göttingen Math.-Physik. Kl. IIa*, **104** (1899); *Pflüger's Arch. ges. Physiol.* **122**, 275 (1908).
3. Nernst, W. und Riesenfeld, E. H.: *Ann. Phys.* (4) **8**, 600 (1902).
4. Riesenfeld, E. H.: *Ann. Phys.* (4) **8**, 609, 616 (1902); Riesenfeld, E. H. und Reinhold, B.: *Z. phys. Chem.* **68**, 459 (1910).
5. Haber, F.: *Ann. Phys.* (4) **26**, 927 (1908).
6. Haber, F. und Klemensiewicz, Z.: *Z. phys. Chem.* **67**, 385 (1909).
7. Beutner, R.: *Biochem. Z.* **47**, 73 (1912); *Z. Elektrochem.* **19**, 319 (1913); *Trans. Am. Electrochem. Soc.* **23**, 401 (1913); *Die Entstehung elektrischer Ströme in lebenden Geweben*, Enke, Stuttgart, 1920; *Physical Chemistry of Living Tissues and Life Processes*, The Williams and Wilkins Co., Baltimore, 1933; *Bioelectricity*, in *Medical Physics*, O. Gasser, Ed., Year Book Publishers, Chicago, 1944, pp. 35–88.
8. Donnan, F. G.: *Z. Elektrochem.* **17**, 572 (1911).
9. Donnan, F. G.: *Chem. Rev.* **1**, 73 (1924); *J. Intern. Soc. Leather Trade Chemists*, April 1933, p. 1; *Z. physik. Chem.* (A), **168**, 369 (1934); *J. Chem. Soc.* p. 707, 1939.
10. Donnan, F. G. and Guggenheim, E. A.: *Z. physik. Chem.* (A), **162**, 346 (1932).
11. Michaelis, L : *J. Gen. Physiol.* **8**, 33 (1925).

12. Michaelis, L. und Fujita, A.: *Biochem. Z.* **161**, 47 (1925).
13. Michaelis, L. und Fujita, A.: *Biochem. Z.* **158**, 28 (1925); **164**, 23 (1925); Michaelis, L. und Dokan, S.: *Biochem. Z.* **162**, 258 (1925); Michaelis, L. and Perlzweig, W. A.: *J. Gen. Physiol.* **10**, 575 (1926–27); Michaelis, L. und Hayashi, K.: *Biochem. Z.* **173**, 411 (1926); Michaelis, L. and McEllsworth, R. and Weech, A. A.: *J. Gen. Physiol.* **10**, 671 (1927); Michaelis, L., and Weech, A. A., and Yamatori, A.: *J. Gen. Physiol.* **10**, 685 (1927); Michaelis, L. and Weech, A. A.: *J. Gen. Physiol.* **11**, 147 (1927).
14. Michaelis, L.: *Bull. Nat. Res. Council* No. **69**, 119 (1929).
15. Michaelis, L.: *Kolloid-Z.* **62**, 2 (1933).
16. Teorell, T.: *Proc. Soc. Exp. Biol. Med.* **33**, 282 (1935); *Proc. Nat. Acad. Sci. U.S.* **21**, 152 (1935); *Trans. Faraday Soc.* **33**, 1054 (1937).
17. Meyer, K. H. et Sievers, J.-F.: *Helv. Chim. Acta* **19**, 649, 665 (1936).
18. Meyer, K. H.: *Trans. Faraday Soc.* **33**, 1073 (1937).
19. Freundlich, H.: *Kapillarchemie*, 2nd Edition, Akademische Verlagsgesellschaft, Leipzig, 1922; 3rd Edition, Leipzig, 1923; *Colloid and Capillary Chemistry* (transl. by H. S. Hatfield), Methuen, London, 1926.
20. Höber, R.: *Physikalische Chemie der Zelle und der Gewebe*, 6th ed., Engelmann, Leipzig, 1926.
21. Loeb, J.: *J. Gen. Physiol.* **2**, 577 (1920); **4**, 463 (1922).
22. Grim, E. and Sollner, K.: *J. Gen. Physiol.* **40**, 887 (1957); **44**, 381 (1960).
23. Sollner, K.: *J. Phys. Chem.* **49**, 47 (1945).
24. Sollner, K.: *J. Macromolec. Sci. – Chem.* **A3**: **1**, (1969).
25. Bethe, A. und Toropoff, T.: *Z. physik. Chem.* **88**, 686 (1914).
26. Bethe, A. und Toropoff, T.: *Z. physik. Chem.* **89**, 597 (1915).
27. Biedermann, W.: *Elektrophysiologie*, G. Fischer, Jena, 1895; *Electrophysiology* (transl. by F. Welby), MacMillan, New York, Vol. I, 1896, Vol. II, 1898.
28. Höber, R.: *Physikalische Chemie der Zelle und der Gewebe*, W. Engelmann, Leipzig, 1902.
29. Verworn, M.: *Allgemeine Physiologie*, 4th Edition, G. Fischer, Jena, 1903, p. 71.
30. Bernstein, J.: *Pflüger's Arch. ges Physiol.* **92**, 521 (1902).
31. Bernstein, J.: *Elektrobiologie*, Friedr. Vieweg und Sohn, Braunschweig, 1912.
32. Bernstein, J.: *Biochem. Z.* **50**, 393 (1913).
33. Dutrochet, R. T. H.: *Ann. Chim. Physique* (2) **35**, 393 (1827); (2) **60**, 337 (1835).
34. Jolly, Ph.: *Ann. Physik (Poggendorff)* **78**, 261 (1849).
35. Graham, Th.: *Phil. Trans. Roy. Soc.* **144**, 177 (1854).
36. Graham, Th.: *Phil. Trans. Roy. Soc.* **151**, 183 (1861).
37. Eckhard, C.: *Ann. Physik* (5) **7**, 61 (1868).
38. Waitz, K.: *Diffusion* in A. Winkelmann, *Handbuch der Physik*, 2nd Edition, J. A. Barth, Leipzig (1908).
39. Wiedemann, G.: *Ann. Physik (Poggendorff)* **87**, 321 (1852); **99**, 177 (1856).
40. Reuss, F. F.: *Mémoires de la Société Imperiale des Naturalistes de Moskou* **2**, 327 (1809).
41. Quincke, G.: *Ann. Physik.* (4) **22**, 513 (1861).
42. Quincke, G.: *Ann. Physik.* (4) **17**, 1 (1859); **20**, 38 (1860).
43. Helmholtz, H.: *Ann. Physik (Wiedemann)* **7**, 337 (1879); *Wissenschaftliche Abhandlungen*, Vol. 1, p. 855, J. A. Barth, Leipzig 1882; Lamb, *Phil. Mag.* (5), **25**, 52 (1888).
44. Smoluchowski, M. v.: *Elektrische Endosmose und Strömungsströme* in *Handbuch der Elektrizität und des Magnetismus* (L. Graetz, Ed.), Vol. II, J. A. Barth, Leipzig 1921, pp. 366–428.
45. Perrin, J.: *J. chim. phys.* **2**, 601 (1904).
46. Gouy, G.: *J. de Physique* (4) **9**, 457 (1910).
47. Wiedemann, G.: *Die Lehre von der Elektrizität*, 2nd Ed., Vol. 1, Friedrich Vieweg und Sohn, Braunschweig, 1893, pp. 982–1023.
48. van 't Hoff, J. H.: *Z. physik. Chem.* **1**, 481 (1887).
49. van 't Hoff, J. H. und Reicher, L. Th.: *Z. physik. Chem.* **2**, 777 (1888); ibid. **3**, 198 (1889).
50. Raoult, F. M.: *Ann. Chim. Phys.* (6) **2**, 66 (1884).
51. Pfeffer, W.: *Osmotische Untersuchungen*, Engelmann, Leipzig, 1877.
52. Traube, M.: *Centralblatt für die med. Wissensch.* 1866, p. 97, 113; *Arch. für Anatomie Physiologie* 1867, pp. 87, 129; *Gesammelte Abhandlungen* Mayer & Müller, Berlin 1899.
53. Kohlrausch, F.: *Ann. Physik (Wiedemann)* **6**, 1, 145 (1879).
54. Kohlrausch, F.: *Gesammelte Abhandlungen*, Vol. II, *Elektrolyte*, J. A. Barth, Leipzig, 1911.

55. Arrhenius, S.: *Z. physik. Chem.* **1**, 631 (1887); **2**, 491 (1888).
56. Ostwald, W.: *Z. physik. Chem.* **2**, 270 (1888).
57. Ostwald, W. und Nernst, W.: *Z. physik. Chem.* **3**, 120 (1889).
58. Nernst, W.: *Z. physik. Chem.* **2**, 613 (1888).
59. Nernst, W.: *Z. physik. Chem.* **4**, 129 (1889).
60. Helmholtz, H.: *Monatsberichte Berlin. Akad. Wissenschaften*, November 26, 1877; *Wissenschaftliche Abhandlungen*, Vol. 1, Barth, Leipzig, 1882, pp. 840–854.
61. Planck, M.: *Ann. Physik. (Wiedemann)* **39**, 161 (1890); **40**, 561 (1890).
62. Henderson, P.: *Z. physik. Chem.* **59**, 118 (1907); **63**, 325 (1908).
63. Becquerel, A. C.: *Compt. rend.* **64**, 919, 1211 (1867); **65**, 51, 720 (1867); **82**, 354 (1876); **84**, 145 (1877); **85**, 169 (1877); *Des Forces Physico-Chimiques*, Typographie Firmin-Didot Frères, Paris, 1875.
64. Freundlich, H. und Sollner, K.: *Z. phys. Chem.* (A) **138**, 349 (1928); 152, 313 (1931).
65. Nernst, W.: *Nachr. Akad. Wiss., Göttingen, Math. Physik. Kl.*, IIa, 1899, p. 104.
66. Nernst, W.: *Pflüger's Arch. ges. Physiol.* **122**, 275 (1908).
67. Donnan, F. G. and Green, G. M.: *Proc. Roy. Soc.* Ser. A, **90**, 450 (1914); Donnan, F. G. and Allmand, A. J.: *J. Chem. Soc.* **105**, 1963 (1914).
68. Donnan, F. G. and Garner, W. E.: *J. Chem. Soc.* **115**, 1313 (1919);
69. Procter, H. R.: *J. Chem. Soc.* **105**, 313 (1914); Procter. H. R. and Wilson, J. A.: *J. Chem. Soc.* **109**, 307. 1327 (1916); *J. Am. Leather Chemists' Assoc.* **11**, 261, 399 (1916); Wilson, J. A. and Wilson, W. H.: *J. Am. Chem. Soc.* **40**, 886 (1918); Wilson, J. A.: *J. Am. Chem. Soc.* **38**, 1982 (1916).
70. Loeb, J.: *J. Am. Chem. Soc.* **44**, 1930 (1922); *Proteins and the Theory of Colloidal Behavior*, McGraw-Hill, New York, 1922.
71. Sörensen, S. P. L.: *Studies on Proteins, Compt. Rend.*, Carlsberg, 1915–1917.
72. Bolam, T. R.: *The Donnan Equilibria and their Application to Chemical, Physiological and Technical Processes*, G. Bell and Sons, Ltd., London, 1932.
73. Collander, R.: *Kolloidchemische Beihefte* **19**, 72 (1924).
74. Collander, R.: *Societas Scientiarum Fennica Commentationes Biologicae* II. (6), Helsingfors 1926, pp. 1–48.
75. Sollner, K. and Beck, P. W.: *J. Gen. Physiol.* **27**, 451 (1944).
76. Fujita, A.: *Biochem. Z.* **170**, 18 (1926).
77. Sollner, K.: *Kolloid-Z.* **62**, 31 (1933).
78. Tammann, G.: *Z. phys. Chem.* **9**, 97 (1892); **10**, 255 (1892).
79. Walden, P.: *Z. phys. Chem.* **10**, 699 (1892).
80. Walden, P.: *Das Leitvermögen der Lösungen*, Parts I and II/III (in two volumes), Akademische Verlagsgesellschaft, Leipzig, 1924.
81. Collander, R.: *Kolloidchemische Beihefte* **20**, 273 (1925).
82. Bethe, A., Bethe, H. und Terada, Y.: *Z. phys. Chem.* **112**, 250 (1924).
83. Biegelow, S. L. and Gamberling, A.: *J. Am. Chem. Soc.* **29**, 1576 (1907)
84. Brown, W.: *Biochem. J.* **9**, 591 (1915).
85. Ostwald, Wo.: *Kolloid-Ztschr.* **23**, 68 (1918); *Grundriss der Kolloidchemie*, Erste Hälfte, Theodor Steinkopf, Dresden and Leipzig, 1923.
86. Bechhold, H.: *Die Kolloide in Biologie und Medizin*, 5th Ed., Theodor Steinkopf, Dresden and Leipzig, 1929.
87. Zsigmondy, R.: *Kolloidchemie*, 3rd Ed., Otto Spamer, Leipzig, 1920; 5th Ed., Part I. Leipzig, 1925; Part II, Leipzig, 1927.
88. Hebler, F.: *Ultrafiltration und Dialyse* in *Kolloidchemische Technologie*, R. E. Liesegang, Ed. Theodor Steinkopf, Dresden and Leipzig, 1927, pp. 70–93.
89. Pauli, W.: *Kolloid-Ztschr.* **31**, 252 (1922); *Biochem. Ztschr.* **152**, 355 (1924).
90. Freundlich, H. und Loeb, L. F.: *Biochem. Ztschr.* **150**, 522 (1924).
91. Bradfield, R. and Bradfield, H. S.: *J. Phys. Chem.* **33**, 1724 (1929).
92. Spiegel-Adolf, M.: *Electrodialyse* in *Abderholdens Handbuch der Biol. Arbeitsmethoden*, Abt. III, Teil B, 1927.
93. Meyer, E.: *Elektroosmose* in *Kolloidchemische Technologie*, R. E. Liesegang, Ed., Theodor Steinkopf, Dresden and Leipzig, 1927, pp. 94–136.
94. Pauli, W. und Valko, E.: *Elektrochemie der Kolloide*, Julius Springer, Wien, 1929.

95. Prausnitz, P. H. und Reitstötter, J.: *Elektrophorese, Elektroosmose, Elektrodialyse in Flüssig-keiten*, Theodor Steinkopf, Dresden and Leipzig, 1931.

96. Michaelis, L.: *Kompendium der Entwicklungsgeschichte des Menschen, mit Berücksichtigung der Wirbeltiere*, Boas and Hesse, Berlin, 1898.

97. Bartell, F. E.: *J. Am. Chem. Soc.* **36**, 646 (1914); Bartell, F. E., and Hocker, C. D.: *J. Am. Chem. Soc.* **38**, 1029, 1036 (1916); Bartell, F. E. and Madison, O. E.: *J. Phys. Chem.* **24**, 444, 593 (1920); Bartell, F. E., and Carpenter, D. C.: *J. Phys. Chem.* **27**, 101, 252, 346 (1923).

98. Bartell, F. E.: *Membrane Potentials and their Relation to Anomalous Osmosis*, in J. H. Mathews (Ed.), Colloid Symposium Monograph, Department of Chemistry, University of Wisconsin, Madison **1**, 120 (1923).

99. Freundlich, H.: *Kolloid-Z.* **18**, 11 (1916).

100. Loeb, J.: *J. Gen. Physiol.* **2**, 387, 659, 673 (1920).

101. Loeb, J.: *J. Gen. Physiol.* **4**, 213, 463 (1922).

102. Loeb, J.: *Am. J. Physiol.* **3**, 135 (1900); *Biochem. Z.* **15**, 254 (1909); *Die chemische Entwicklungs-erregung des tierischen Eies*, Julius Springer, Berlin, 1909.

103. Höber, R.: *Physikalische Chemie der Zelle und Gewebe*, 3rd Ed., Wilhelm Engelmann, Leipzig, 1911; 4th Ed., 1914.

104. Hittorf, W.: *Z. physik. Chem.* **39**, 613 (1902); **43**, 239 (1903); *Z. Elektrochem.* **8**, 482 (1902).

105. Du Bois-Reymond, E.: *Berl. Monatsberichte*, 1859, p. 405.

106. Bethe, A.: *Zentralblatt f. Physiol.* **23**, 1909, No. 9; Internat. Physiol. Kongress, Wien 1910; *Münch. medizin. Wochenschr.* 1911, No. 3 (Quoted after Bethe and Toropoff).

107. Cremer, M.: *Z. Biologie* **47**, 1 (1908).

108. Perrin, J.: *Ann. Chim. Phys.* **2**, 661 (1904).

109. Girard, C. r. Acad. Franc. Feb. 19th and May 3rd, 1909; July 4, 1910 (Quoted after Bethe and Toropoff)

110. Cybulski, *Bull. de l'Acad. des sciences de Cracovie* **6**, 662 (1903); Cybulski et Borkowski, ibid. **12**, 660 (1909) (Quoted after Bethe and Toropoff).

111. Freundlich, H.: *Kapillarchemie*, Akademische Verlagsgesellschaft, Leipzig, 1909; *Z. phys. Chem.* **79**, 385 (1912).

112. Pauli, W.: *Pflüger's Arch. ges. Physiol.* **136**, 488 (1910).

113. Wilson, J. R. (ed.): *Demineralization by Electrodialysis*, South African Council for Scientific and Industrial Research, Pretoria, South Africa; Butterworths Scientific Publication, London, 1960.

114. Spiegler, K. S. (ed.): *Principles of Desalination*, Academic Press, New York, 1966.

115. *Saline Water Conversion Reports* for 1963–1972, U. S. Dept. of the Interior, Office of Saline Water, Washington, D. C., 1963–1972.

116. Smith, J. D. and Eisenmann, J. L.: *Electrodialysis in Advanced Waste Treatment*, U. S. Dept. of the Interior, Cincinnati, Ohio, Feb. 1967.

117. Sollner, K.: *Z. Elektrochem.* **36**, 36 (1930).

118. Michaelis, L.: *Die Wasserstoffionenkonzentration*, 2nd Ed., Julius Springer, Berlin 1922; *Hydrogen Ion Concentration*, Williams and Wilkins Company, Baltimore, 1926.

119. Loeb, J. und Beutner, R.: *Biochem. Z.* **41**, 1 (1912).

120. Schwalbe, C. G.: *Die Chemie der Cellulose*, Bornträger, Berlin, 1910/11 and 1918.

121. Hess, J.: *Die Chemie der Zellulose, und ihrer Begleiter*, Akademische Verlagsgesellschaft, Leipzig, 1928.

122. Sollner, K.: *J. Phys. Chem.* **53**, 1211, 1226 (1949).

123. Mond, R. und Hoffmann, F.: *Pflüger's Arch. ges. Physiol.* **220**, 194 (1928).

124. Bjerrum, N. und Manegold, E.: *Kolloid-Z.* **42**, 97 (1927); **43**, 5 (1927).

125. Manegold, E.: *Kolloid-Z.* **61**, 140 (1932); *Trans. Faraday Soc.* **33**, 1088 (1937); *Kapillar Systeme*, Vol. 1, Strassenbau, Chemie und Technik Verlagsgesellschaft m.b.H., Heidelberg, 1955.

126. Sollner, K.: *Z. Elektrochem.* **35**, 789 (1929).

127. Grotthuss, C. J. D. von: *Ann. Physik. (Gilbert)* **61**, 65 (1819).

128. Braun, F.: *Ann. Physik. (Wiedemann)* **42**, 450 (1891); **44**, 473 (1891).

129. Coehn, A.: *Z. phys. Chem.* **25**, 651 (1898); *Z. Elektrochem.* **4**, 501 (1898).

130. Sollner, K.: *Z. Elektrochem.* **36**, 234 (1930).

131. Grollman, A.: *Improvements in the Mode of Measurement of Osmotic Pressure and their Applica-tion to a Study of aqueous Phenol Solutions at 30°, Ph. D. Thesis*, Johns Hopkins University, 1923.

132. Grollman, A. and Frazer, J. C. W.: *J. Am. Chem Soc.* **45**, 1710 (1923).

133. Grollman, A.: *J. Gen. Physiol.* **9**, 813 (1926).
134. Morse, H. N.: *The Osmotic Pressure of Aqueous Solutions, Report on Investigations made in the Chemical Laboratory of the Johns Hopkins University during the Years 1899–1913*, The Carnegie Institution of Washington, Washington, D.C., 1914.
135. Grollman, A. and Sollner, K.: *Trans. Electrochem. Soc.* **61**, 477 (1932).
136. Sollner, K. und Grollman, A.: *Z. Elektrochem.* **38**, 274 (1932).
137. Grollman, A. and Sollner, K.: *Trans. Electrochem. Soc.* **61**, 487 (1932).
138. Sollner, K.: *J. Phys. Chem.* **49**, 47, 171, 265 (1945); *J. Electrochem. Soc.* **97**, 139 C (1950); *The Annals of the N.Y. Acad. Sci.* **57**, 177 (1953); *Macromol. Sci. – Chem. A. 3* 1 (1969); etc.
139. Carr, C. W. and Sollner, K.: *Nature* **204**, 878 (1964).
140. Lippmann, G.: *Compt. rend.* **145**, 104 (1907).
141. Aubert, M.: *Ann. chim. phys.* (8), **26**, 145, 551 (1912).
142. Carr, C. W. and Sollner, K.: *J. Electrochem. Soc.* **109**, 616 (1962).
143. Höber, R. und Hoffmann, F.: *Pflüger's Arch. ges. Physiol.* 220, 558 (1924).
144. Sollner, K.: *Biochem. Z.* **244**, 370 (1932).
145. Neihof, R. and Sollner, K.: *J. Phys. Colloid Chem.* **54**, 157 (1950); *J. Gen. Physiol.* **38**, 613 (1955).
146. Sollner, K.: *Arch. Biochem. Biophys.* **54**, 129 (1955).
147. Weinstein, J. N. Bunow, B. J. and Caplan, S. R.: *Desalination* **11**, 341 (1972).
148. Botrè, C. and Scibona, G.: *Ann. Chim. (Rome)* **52**, 1199 (1962).
149. Sollner, K. and Shean, G. M.: *J. Am. Chem. Soc.* **86**, 1901 (1964); Shean, G. M. and Sollner, K.: *Ann. N.Y. Acad. Sci.* **137**, 759 (1966).
150. Sollner, K.: *The Basic Electrochemistry of Liquid Membranes*, in *Diffusion Processes, Proceedings of the Thomas Graham Memorial Symposium, University of Strathclyde*, Vol. 2, Sherwood, J. N., Chadwick, A. V., Muir, W. M., Swinton, F. L. (eds.), Gordon & Breach, London and New York, 1971, pp. 655–730.

THE DEVELOPMENT OF THE
MODERN MEMBRANE CONCEPTS AND THE RELATIONS
TO BIOLOGICAL PHENOMENA

TORSTEN TEORELL

Institute of Physiology and Medical Biophysics, Biomedical Center, University of Uppsala, Uppsala, Sweden

Mr Chairman, Ladies and Gentlemen.

During the negotiations for this talk the Chairman indicated a wish that I should cover the period after about 1930, which was supposed to be the end of Dr Sollner's historical review of the early work. However, since the thirties, when both Dr Sollner and I entered the membrane field, an enormous development has taken place. It is an impossible task to give a just and well-balanced account of the evergrowing contributions from the vast number of research workers. Realizing this difficulty, I thought it was more advantageous for me and for you to attempt another type of historical review. I will try to outline the *conceptual* developments. Regrettably, that implies that many names worthy of mentioning have to be omitted. Likewise, hardly any mathematical formulations will be presented. *Einstein* once said, that "thoughts and ideas not formulas, are the beginning of every physical theory". I also agree with *Willard Gibbs*, that "an important principle in any department of knowledge is to find the approach from which a problem appears in its greatest simplicity". Therefore this presentation will be highly simplified and sweeping.

1. The Background

First, I will try to recollect what sort of ideas were prevailing about 50 years ago, which inspired my generation to go into research. At that time I was a student of biochemistry and physiology. It was a time of strong belief in the utility of physico-chemical principles for biological research. It was also a time of cross-fertilisation between the branches of natural sciences. I present in Table I a kind of updated chronological table indicating the development of new concepts or methods. At first sight, it may appear to be a very confusing picture. Some items are merely the development of tools, others are formations of abstract concepts and the majority of the achievements may consist of probing explorations. A closer examination, however, reveals that the sequential developments are quite logical, one thing leads to another, although not always in a straight line, but rather like the growth of branches on a developing tree. Diversified as the individual achievements may appear, it is yet possible to discern a common, ultimate objective in the strivings of all 'membranologists'. Their concern focuses on the transport phenomena, whether the membranes are artificial and dead, or whether they are the site of life processes, as in cells and

TABLE I

A 'chronology' of membrane research

Year	Physicochemists Biophysicists	Biologists	Micromorphology	Physiologists Biochemists
1850–1900	Osmosis, El-diffusion	Water and salt transport Cell-membrane concept	Staining process, cell details	Chem. tissue analyses 'Secretion'
1900–1940	Donnan effect, Surface chem. 'Fixed Charge' concept	Lipoid solubility, Ion 'sieving' (K^+ select.) Cell pot./ 'Active transp.'	Polarization-MS, UV-microscopy, Phase contrast MS	Oedema, red cells (400 Å) Macro-molecules
1940–1960	Isotopes, ion exchange, Irrev. thermodynamics, Membrane oscill.	Ion pumps H. H. Excitability (Na^+/ K^+) 'Pore' size ($0 \lesssim 10$ Å) Carrier transport	EMS, XRMS, ARG, Intern. structures 'Unit membrane', XR-crystallography	Lipid chemistry Mitochond./bact. enzymes (ATP) K^+-Na^+-ATP-ases
1960–	Desalination Solid/lig. ion exchangers 'Fuel cells' (chem.→ electr.)	K^+/Na^+ selectivity, Bilayer models ('Valinomycin') Mitochondria [ion-H_2O pump]	'Multiple mem-branes' 'Membrane mobile'	Mitoch. (H^+, K^+, H_2O) Rhythmical-chem. reactions (=biol. 'Fuel cells')

biological tissues. One might even venture to condense the whole science of membranology into a single formula, codified as follows:

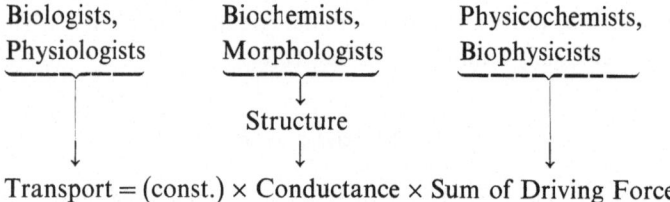

$$\text{Transport} = (\text{const.}) \times \text{Conductance} \times \text{Sum of Driving Forces}$$

The permeability problems of cell and tissue membranes round about 1930 were an inheritance from the earlier decades. The features of the cell membranes were mainly studied by chemical analysis. For example, in the red blood cells one was aware of a striking ion distribution. The inside of the cell was rich in potassium and poor in sodium. In the surrounding blood plasma the situation was the reverse. Although one had a gross appreciation of the chemical composition of the red cell membrane, one had no picture of its ultrastructure. Information about this was difficult to obtain, though the 'micromorphologists' in this period started to develop new, powerful forms of microscopy. The dominating permeability theories of the time focused on the presence of lipoids in the cell membranes. The peculiar ion distributions observed were ascribed to a preferential solubility, say of K in the lipoid phase of the cell membrane. The protein content of the membranes was, however, not entirely neglected. *Michaelis* had reemphasized the importance of 'geladene Membranen' and during the twenties he had published well appreciated papers of work on dried collodion. *Mond* in Germany was probably the first one to suggest that the red cell membrane could be similar to Michaelis's model. However, by and large, the lipoid solubility theory had the monopoly for many years ahead. At this time *Beutner* and *Osterhout* were running experiments with 'liquid membrane' models. In Osterhout's laboratory at the Rockefeller Institute of New York, a Mecca for the bio-scientists of its time, very ingenious experiments were run with guaiacol, which easily transported K-ions, but not Na-ions. Beutner preferred to measure the phase-boundary potentials of similar systems. These phase-boundary potentials, which have had a long tradition, as mentioned by Dr Sollner, greatly attracted the interest of the physiologists, who had long known that there existed electrical currents and potentials in living material. It was quite understandable that the cell potentials now were interpreted as altered mobility in the membrane, due to a differential solubility. Both Michaelis and Osterhout inserted their potential data in the *Nernst* equation for the diffusion potential. Of course, they came out with mobility ratios for cations/anions, which were far different from those the electrochemists had calculated for purely aqueous solutions. Beutner and Michaelis were much concerned with the 'concentration effect', namely that the potentials of their model systems were highly concentration dependent. However, I will revert to this later in this talk.

All biologists agreed that the cell membrane existed, it was not simply an open boundary between the protoplasmic phase and the surrounding tissue fluid. The dimension of the membrane structure proper became estimated first in the twenties by *Fricke*, who applied high frequency conductance measurements in red cell suspensions. (Incidentally, already in 1910 *Höber* had used more primitive such measurements and demonstrated that the electrical conductance inside the cells was quite high.) *Fricke* gave an estimate of about 400 Ångström for the membrane thickness.

Turning to the 'physical chemists' at the corresponding time after World War I, they were very much concerned with the properties of colloidal solutions. Osmotic pressure measurements of blood plasma and their components were performed. The word 'macromolecules', I think, was coined much later by The Svedberg in Uppsala, who, by aid of his famous ultracentrifuge, gave the first reliable figures for the molecular weights of the proteins. The concept 'colloid osmotic pressures' was quickly incorporated into physiology, since *Schade* in Germany and *Starling* in England had ascribed certain cases of oedema to changes of the protein content and the blood pressure conditions.

The most useful and fruitful concept was the *Donnan* effect. It was published already in 1911, it was featured by the leather chemists *Procter* and *Wilson* in 1915, but it was not until after World War I that the biologist *Jacques Loeb* of the Rockefeller Institute made it available to the biologists, through his famous book 'The Proteins and the Theory of Colloidal Behaviour' (1921, 1924). I witnessed personally the progress of the Donnan applications during the late twenties. Together with the not too old pH concept of *Sörensen* and the extensive formulation of the isoelectric point concept by *Michaelis*, it contributed very strongly to create a strong belief in physical chemistry as instrumental in biological research.

2. The 1930s and 1940s

I have dwelled at some length on the years immediately before 1930 and have tried to characterize the scientific atmosphere of that particular time, which of course influenced us, who then went into membrane research, as Dr Sollner and myself. I had just concluded a work on the secretion of gastric acid of the stomach. This was – and still is – a most startling 'ion accumulation' problem, as the term went in those days. The pure gastric acid has a pH of about 1, while the acid producing cells in the stomach have a pH of about 7.4, which means an accumulation of more than one million times! I wanted to study more fundamentals and got a fellowship to study with Osterhout at the Rockefeller Institute, where Michaelis was also a member. I learned fundamentals of electrochemistry from Osterhout's collaborators in the *McInnes* group and chose to determine 'transference numbers' in membrane models. I used an old sample of cellophane and found that the transference numbers of various chlorides were clearly concentration dependent, in fact, Beutner's 'concentration effect' was present.

A first attempt was to interpret my results in terms of phase-boundary potentials. A similar approach had been made by *Wilbrandt*, who was working with Michaelis at the same time as I was with Osterhout. I failed, and turned to a sort of phase-boundary absorption idea and finally elected to choose the simplest possible phase-boundary distribution scheme, namely that of Donnan. This meant that the cellophane membrane was conceived of as a 'colloid electrolyte' in the sense of Loeb. The total membrane potential was taken as the sum of two boundary (Donnan) potentials acting in opposition. However, the results using my own data, as well as comparisons with other concentration effect data, were not yet satisfactory. Something was missing! Maybe there was a 'forgotten' potential residing *inside* the membrane? This inside was then treated as of 'mixed' electrolytes, consisting of the bound membrane ionic groups of the colloid (the cellophane) and contributions of the free ions from the surrounding solutions, which were calculated according to the Donnan principles. The easiest way to handle the intramembrane potential was to apply the *Henderson* formula. Now everything came up seemingly satisfactory and I rushed to publish a condensed paper in the fall of 1935, which perhaps was the first quantitative fixed charge membrane theory. About half a year later *Meyer* and *Sievers* in Switzerland published very similar concepts at a much greater length. It may be of some interest to reproduce here, in Table II and Figure 1, the original table describing the course of the concentration effect, and the first explicit diagram of the concentration profiles at a fixed charge membrane, which I presented in the discussions of the first Faraday Society Meeting on membranes (London 1937).

As you may understand the new formula was the result of a fusion of ideas, which were 'on the air' at that time. All the ingredients, the Donnan-distribution, the potentials formulas were there long before – so to speak, in a bottle. But somehow, "the wine had to ripen". It took some time and it was my luck and also Meyers' and Sievers' to uncork the bottle. We used a very simple trick: to combine the Donnan-distribu-

TABLE II

X or membrane 'activity' $=1$. Membrane negative. Mobility relation $u:v$ in the membrane the same as in water. (Signs refer to the dilute solution in the external circuit.)

a_1	a_2	Partial E.M.F.		Total E.M.F. mv.
		Boundary mv.	Diffusion mv.	
100	10	+ 1.1	−13.2	−12.1
10	1	+10.9	−12.1	− 1.3
5	0.5	+20.5	−12.0	+ 8.5
1	0.1	+46.2	− 5.4	+41.8

[b] From Teorell *Proc. Soc. Exp. Biol. a. Med.*, 1935.
[a] The potential summation at a fixed charge membrane revealing the 'concentrations effect'. (*From Proc. Soc. exp. Biol. and Med.*, 1935)

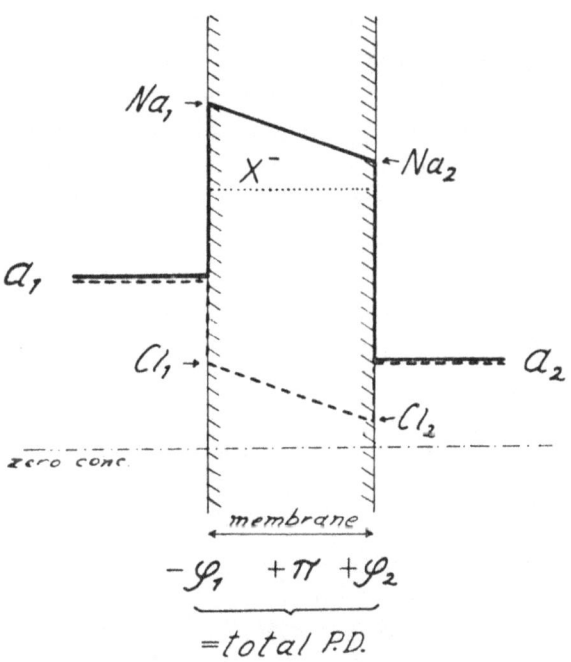

Fig. 1.   The first diagram of 'concentration profiles' at a fixed charge membrane (X⁻ is the fixed ion). (From *Disc. Farad. Soc.*, 1937.)

tion with modified diffusion equations. I couldn't imagine in the beginning that this was especially startling. Somehow, it attracted attention, witnessed also by a great deal of criticism. At the Faraday Society Meeting there was, however, an encouraging statement by *Manegold*, which deserves to be cited in extenso from the transaction publication: "– *Prof. E. Manegold* (Dresden) said: The Meyer-Teorell quantitative theory of permeability can be very well applied (so far as concerns the sieve-effect) to glass membranes, which were considered by Michaelis in connection with the potential of the glass electrode. Brittle glass contains a fixed $-0-Si-0-$ lattice of which the negative charges can be compensated by immobile surplus cations and mobile alkali ions. -'' Later we became aware of the Russian *Nicolsky*'s similar opinions about the mechanisms of the pH glass-electrode. Another kind of ion specific electrode was introduced by *Marshall* which allowed measurement of free calcium ion concentrations. I will revert to ion specific electrodes later on. Over the years I have had many misgivings about my 'Attempt to Formulate a Quantitative Theory for Membrane Permeability', the title of the first two-page publication. Various reasons and World War II prevented me from improving upon it until after the war.

Let us still remain in the last five years before the war and look at the activities in other laboratories. *Sollner* had started his pioneer work on 'permselective' membranes, which has furnished a solid basis for all further developments. While we were at the Rockefeller Institute, Dr *Wilbrandt* and I paid a visit to Princeton to meet a contemporary young man from England, *I. Danielli*. He demonstrated with pride that he had succeeded in making kind of stable, very thin bubbles of oil and proteins, which he claimed to be just two monolayers of orientated molecules, fused back to back. I wish to remind you that surface chemistry was fashionable both in England and in the U.S. at that time. Danielli's bimolecular layers, nowadays simply called 'bilayers', became immensely popular right from the start. The electron microscope, which came into use during the war years or immediately afterwards, revealed that the cell membranes, as I pointed out before, were extremely thin. They could now be estimated to be about 80 Å in thickness. Danielli's picture of the cell membranes as stable bimolecular layers of orientated proteins and lipids became highly probable (the 'unit membrane'). These views gave rise to one of the earliest objections against the 'TMS'-theory (as Meyer and I later agreed to abbreviate it to): – Was the application of the classical Nernst-Planck diffusion equations permissible on such small dimensions? During discussions I usually retorted that *Planck* had found the lower limit to be about 100 Å. Incidentally, Planck was – and still is – my 'hero' in the Science. Well, the membrane thickness offers certainly difficulties even today.

There were other failures for the original TMS concepts. *Sollner* and his collaborators had started to study 'mosaic membranes' with positive and negative parts in juxtaposition, models of a considerable biological interest. In these configurations local currents of electricity could flow and many odd ionic distribution effects were observed, which were not easy to put into the TMS mathematical framework. A 'sandwich-type' of fixed charged membranes could also be envisaged. They were materialized considerably later (*Mauro* and others) and were then found to have peculiar rectification properties, related to the technological 'transistors'. In due time it turned out that another method of theoretical approach than the kinetical was, by far, more advantageous in handling the heterogeneous membranes. This was the method of irreversible thermodynamics, which began in Belgium-Holland (*Prigogine*, of the Dutch we have present here Professor *Staverman*). This new science came into maturity after the war, I will mention it later.

Perhaps the greatest advancement in permeability research in the late thirties was the introduction of radio-isotopes, particularly useful for the biologists. It was clearly demonstrated that very rapid ion exchanges could take place at seemingly stable membrane-environment systems. The peculiar K/Na situation, for instance in the red blood cell, alluded to earlier, could be resolved as a dynamic equilibrium, or rather steady state, where the influx and the outflux of ions balanced each other quite rapidly. However, it also became evident that a single electro-chemical potential gradient was not the only driving force. One had to invoke new terms to describe the discrepancies between the actual observations and the theoretical calculations based on the Nernst electro-diffusion potential or the Donnan distributions. I think it was the great physiologist. *A. Krogh* in Copenhagen, who introduced the term 'active transport'. His associate *Ussing* defined the required extra force in terms of a specific ion potential, for instance, the 'sodium potential'. Their studies were performed on frog skins. These studies on other biological material, including isolated giant nerves from squids by *Hodgkin, Huxley* and *Keynes*, round about 1950 furnished more material for the concept of specific 'ion pumps', which were assumed to be driven by metabolic processes. According to the final 'H – H' theory the action potentials (being AC-like pulses of about 1/10 volt) arise from alternating shifts in Na- and K- 'permeability'. The first inward rush of the Na ions into the axoplasma should be reversed ('pumped out'). Figure 2 depicts a typical action potential picture of the heart, first recorded by *Weidmann* at this time. The studies on plant cells (algae) and the nerve axon also indicated the presence of electrical rectification and something called 'negative electrical

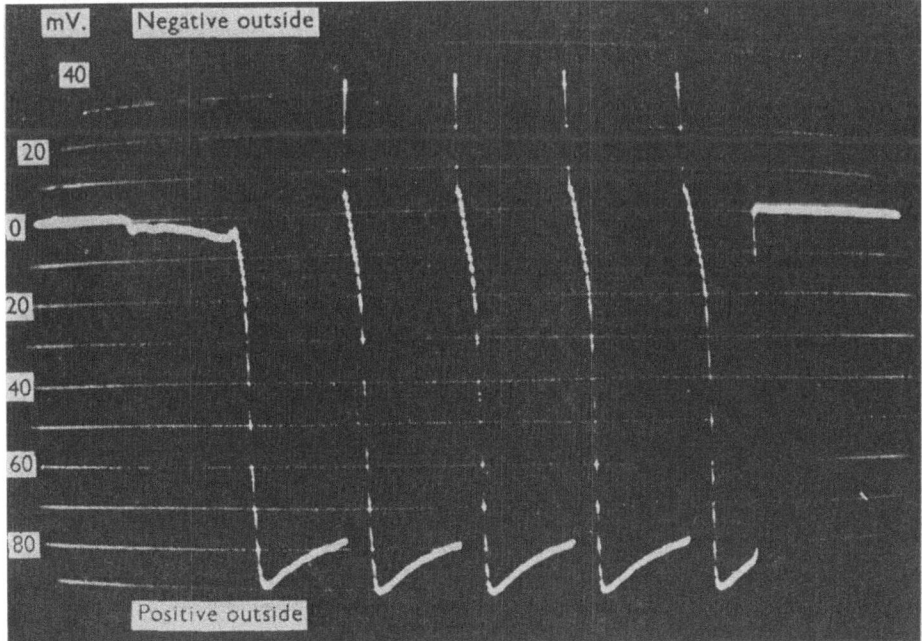

Fig. 2.   An early record of an action potential from the heart. (From Weidmann, 1951.)

conductance' (*K. S. Cole*, 1939, and others). That rectification could be a property also of artificial membranes had been known and had been elaborated at some length by *Goldman* in a fine paper published during the war (1943) (which I did not become aware of until long after the war).

3. The Years after 1950

An extension of the original TMS-theory including the electrical rectification properties, based on appropriate ionic flux equations (modified *Planck* equations) were worked out for homogeneous fixed charge membranes and I presented it at a meeting in Göttingen in 1951, organized by *Bonhoeffer*, assisted by his young associates *Schlögl*, *Hellferich*, *Manecke* and *U. Frank*. From the ensuing publication in the *Zeitschrift für Elektrochemie* I reproduce here a picture being a comprehensive presentation of the events in a fixed charge membrane, subject to a current flow, i.e. membrane electrophoreses (cf. Figure 3). Later I wrote a review article (1953) with further extensions and tried to launch the name 'Ionic membranes' for the fixed charged membranes, as suggested earlier by *Dean*. In fact, 'fixed charge' is a misnomer, because 'charge' may refer to *space* charge and usually we mean *fixed ionic groups*. Well, the term 'fixed charge' has been established, so there is nothing to do about it. The progress during the fifties was very significant. A most important and thorough treatment of membranes was given by *G. Schmid*, who successfully reconciled the old zeta-potentials of

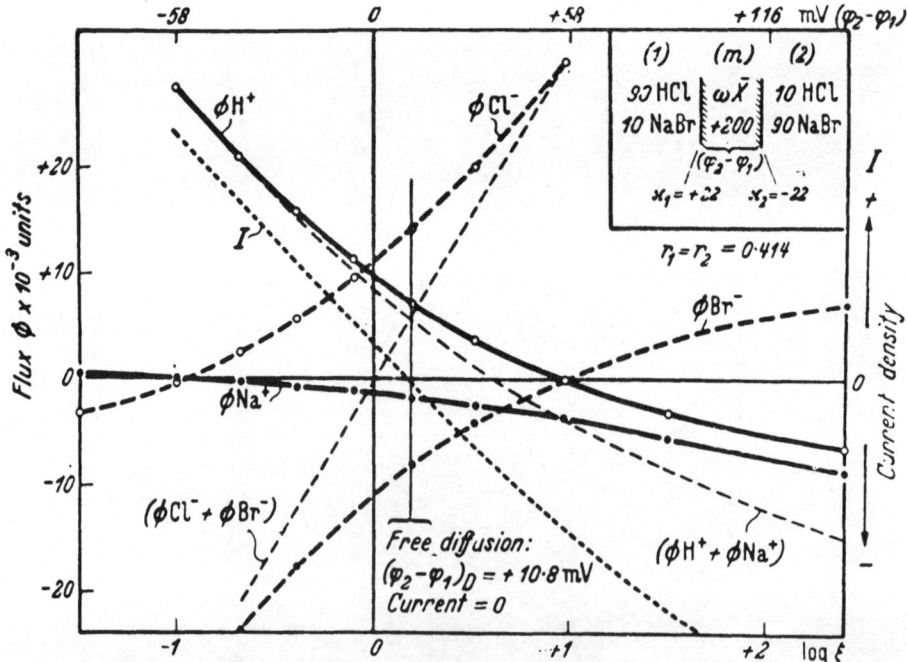

Fig. 3. A comprehensive diagram of the ion fluxes at a fixed charge membrane subject to electrical current flow. (*Zeitschr. f. El. chemie*, 1951. Reproduction from *Progr. Biophys.*, 1953.)

the colloid chemists with the TMS concepts. Schmid also introduced the effects of water permeability, which included electro-osmosis. We should remember that the original TMS formulas assumed the membranes to be 'water tight'. *Schlögl* and his collaborators published a beautiful series of extensions of the fixed charge theory, which were incorporatedin *Hellferich's* excellent book *Die Ionenaustauscher*. Incidentally, Hellferich had worked with the fine physical chemist *Scatchard*, who was much interested in membranes and the influences of the 'unstirred layers' were now formalized. Another name, in Germany, worthy of mentioning for his very original membrane research, was *R. Kuhn* with his 'permutierende Membranen'. At about this time *Meares* in Scotland got his devotion to fixed charge membranes, leading to so many important contributions over the following years to come. Both Schlögl and Meares were familiar with the irreversible thermodynamics, which, by degrees, came into use after the war. They could combine kinetics with the new ideas of 'conjugated forces and fluxes', which gave them much more theoretical power.

It should be pointed out that new materials for membranes appeared on the market during the fifties. The companies manufacturing material for water softening offered the new ion exchange resins as membranes. By the use of polymerisation chemistry membranes could be 'tailor made'. This implied that the theories could be immediately tested on a suitable membrane. This may illustrate again the importance of feed back between the various branches of sciences and technology.

For my own part I was very intrigued by the 'negative conductance', necessary for the oscillatory action potentials in the nerves and of the heart (cf. Figure 2), which during this period were intensively studied by numerous neurophysiologists, who came in a wake of *Hodgkin-Huxley* (Nobelprize 1954) and of *Tasaki*. I still had a belief that straight physical chemistry was not yet exhausted and that it was not yet necessary to invoke some vague 'active transport' or 'ion pumps' to achieve oscillations. I reasoned, if nerve can be stimulated to action by external electric current flow, what would happen if you drove a current through a fixed charge membrane? Of course, we did know of a slight rectification, which depended on strength and/or the direction of the current. But, what were the requirements for possible *oscillations* of the potential and resistance? After a great deal of experimentation we stumbled upon a most simple membrane, a sintered glass filter, used as a filter disc. This had a weak charge due to the silicic acid. It showed a quite marked electro-osmosis and, above all, it gave regularly damped oscillation, and under somewhat more narrow conditions; alas there occurred beautiful sustained, undamped oscillatory excursions of potential and resistance, which approximately mimicked the biological nerve events (cf. Figure 4). What was later to be called the 'membrane oscillator' had been found and we published a reasonable theory in terms of a set of differential equations in the late fifties.

Important experimental extensions and theoretical amplifications of the membrane oscillator were later made by *U. Franck*. He also succeeded in making an experimental arrangement which could demonstrate the *propagation* of potential waves along the membrane – simulating another striking feature of the nerve signals, that they can spread. I wish to emphasize that our artificial nerve model is no competitor to the

widely accepted Hodgkin-Huxley formalism, which contains a quite different set-up
of differential equations. In recent years I have tried to elaborate a biophysical formalism of a somewhat extended model, based on a membrane oscillator, which has
been denoted 'the electro-hydraulic excitability analogue'. The desire has been to be

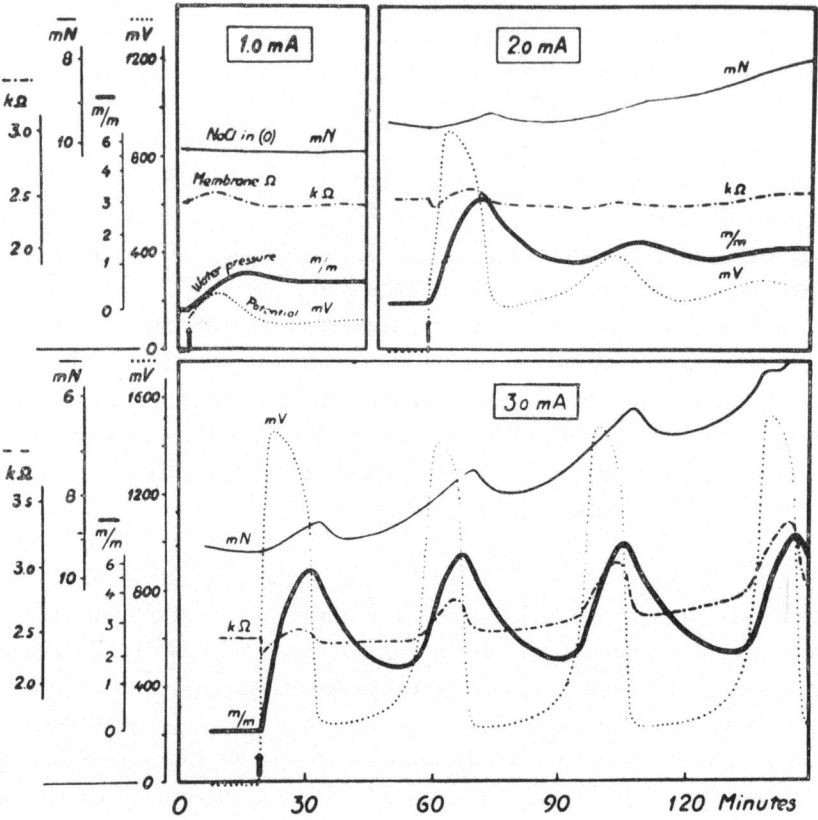

Fig. 4. Oscillations displayed at a fixed charge membrane subject to a steady electrical current (the
'membrane oscillator'). (From *Exp. Cell Research*, Suppl. 3, 1955.)

able to get some background to the facts that the nerve 'signal code' implies a 'frequency modulation' and that many nervous structures are sensitive to mechanical
pressures (compare the reception of sound in the ear!). I will dwell briefly on these
interests in another talk at this meeting. Dr *Tasaki*, with whom I soon established a
profitable cooperation about 15 years ago, accepted the fixed charge theory for nerves,
but has revised it very thoroughly to what may be called the 'conformation change
hypothesis'. I leave the details to Dr Tasaki's forthcoming talk.

As you may understand from my recitation, in the search for an understanding of
the nature of biological transport and excitation phenomena many areas of what is
now called 'membranology' have been covered. Almost 20 years after the first Faraday

Society Membrane Meeting a new Faraday Discussion was held in Nottingham, in 1956. For that occasion I had compiled a sort of 'family tree' of the developments up to that time (Figure 5). You recognize here many names of the older generations. Since then many new branches and names have been added on the tree. *Eisenman*

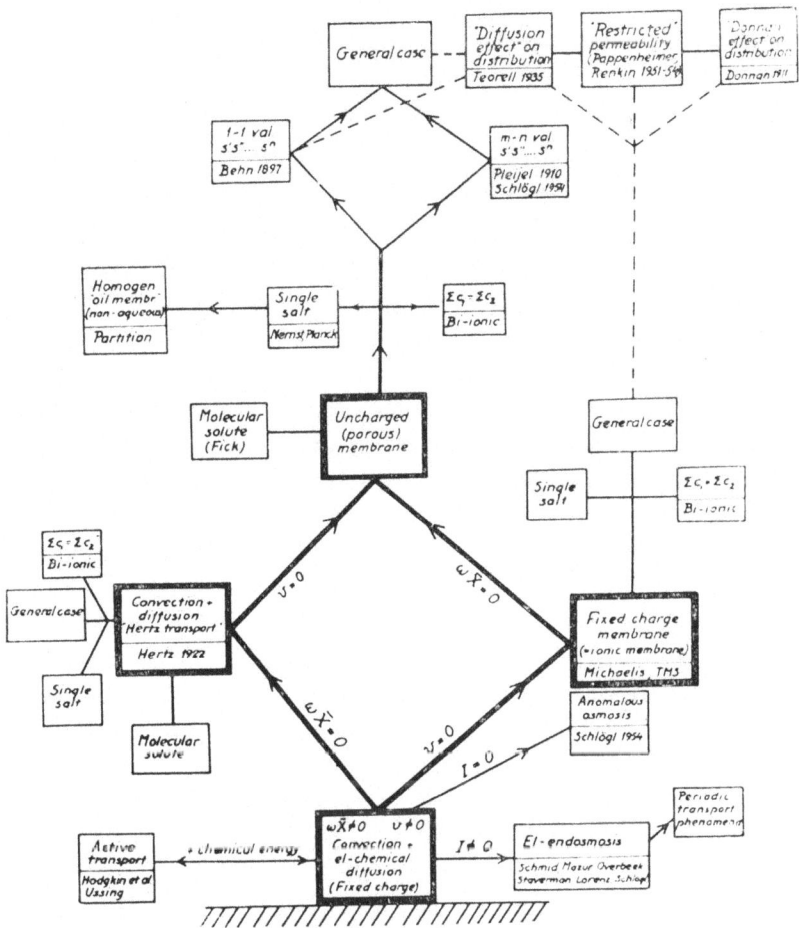

Fig. 5.   The developments of the membrane research as seen in 1956. (From *Disc. Farad. Soc.*, 1956.)

published some years ago a similar picture, which, among other things, incorporated also 'liquid fixed charge' and 'mobile site' membranes. If anybody should attempt to draw the family tree of 1973 it would be a colossal contraption in many dimensions. The output of papers dealing with membranes was summarized a few years ago in a book by *Lakshminarayanaiah;* for 1971 alone the output of papers was round about 700! For me it is impossible to give an orderly account of more recent progress, even in the fields of my personal interest. I will therefore rest at this point and confine myself to mentioning a few modern developments only in those areas which, I think, point to the future.

4. Selected Contemporary Membrane Research

The membranes are becoming 'community oriented' in recent times. *Desalination of water* and *treatment of sewage effluents* are new, important areas of applications of membranes *K. S. Spiegler* and *Gregor* of this meeting are well known names, among others, in this field.

The *ion-selectivity problem,* so important already in the early days, will probably get elucidation from the work in the new field of membranology, dealing with the bimolecular *'bilayers'*. The study of bilayers is immensely popular in our time. The interesting thing is that one can 'condition', 'dope', these films with small additions of other substances, which can drastically change the permeability. For instance, the addition of valinomycin (essentially an antibiotic) makes certain films exclusively permeable for K-ions. The bilayer membrane now behaves like a potassium glass electrode, one of the ion specific electrodes so profitably studied by *Eisenman* and collaborators. It is not astonishing that *Eisenman* in recent years has shifted his activities from glass and is now busy elaborating selectivity theories for the bilayers! In a way, the bilayer work is a rehabilitation of the classical theories dealing with 'partition' or 'solubility'. It now seems possible, just as in glass, to interrelate the membrane molecular structure with the atomic structure of the penetrating ion. Perhaps we can expect that the riddle of the biological K/Na distribution soon can be better understood. A remarkable discovery was made by *Müller* and *Rudin* who showed that certain kinds of conditioned bilayers could exhibit periodic, rather high-frequent potential changes. These were in many respects very similar to the real nerve action pictures. Interesting are also some experiments by *A. Monnier*, who got oscillations on oxidized films of linseed oil. I have heard of many reports of other kinds of model films, which displayed rhythmicity. I would venture a good guess that *oscillatory membrane events* will be a subject for the membranologists of the future.

Another dominating interest in contemporary research has been concerned with the terms *carrier transport, facilitated transport* and *'chemio-membrane' transport*. With 'chemio-transport' I refer to the influence of chemical reactions, occurring within membranes, on the transport events and the ensuing steady-state distributions of ions and molecules (*Mitchell* and others). In these domains the irreversible thermodynamics has excelled. It is, and will remain, a much more powerful tool than 'straight' kinetics or any 'activated state' theory. In this context, I wish to recall the important contributions of our late friend *Aron Katchalsky* to the bio-thermodynamics. Together with his many collaborators in Israel and the U.S., he has erected a lasting monument in honor of himself.

Among those, we have present here, Dr Ora *Kedem* and Dr *Caplan* and others, who will certainly direct the future research.

Let us proceed and discuss those membranes which are, so to speak, *'energized membranes'*, those, where the membrane matrix is the site for chemical reactions. Several workers have incorporated enzymes and other catalysts in membranes with interesting effects. Our host here, Professor *Sélégny*, and his collaborators have developed efficient 'oxygenator' membranes by the incorporation of catalase and carbonic anhydrase in suitable polymers, which facilitate the diffusion of oxygen and CO_2. Some people have even been able to make 'enzyme electrodes', which are directed towards the glucose in blood, i.e. a 'glucose electrode'. In this particular area we may expect the most rewarding technical and medical applications. The new discoveries of cyclical events in test tube solutions or cell suspensions made by *B. Hess* in Germany and the *Britton Chance* group and others in the U.S. have clearly demonstrated that chemical reactions can be periodical on the time scale. What could one not expect then from chemical reactions within structured membranes? The riddle of the biological excitation phenomena might find its final answers after studies of charged membranes with 'mobile sites', where chemical reactions take place, controlled by enzymes, fixed to the groups of the membrane matrix. Probably this kind of system would be highly labile and would easily become oscillatory. If this happens, many pertinent questions will arise: What sort of energy is available to drive the oscillations? What sort of detailed coupling exists between the chemical forces and the transport forces? What can be the role of the membrane structure in such events?

5. Looking Towards the Future

As you may observe, I have started to make predictions for the future. Nothing is more difficult, nothing is more futile – a human being can only make extrapolations from observations of concepts and events in his contemporary time. But, nevertheless I will speculate about the coming time. How would the everlasting riddle of the 'origin of life' affect the membranology in the future? In a sweeping statement one might venture to say, that ordered structures of macromolecules was the first requirement for life and the first organism deserving the name of a 'living cell'. Taking this for granted, the pertinent question would arise: What kind of forces are arranging the high molecular materials into membranes? The proposition might be seen as very remote, but, in fact, the problem has already been posed, not in philosophical but in highly refined mathematical terms, by two leaders in the field of irreversible thermodynamics. Both *Prigogine* in Belgium and the late *Aron Katchalsky* of Israel started to attack this most important problem of '*structure formation*'. Here is a challenge for the future.

I am now at the end of my presentation of some developments within membranology over the last 50 years, which I have been able to witness. What wish, or advice, for the future research could I come up with? My personal answer is this – Look for, or make membranes, which have properties of '*fuel cells*'! We know that fuel cells in modern technology can convert chemical energy directly into electrical energy. We know that the background of biological excitability is electrical and energy requiring. Why should not Nature have local electrical current sources of fuel cell type incorporated in the living structures? A revelation of a scheme for the coupling between metabolic energy and electrical ionic current may be a dream – not entirely a dream – rather a vision, which could be realized by the next generations of membrane students.

II

EQUILIBRIA

THE NATURE OF THE SELECTIVE BINDING OF IONS BY POLYELECTROLYTE GELS: VOLUME AND ENTROPY CHANGE CRITERIA*

GEORGE E. BOYD

*Oak Ridge National Laboratory, Oak Ridge, Tenn. 37830, U.S.A.***

Abstract. The volume, ΔV^0, and entropy, ΔS^0, changes which accompany the selective binding of ions in ion exchange reactions at 25° between aqueous electrolyte mixtures and polyelectrolyte gels are considered in a discussion of the range of applicability of the Katchalsky and Rice-Harris theories. Experimental methods for the estimation of ΔV^0 are reviewed, and dilatometer measurements with soft and hard cationic and anionic gels are reported. Apparent molal volume determinations with aqueous model compound and linear polyeletrolyte-analog solutions also were conducted to elucidate the nature of the hydration of polyelectrolyte gels.

Four types of ion-binding processes were distinguished based on ΔV^0 and ΔS^0: (a) 'Field' binding; (b) Site-binding via ion-pair formation; (c) 'structure-enforced' binding; and, (d) charge-transfer binding via complex formation. Multiply charged cations, except possibly Cr^{+++}, were bound as ion-pairs in all gels. Field binding was inferred only for the uptake of singly-charged cations in lightly cross-linked gels. The larger, symmetric tetra-n-alkylammonium cations showed structure-enforced binding; charge-transfer complex formation appeared to be involved in the binding of the heavier halides.

1. Introduction

Polyelectrolyte gels when placed in aqueous mixtures of simple electrolytes at room temperature frequently undergo perceptible changes in their external volume. At the same time, a selective binding of some of the ions in solution occurs with a lowering of the free energy, and with changes in the enthalpy, entropy and heat capacity of the system. The volume of the system frequently appears to change only slightly suggesting that ΔV is small. Speculations as to the origins of these thermodynamic property changes and their bearing on the nature of the binding of ions by polyelectrolytes have continued to excite interest. Electrostatic forces doubtless are usually of first importance, and the quantitative theories of Katchalsky and coworkers [1, 2] and of Rice and Harris [3] attempt to account for the equilibrium behavior of ion exchange gels in terms of the interaction of the counter-ions with the fixed charge.

The theory of Katchalsky, which is based on a cell model, should apply only to lightly cross-linked gels, although it has been extended [4] to heavily cross-linked preparations. In this theory the interaction between all groups on one and the same chain in the gel network is calculated ignoring the interactions with neighboring chains. No chemical interactions between the counter-ions and fixed charges are assumed (i.e., no ion-pair formation). Rather, the counter ions are constrained by the electrostatic field (i.e., 'field binding'). Thermodynamic property changes accompanying the

---

* Presented at the 1973 NATO Advanced Study Institute on Charged and Reactive Polymers, Forges les Eaux, Normandy, France, September 15–23, 1973.

Eric Sélégny (ed.), Charged Gels and Membranes I, 73–89. All rights reserved.

replacement of one type of counter ion by another therefore should be approximately the same as those observed in the mixing of simple electrolytes (i.e., $\Delta V \leqslant 1$ ml mole^{-1}).

The Rice-Harris theory, which is based on a lattice model, applies to moderately and heavily cross-linked polyelectrolyte gels. Ion-pairs are assumed, and, in fact, these are the sole cause of the selective binding. Volume changes of the same order of magnitude as for the formation of ion-pairs in electrolyte solutions may be expected (i.e., 3 to 15 ml mole^{-1}) depending on whether outer sphere (solvent separated ion-pairs) or inner sphere (contact ion-pairs) complexes are formed. Recently Hemmes [5] has shown that the volume change, ΔV_F, for the creation of a 'Fuoss ion-pair' is given by:

$$\Delta V_F = \frac{|z_A z_B| \, e_0^2 N}{aD} \left(\frac{d \ln D}{dP} \right)_T - RT\beta \tag{1}$$

where a is the distance of closest approach of the ions, D is the dielectric constant, e_0 is the electronic charge, β is the solution compressibility, and the other symbols have their usual meanings. The first term on the righthand side of Equation (1) is greater than the second, hence, the predicted volume change is positive. (A similar conclusion can be deduced for 'Bjerrum ion-pairs'.) Further, Equation (1) shows that ΔV_F increases as the distance of closest approach decreases. Hence, a criterion for site binding of ions by polyelectrolyte gels when ion association occurs is that ΔV will be positive and will increase as the strength of binding increases.

Enthalpy, entropy and heat capacity changes also reflect the interactions involved in selective ion binding reactions. Thermochemical measurements [6], for example, have shown that in mixtures of like, equally charged ions selectivity is determined by the enthalpy lowering, whereas with heterovalent ion exchange reactions the entropy increase is governing. Heat capacity changes are especially revealing [7], but, unfortunately, very few determinations of this kind have been reported.

Other types of interactions (i.e., formation of charge-transfer complexes or 'structure-enforced ion-pairs') of counterions in polyelectrolyte gels may determine selective binding, and the sign and magnitude of the volume or entropy change for these is not readily predictable. Non-electrostatic effects on the hydrogen-binding of water must always be present and sometimes may be dominant.

2. Methods for the Experimental Estimation of Volume Changes in Polyelectrolyte Gel systems

Definitions: The volume change in the selective binding of an ion in an ion exchange reaction such as, for example, in the reaction between singly-charged cations

$$M_1 R(a = 1 \text{ equil. } 0.2mM_1\text{Cl}) + M_2\text{Cl}(0.2 \text{ m}) \rightarrow$$
$$[M_1 + M_2] R(x_{M_2} \simeq 0.05, \text{ equil. } 0.2 \text{ m } M, \text{ Cl}) + M_1\text{Cl}(0.2m) \tag{2}$$

may be estimated in several ways. The ΔV in the reaction is

$$\Delta V = [\phi_v^0(M_2 R) - \phi_v^0(M_1 R)] - [\phi_v(M_2\text{Cl}) - \phi_v(M_1\text{Cl})] \tag{3}$$

where $\phi_v^0(M_1R)$ and $\phi_v^0(M_2R)$ are the apparent molar volumes of the respective homo-ionic forms in their 'standard states' defined as the gel in equilibrium with 0.2 m aqueous chloride solution containing the same cation as in the gel. The volume change is therefore seen to be equal to a 'difference in differences' between apparent molar volumes. The correction, $\Delta\phi_v$, for the electrolyte solutions to a standard state of infinite dilution is small, hence the ΔV of Equation (3) approximates ΔV^0, the volume change in the standard state reaction:

$$M_1R(a=1, \text{equil. } 0.2 \text{ m } M_1Cl) + M_2Cl(a=1) \rightarrow$$
$$M_2R(a=1, \text{equil. } 0.2 \text{ m } M_2Cl) + M_1Cl(a=1) \tag{4}$$

Pressure Coefficient Measurements: The well-known equation

$$(d \ln K_a/dP)_{T, x} = -\Delta V^0/RT \tag{5}$$

for the effect of pressure on the thermodynamic equilibrium constant, K_a, at constant temperature and composition may be employed to estimate ΔV^0, and, in principle, values more accurate than by any other method can be obtained. Only two applications of Equation (5) to ion exchange equilibria have been reported [8, 9] and the results given in one report [8] are puzzling: hydrostatic pressures approaching 6670 bar had no effect on the potassium-hydrogen ion exchange equilibrium with cross-linked polystyrenesulfonate gels, while with the strontium-hydrogen equilibrium K_a increased with pressure (i.e., ΔV^0 for the selective binding of Sr^{++} must be negative.)

Density Measurements on Hydrated, Cross-Linked Gels: Apparent molar volume measurements on the homo-ionic gels in equilibrium with dilute aqueous electrolyte solutions may be employed together with Equation (3) and the published measurements [10] of ϕ_v values for electrolyte solutions to estimate ΔV. The apparent equivalent volume of MR is defined by,

$$\phi_v(MR) = \frac{V - n_w V_w^0}{n_2} = V_e - x_w V_w^0 \tag{6}$$

where V_e is the equivalent volume of the gel (ml equiv^{-1}), x_w is the equivalent water content (moles H_2O/equiv. or g·H_2O/equiv.) and V_w^0 is the molar volume (18.069 ml) of pure water. Measurements of $\phi_v(MR)$ of high accuracy are required as relatively small differences (cf. Equation (3)) between large quantities give ΔV. Thus far, because of experimental difficulties it has been almost impossible to measure the apparent volumes of lightly cross-linked gels.

Dilatometric Measurements: Direct measurements of the volume change at room temperature and one atmosphere pressure may be performed by conducting reaction (2) in a dilatometer. A convenient instrument is the inverted 'V' type, two-bulb instrument described by Linderstrøm-Lang and Lang [11]. A few measurements of volume changes occurring in the selective binding of ions by cross-linked polystyrenesulfonic

acid gels were carried out by Strauss and Leung [12] during the course of their pioneering investigations with polyelectrolytes in aqueous solutions. Additional systematic measurements are desirable because of their relevance to the theories of Katchalsky and of Rice and Harris; these measurements are described in this paper.

Density Measurements Related to the Hydration of Polyelectrolyte Gels: The nature of
the water in polyelectrolyte gels is closely related to their apparent molal volumes and
hence to ΔV^0. Numerous NMR (proton resonance) investigations on hydrated ion
exchangers have emphasized the similarities of the gel solution with concentrated
aqueous electrolyte solutions, while other types of researches have revealed unique
properties not shared by electrolytes. [13] Density measurements with aqueous solutions of model compounds and of linear polyelectrolyte analogs therefore were performed to determine the nature of the hydration of the molecular network of the
charged gel.

3. Experimental

Materials: Strong-acid type cation exchange gels (Dowex-50W) based on cross-linked
sulfonated polystyrene were utilized. Their cross-linkings in nominal percent divinylbenzene (DVB) were 0.5%, 8%, and 24%; their exchange capacities were 5.46, 5.11,
and 4.30 mequiv. per g, dry hydrogen-form, respectively. A strong-base anion exchange
gel (Dowex-1) of 0.5% nominal DVB cross-linking and 4.34 mequiv. per g dry chloride-
form capacity was employed for measurements of the anion selective binding reactions.
Determinations of the weak base capacity indicated less than 2% of the capacity was
attributable to this source.

Apparatus and Procedure: Pycnometric methods for the measurement of the equivalent
volume, V_e, and the procedure for the estimation of the equivalent water content, x_w,
of cross-linked ion exchange gels have been described [14, 15]. The pycnometer liquid
used in measurements with gels in which x_w was varied was *n*-octane; isopiestic vapor
pressure equilibrations with saturated salt solutions were conducted at constant temperature in a humidistat to adjust the equivalent water contents.

The procedure followed in the dilatometer measurements of ΔV in ion selective
binding has been reported in detail [16]. Several precautions essential to the successful use of this simple but elegant technique should be observed: (a) The temperature
of the bath in which the dilatometer is immersed should be constant to $+0.001\ °C$ for
at least 30 min; (b) The solutions in the dilatometer should be degassed before the
capillary stem is attached to the bulb system; (c) The standard taper ground glass
joint connecting the capillary to the bulb system must be lubricated carefully so that it
is leak-free. Fluorocarbon greases (i.e., KEL-F #90, 3MCo., or Fluoroglide, Chem-
plast, Inc.) are believed to be superior to glycerol, phosphoric acid and other recommended compounds [17].

Density measurements on aqueous polyelectrolyte and model compound solutions
were conducted at 25.000 °C with a digital precision densimeter [18]. Instrument cali-

brations were made with sodium chloride solutions of accurately predetermined concentrations. Details on the source and purification of the compounds, preparation of solutions, etc. are given elsewhere [19].

4. Experimental Results and Discussion

Model Compound Solutions: Density measurements on aqueous solutions of *p*-ethylbenzenesulfonic acid (HpEBS) and three of its salts were employed to compute the apparent molal volumes of these model compounds for the strong-acid ion exchanger, Dowex-50. These data, shown in Figure 1, were fitted to the Redlich-Meyer [20] equation

$$\phi_v = \phi_v^0 + S_v\sqrt{c} + b_v c \tag{7}$$

where ϕ_v^0 is the apparent molal volume at infinite dilution ($\phi_v^0 = \bar{V}_2^0$), S_v is the theoretical limiting law slope, equal to 1.868 for $1-1$ electrolytes at 25 °C, and b_v is an empirical measure of the deviation from the limiting law. The concentration dependence of ϕ_v for two quaternary ammonium halides, trimethylbenzylammonium chlo-

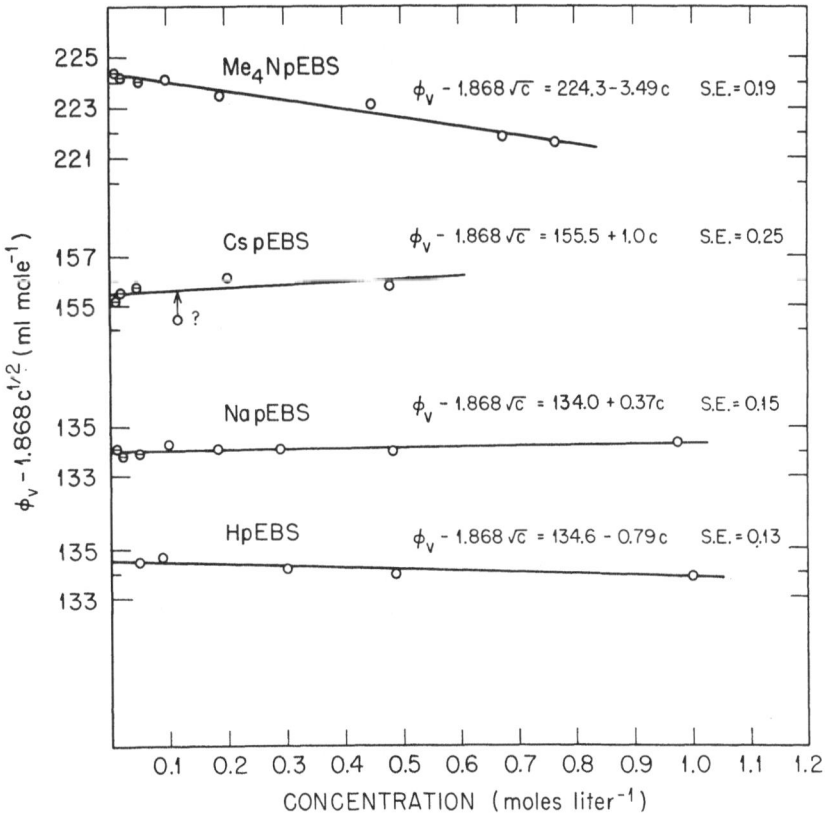

Fig. 1. Concentration dependence of the apparent molal volume of *p*-ethylbenzenesulfonic acid and its salts at 25°.

ride, which is a model compound for the chloride form of the strong-base anion exchanger, Dowex-1, and dimethylethanol-benzyl-ammonium bromide, which is the model compound for Dowex-2, is shown in Figure 2. The conventional partial molal volume of the pEBS$^-$ ion at 25° may be derived from these measurements using values [10] for $\bar{V}^0 (= \phi_v^0)$ for H$^+$, Na$^+$, Cs$^+$, and Me$_4$N$^+$. Thus, $\phi_v(p\text{EBS}^-) = 134.7 \pm 0.2$ ml mole^{-1} where the uncertainty is the standard error. Values for the cations Me$_3$BzN$^+$ and Me$_2$EtOHBzN$^+$ are 151.3 and 166.9 ml mole^{-1}, respectively. Interestingly, the

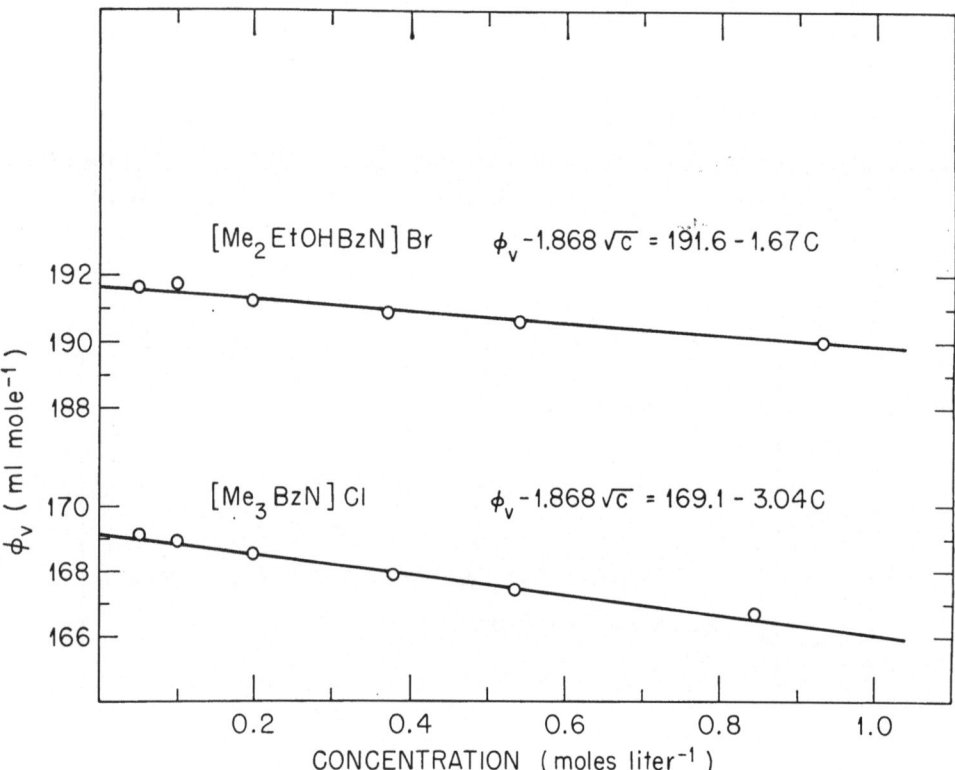

Fig. 2. Concentration dependence of apparent molal volumes of quaternary ammonium salts at 25°.

value of b_v for Me$_3$BzNCl(-3.0) is close to that observed [21] for Me$_4$NCl(-2.5). The deviations from the limiting law shown by Et$_4$NCl(-16.0), n-Pr$_4$NCl(-23.8) and n-Bu$_4$NCl(-35.5) are much larger [21].

Polyelectrolyte Solutions: Apparent molal volumes computed from density measurements on aqueous solutions of salts of polystyrene-sulfonic acid at 25° are plotted in Figure 3 as a function of the logarithm of the concentration. Also shown are the ϕ_v values (ml monomole^{-1}) for the cationic polyelectrolyte, polyvinylbenzyltrimethyl-ammonium chloride. A linear dependence of ϕ_v on log c for concentrations below

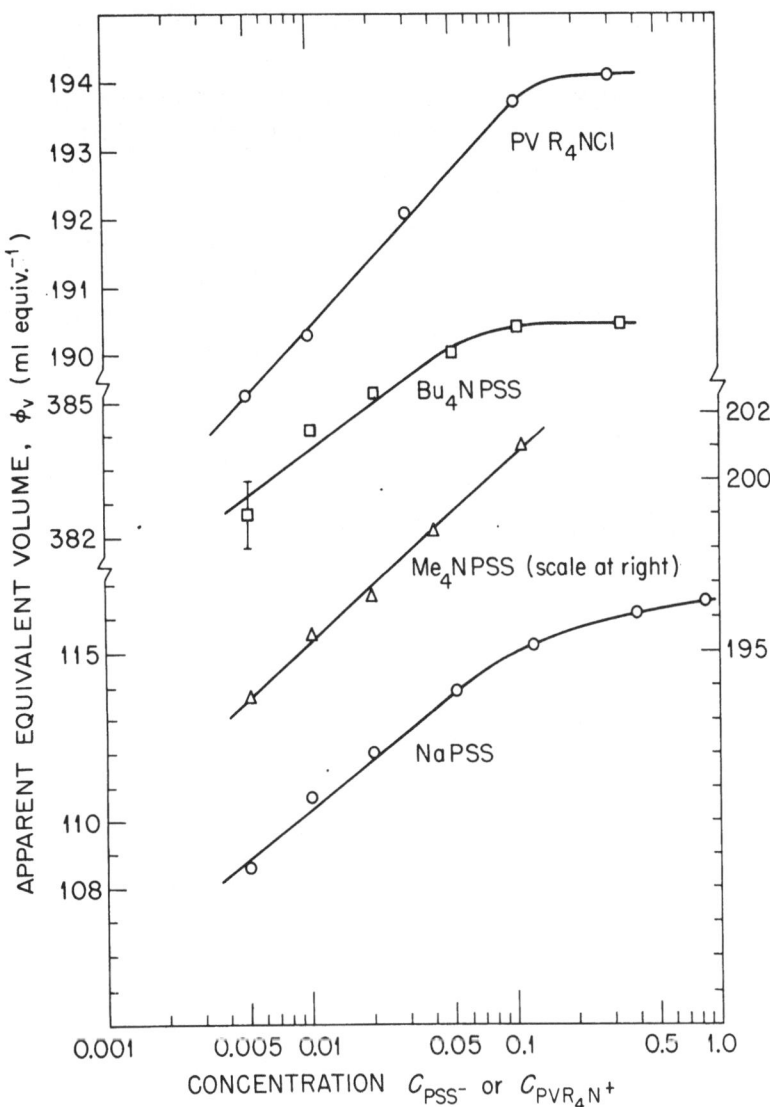

Fig. 3.   Concentration dependence of the apparent equivalent volumes of several polystyrenesulfonates at 25 °C.

0.1 M was observed in agreement with polyelectrolyte theory and with the recent
report of Skerjanc [22], but not with an earlier report [23] on apparent molal volumes
of polyelectrolytes in aqueous solutions. Because of the logarithmic concentration
dependence of ϕ_v it is not possible to extrapolate to infinite dilution to obtain conventional partial molal volumes per monomole of polyelectrolyte. None-the-less, it is clear
that ϕ_v for polystyrenesulfonate is substantially smaller than that for pEBS$^-$ ion. The
difference in the case of the sodium salts amounts to approximately 23 ml per monomole showing that polymerization produces a substantial decrease in ϕ_v in agreement
with Tondre and Zana [23]. This effect has been interpreted by Ikegami [24] and

by Oosawa [25] as being caused by the cooperative electrostrictive action of the charged groups of the polyelectrolyte on the water structure. The water adjacent to the polyelectrolyte thus belongs to two regions: that associated with the intrinsic hydration spheres of the ionogenic groups and that which lies in the cylindrical hydration sheath created by the overlap of the spherical regions.

Apparent Molal Volumes of Polyelectrolyte Gels: A summary of $\bar{\phi}_v$ for various salt forms of a moderately cross-linked (nominal 8% DVB content) polystyrenesulfonate gel is given in Table I together with experimentally determined values [15, 26] of the

TABLE I

Typical values for V_e, x_w and $\bar{\phi}_v$ for salt-forms of moderately cross-linked polystyrenesulfonate in equilibrium with water at 25 °C

Salt-form	V_e (ml equiv.$^{-1}$)	x_w (moles H$_2$O equiv.$^{-1}$)	$\bar{\phi}_v$ (ml equiv.$^{-1}$)
H$^+$	317.5	10.2	118.5
Li$^+$	313.2	10.1	117.2
Na$^+$	295.4	8.8	122
K$^+$	282.7	7.6	131
Rb$^+$	285.0	7.5	137.5
Cs$^+$	289.8	7.5	139.9
NH$_4$$^+$	293.5	7.8	136.9
Ag$^+$	251.9	6.7	120.6
Mg^{++}	297.1	9.8	(119.8)
Ca^{++}	280.0	8.6	124.6
Sr^{++}	274.5	8.3	125.3
Ba^{++}	258.0	7.0	132
H$^+$	320	11.2	115.9
NH$_4$$^+$	294	8.6	136.0
Me$_4$N$^+$	360	8.7	200.2
Et$_4$N$^+$	390	7.5	251.4
n-Pr$_4$N$^+$	429	6.3	311.5
n-Bu$_4$N$^+$	457	4.2	378.0

equivalent volume (ml equiv^{-1}) and water content (moles H$_2$O equiv^{-1}) of the gel in equilibrium with pure water at room temperature. Three points may be noted: (a) V_e and $\bar{\phi}_v$ increase with the atomic weight, A, for singly-charged cations and decrease with A for divalent cations; (b) x_w decreases with increasing A for both uni- and divalent cations; (c) V_e and ϕ_v for the *n*-tetra-alkylammonium ions are large and increase with the size of the organic ion while x_w decreases.

The limited utility of the $\bar{\phi}_v$ values in Table I for the estimation of ΔV with Equation (3) is illustrated by the entries in Table II. The second column of Table II gives $\Delta \bar{\phi}_v$ computed from the $\bar{\phi}_v$ values of column 4 of Table I. Column 3 of Table II was computed for 0.1 N aqueous solutions with data tabulated in the literature [10]. A comparison of the ΔV's in columns 4 and 5 with those in column 6 confirms the fact that

TABLE II

Comparisons of ΔV values (ml equiv.$^{-1}$) for reactions with H-form of nominal 8% DVB cross-linked polystyrenesulfonate derived from apparent equivalent volume, $\bar{\phi}_v$, and dilatometer measurements

Cation	$\Delta\bar{\phi}_v$	$\Delta\phi_v$	ΔV^a	ΔV^b	ΔV^c
H$^+$	0.0	0.0	0.0	0.0	0.0
Li$^+$	−1.9	−0.7	−	−1.2	0.8
NH$_4$$^+$	20.3	18.2	2.1	1.2	−
Na$^+$	2.5	−1.0	−	1.5	4.1
Cs$^+$	20.5	21.6	−	−1.1	4.8
Ag$^+$	−0.8	0.3	−	−1.1	3.1
Me$_4$N$^+$	84.9	89.3	−4.4	−6.2	−4.0
Et$_4$N$^+$	136.2	146.2	−10.0	−6.9	−
Pr$_4$N$^+$	196.5	210.5	−14.0	−	−
Bu$_4$N$^+$	263.2	269.6	−6.4	−	−
Ca^{++}	−6.3	2.3	−	−8.6	9.0
Ba^{++}	1.2	7.6	−	−5.4	7.7

[a] ϕ_v: ORNL (1956)
[b] ϕ_v: Gregor et al. (1951–52)
[c] Dilatometer measurements: ORNL

when the volume change is small values estimated with Equation (3) may be seriously in error and even show an incorrect sign. The measurement of $\phi_v^0(MR)$ values to better than 0.3% would require great care. The direct measurement of ΔV with a dilatometer is clearly superior, and, if $\phi^0(MR)$ of one salt form is determined with high accuracy, reliable apparent molal volumes for other salt forms can be obtained with the ΔV's determined directly.

Solvation of Polyelectrolyte Gels: Measurements of V_e as a function of the hydration of the ion exchange gel (Figures 4 and 5) are of interest in that they show that the partial molal volume, $\bar{v}_w$, of the water taken up initially is much smaller than the normal molar volume of 18.069 ml. Thus, from Figure 4, and the definition,

$$\bar{v}_w = (\partial V_e / \partial x_w),$$ (8)

the first water absorbed by the sodium form ($x_{zn} = 0.0$) possesses $\bar{v}_w = 12.44$ ml mole^{-1}. As might be expected, there appears to be much less electrostriction of the water in anion exchangers (cf. Figure 5). Only with the fluoride form is V_e non-linear at small x_w values.

Frequently, V_e and x_w are observed to vary almost linearly with the ionic composition of the polyelectrolyte gel. The data plotted in Figure 6, which exhibit a strongly non-linear dependence, are of interest because of the linear behavior of the apparent molal volume. Additivity of ϕ_v is observed frequently in electrolyte mixtures to a high order of precision, therefore, it is perhaps not unexpected with polyelectrolyte solutions and gels.

GEORGE E. BOYD

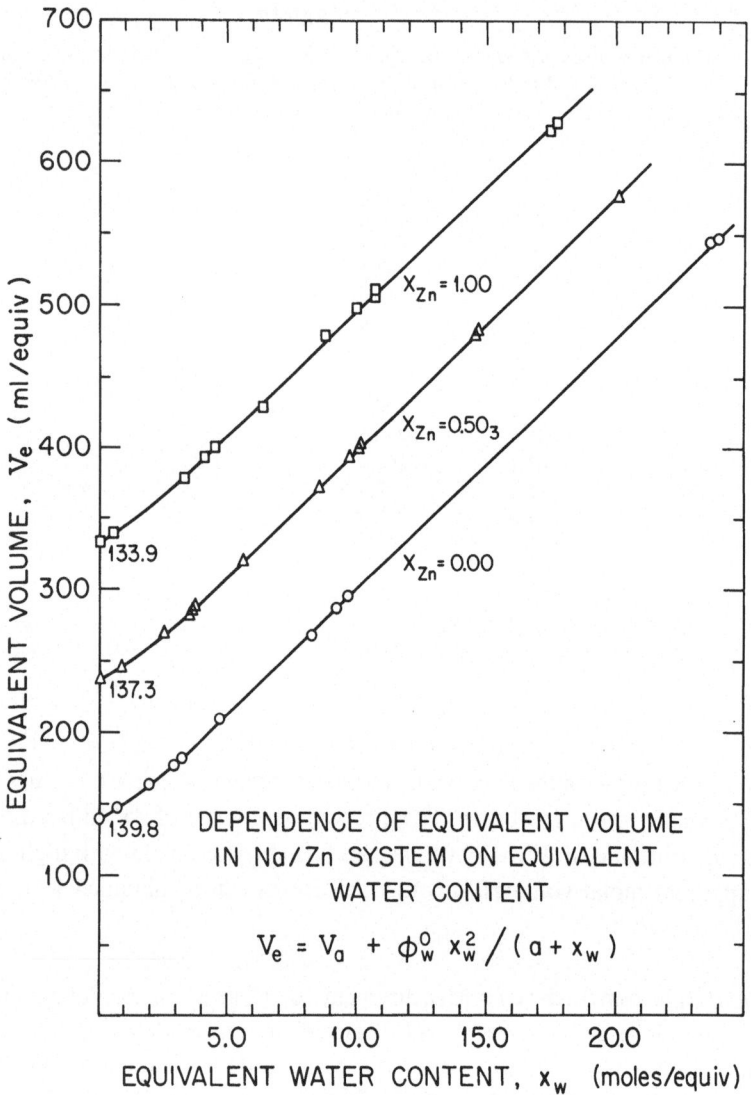

Fig. 4. Dependence of equivalent volume in Na/Zn system on equivalent water content.

Dilatometer Measurements of ΔV for Se lective Binding of Ions: An important observation made during the conduct of the dilatometer measurements was that the volume changes in the selective binding reactions were additive within experimental error. For example, with a lightly cross-linked cation exchange gel the ΔV for the exchange of tetramethylammonium with NH_4^+ ion in the gel was -0.94 ml/equiv. The *sum* of the ΔV's for the reaction of Me_4N^+ with Na^+ ion (-1.23 ml/equiv.), of Na^+ with Cs^+ ion (-0.66 ml/equiv.) and of Cs^+ with NH_4^+ ion (0.92 ml/equiv.) was -0.97 ml/equiv., which is considered to be in agreement with -0.94 ml/equiv. Because of additivity the ΔV values in Tables III–V are given relative to the replacement of H^+

ion by cations, and relative to the replacement of F^- ion by anions. Hydrogen ion is the least strongly bound ion, except for Li^+, by polystyrenesulfonate in solution or in gels. Raman [26], IR [27] and NMR [28] measurements are in agreement that polystyrenesulfonic acid is highly dissociated, hence the binding of H^+ may be assumed

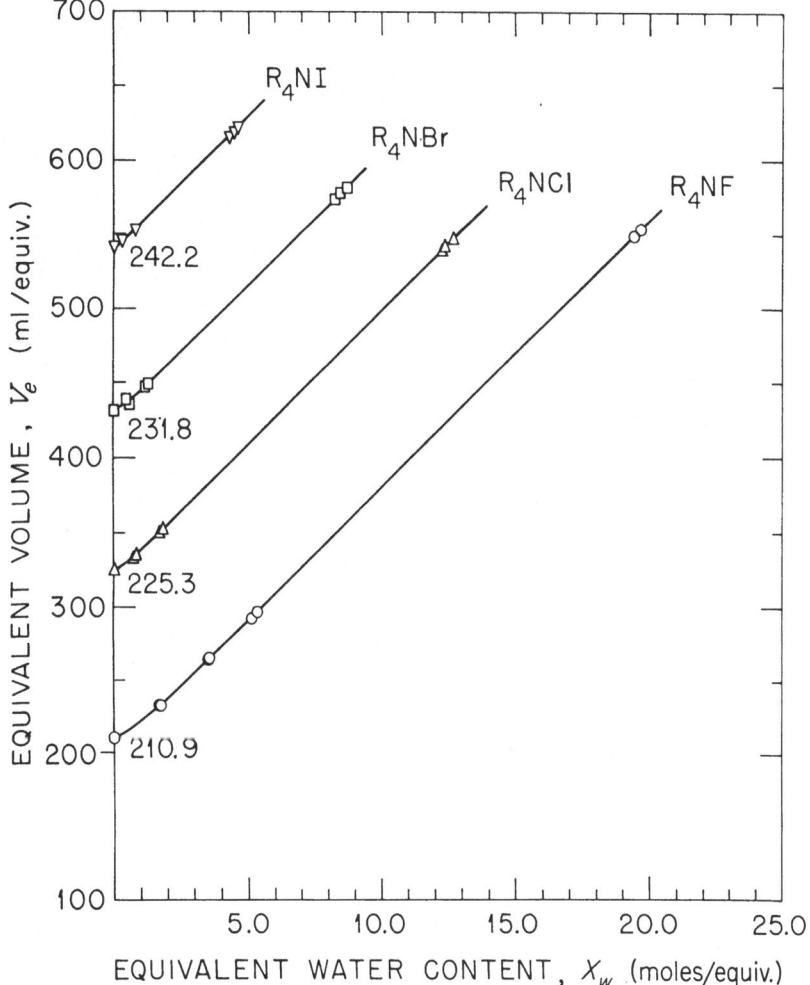

Fig. 5.   Equivalent volumes of halides of Dowex-2X2 as a function of their hydration.

to be non-localized. Fluoride is the anion least strongly bound by polyvinyl-trimethyl-ammonium type ion exchange gels.

Selective Binding of Ions in Lightly Cross-Linked Cation and Anion Exchange Gels: The volume changes observed with singly-charged cations were small and were either positive or negative. (Table III). Selective binding of the alkali-metal cations and NH_4^+ ion was accompanied by a volume increase and a negative ΔS^0. Field binding of these

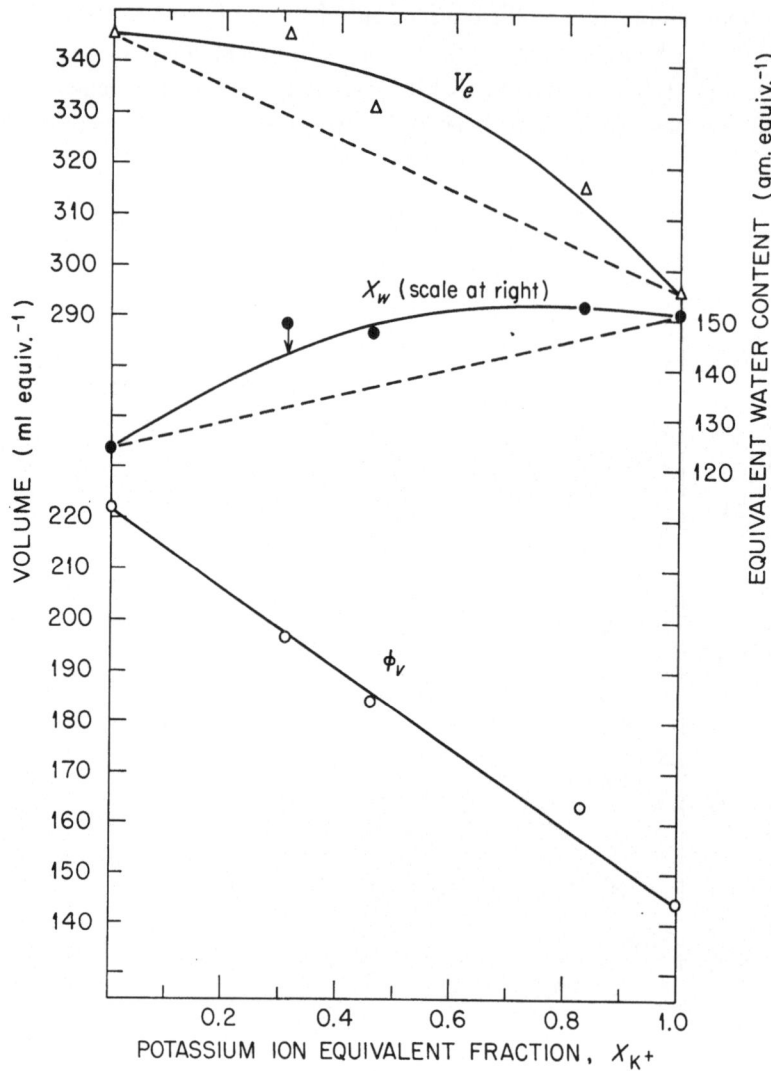

Fig. 6. Additivity of apparent equivalent volume, ϕ_v, in the $Me_4N^+-K^+$ ion exchange reaction with DVB10 polystyrenesulfonate. (Computed from Data of Gregor, Gutoff & Bregman, 1951).

ions may be inferred and the small entropy decrease may be attributed to a net in-increase in the ordering of the solvent molecules in the system.

The volume *decreases* which accompanied the selective binding of the *n*-alkylam-monium cations were significant, as electrostatic ion-pair formation must be excluded. A substantial entropy increase and the absorption of heat also accompanied the uptake of the largest of these ions. The process may be described as 'structure-enforced' binding [24] because of the apparent importance of ion-solvent and the solvent-solvent rather than ion-ion interactions. Since the researches of Frank and Wen [30] it has been recognized that large, organic cations interact in a special way with hy-

TABLE III

Volume, standard free energy, enthalpy, and entropy changes for the replacement of H$^+$ ion from a lightly cross-linked polystyrenesulfonate gel

Cation	ΔV (ml mole^{-1})	ΔG^0 (kcal mole^{-1})	ΔH^0 (kcal mole^{-1})	ΔS^0 (cal deg^{-1} mole^{-1})
Li$^+$	0.1	−0.05	0.07	0.0
NH$_4^+$	0.2	−0.06	−0.28	−0.8
Na$^+$	0.4	−0.01	−0.26	−0.9
K$^+$	0.6	−0.13	−0.49	−1.2
Cs$^+$	1.1	−0.24	−0.62	−1.3
Ag$^+$	1.7	−0.23	−0.18	0.1
Me$_4$N$^+$	−0.8	−0.29	−0.82	−1.9
Et$_4$N$^+$	−2.5	−0.36	−0.76	−0.7
n-Pr$_4$N$^+$	−3.5	−0.40	+0.30	+2.2
n-Bu$_4$N$^+$	−2.0	−0.49	+1.95	+8.1
Me$_3$BzN$^+$	−1.2	–	–	–
Mg^{++}	2.6	–	–	–
Ca^{++}	4.2	–	–	–
Ba^{++}	6.2	–	–	–
Zn^{++}	4.2	–	–	–
Cr^{+++}	2.4	–	–	–
La^{+++}	12.3	–	–	–
Th^{4+}	15.3	–	–	–

drogen-bonded solvents. Recent X-ray diffraction measurements [31] with tetra-n-butylammonium fluoride solutions have supplied direct, quantitative evidence that such ions promote short-range order in water:

(1) The fraction of cavity sites occupied by water molecules in the expanded ice-I 'structure' of liquid water decreases from 0.5 for pure water to zero for the solution.

(2) The nearest neighbor network distances between water molecules are shortened.

(3) The 'root-mean-square' displacements associated with the network distances are significantly smaller than in pure water (i.e., the interaction is sharper).

Changes in the short-range order caused by large, organic cations are reflected also by the increased partial molal volume, $\bar{v}_w$, of water in the solution, by an increase in its heat capacity, and by a decrease in its entropy. When, because of selective binding, R$_4$N$^+$ ions leave the aqueous phase, the cavities in the structure partially refill with water molecules and $\bar{v}_w$ decreases and approaches 18.068 ml mole^{-1}; hence, ΔV is negative and ΔS^0 is positive. The volume change with tetra-n-butylammonium ion appears (Table III) to be irregular in that its ΔV was smaller than with tetra-n-propyl-ammonium ion. This behavior apparently arises because of cation-cation interactions and interactions between the structural effects around the tetra-n-butylammonium counter-ions in the gel solution [32].

The large difference in the ΔV values for the two trivalent cations, Cr^{+++} and La^{+++}, deserves special comment as it would appear that the latter is site-bound whereas the former is perhaps only field-bound. Both cations are strongly hydrated in aqueous solutions. Chromic ion is known from H$_2^{18}$O exchange experiments [34] to

TABLE IV

Volume, standard free energy, enthalpy and entropy changes for the replacement of F^- ion from a lightly cross-linked polyvinyl-trimethylbenzylammonium anion exchange gel

Anion	ΔV (ml mole^{-1})	ΔG^0 (kcal mole^{-1})	ΔH^0 (kcal mole^{-1})	ΔS^0 (cal deg^{-1} mole^{-1})
F^-	0.00	0.0	0.0	0.0
Cl^-	0.68	−0.82	−1.61	−2.6
Br^-	0.73	−1.15	−2.79	−5.5
I^-	0.75	−1.94	−5.04	−10.4
p-EtBzSO$_3^-$	−0.58	–	–	–

form an exceptionally stable hexa-aquo species, $Cr(H_2O)_6^{+++}$, which keeps its identity when it enters the gel solution of cross-linked cation exchangers [35]. It may be concluded that the hexa-aquo ion does not readily form solvent-separated ion-pairs with the sulfonate groups of the polyelectrolyte gel because of the slow-exchange of its water. Lanthanum ion, which is believed to occur in dilute solutions as $La(H_2O)_9^{+++}$, shows rapid water exchange, and hence can be site-bound via ion-pair formation.

Perceptibly larger volume increases were observed (Table III) for the selective binding of multiply-charged cations by the lightly cross-linked gel. Site-binding appears to occur in all cases; with Ba^{++}, La^{+++}, and Th^{4+} ions contact ion-pairs may be formed.

The volume change criterion for the selective uptake of halide ions by a lightly cross-linked cationic gel based on polystyrene is not obeyed (Table IV). The magnitude of the volume increase suggests that only field binding is involved, yet, the thermodynamic data, especially the changes in ΔH^0, show that strong interactions are present, particularly with I^- ion. Other types of forces than purely electrostatic must be important, and, in view of the aromatic character of the polyelectrolyte gel, the possibility of binding via the formation of 'charge transfer' complexes must be considered. Earlier studies [33] have shown that the concept of charge-transfer interactions of halide ions with ion exchange gels and permselective membranes containing aromatic groups is consistent with their properties. Thus, I^- ion with the greatest charge-transfer (c.t.) interaction has a larger affinity than Cl^- ion which has the smallest c.t. interaction. Gels not containing aromatic groups, but otherwise similar in properties, bind I^- and Cl^- ion about equally strongly.

The ΔV for the selective binding of the p-ethylbenzene-sulfonate ion was negative, suggesting that this species may induce short-range order in the aqueous phase. However, because of the polar sulfonate group in the anion, the volume decrease is appreciably smaller than was observed with the more hydrophobic trimethylbenzyl-ammonium cation (cf. Table III).

Selective Binding of Ions in Moderately and Heavily Cross-Linked Polyelectrolyte Gels:
Increasing the cross-linking of a polyelectrolyte gel reduces the amount of water its salt-forms will hold when they are in equilibrium with dilute aqueous electrolyte

TABLE V

Volume, standard free energy, enthalpy, and entropy changes for the replacement of H^+ ion from variously cross-linked polystyrenesulfonates[a]

Cation		0.5% DVB	8% DVB	24% DVB
		(A) Single-charged cations		
Li^+	ΔV	0.1	0.8	1.5
	ΔG^0	0.05	0.13	0.08
	ΔH^0	0.07	0.26	0.43
	ΔS^0	0.0	0.6	1.1
Na^+	ΔV	0.5	4.1	6.5
	ΔG^0	−0.01	−0.23	−0.48
	ΔH^0	−0.26	−1.18	−1.40
	ΔS^0	−0.9	−3.0	−3.1
NH_4^+	ΔV	0.2	5.1	−
	ΔG^0	−0.06	−	−
	ΔH^0	−0.28	−	−
	ΔS^0	−0.8	−	−
Cs^+	ΔV	1.2	4.8	5.7
	ΔG^0	−0.24	−0.56	−0.93
	ΔH^0	−0.62	−1.95	−2.88
	ΔS^0	−1.3	−4.5	−6.6
Ag^+	ΔV	1.8	3.1	6.4
	ΔG^0	−0.23	−1.06	−1.95
	ΔH^0	−0.18	−1.07	−1.74
	ΔS^0	0.1	0.1	0.7
Me_4N^+	ΔV	−0.7	−4.0	−
	ΔG^0	−0.29	0.04	−
	ΔH^0	−0.82	−1.14	−
	ΔS^0	−1.9	−3.8	−
		(B) Multiply-charged cations		
Ca^{++}	ΔV	5.8	9.8	−
	ΔG^0	−	−1.16	−
	ΔH^0	−	0.48	−
	ΔS^0	−	5.8	−
Ba^{++}	ΔV	6.2	15.4	19.6
	ΔG^0	−	−2.04	−
	ΔH^0	−	−0.86	−
	ΔS^0	−	4.4	−
La^{+++}	ΔV	12.3	16.5	27.3
	ΔG^0	−	−3.36	−
	ΔH^0	−	0.60	−
	ΔS^0	−	13.8	−
Th^{4+}	ΔV	15.3	24.0	34.0
	ΔG^0	−	−	−
	ΔH^0	−	−	−
	ΔS^0	−	−	−

[a] ΔV in ml mole^{-1}; ΔG^0 and ΔH^0 in kcal mole^{-1}; ΔS^0 in e.u. mole^{-1}.

solutions (i.e., reduced swelling). The extent of ion-solvent interaction in the gel is diminished, ion-ion interactions are increased, and electrostatic ion-pair formation is facilitated. As may be seen in Table V, ΔV increases with increased cross-linking and with ionic charge. Site-binding occurs for nearly all ions, except H^+ and Li^+, and, with the heavily cross-linked gel, it seems likely that contact ion-pairs are formed by the multiply-charged cations while the singly-charged species give solvent separated pairs (i.e., 'outer sphere' complexes).

The selective binding of multiply-charged ions is also accompanied by substantial entropy increases suggesting that ion dehydration and water structure changes must occur. However, as has been pointed out by Hemmes [5] the volume increase accompanying the release of water molecules is small (i.e., 1 to 5 ml $mole^{-1}$) relative to the increase consequent to the destruction of the ordering of the solvent when ion association occurs. It is not possible to distinguish between solvent-separated and contact ion-pair formation from the data in Table V. Results from studies on aqueous solutions indicate that ΔV for the conversion of outer sphere to inner sphere complexes is small and positive. [5] Some of the ΔV's in the fifth column of Table V are so much greater than typical values from solutions that the assumption of the formation of at least some inner sphere complexes must be made. For example, the formation of $LaFe(CN)_6$ pairs in water [34] where only an outer sphere complex is believed to exist is $+8.0$ ml/mole. The formation of solvent-separated pairs by $MgSO_4$ and $MnSO_4$ in water have been measured [35] as 8.3 and 5.4 ml/mole, respectively.

5. Conclusions

Our researches on the volume and entropy changes which accompany the selective binding of ions by polyelectrolyte gels have led us to the following tentative conclusions:

(1) The binding of singly charged counter-ions by lightly cross-linked polystyrenesulfonate gels is non-localized (i.e., no site-binding) except possibly for Ag^+ ion. Alkaline-earth, rare earth and thorium ions appear to be site-bound; solvent-separated ion pairs may be formed.

(2) With moderately and heavily cross-linked gels site-binding of all cations except H^+ and Li^+ appears to occur. The magnitude of ΔV with singly-charged cations is such that solvent-separated ion-pairs are implied. Ba^{++}, La^{+++}, and Th^{4+} ions, however, appear to form contact ion pairs. The release of water of coordination and changes in water structure are indicated by the positive entropy changes observed.

(3) The selective uptake of the larger tetra-alkylammonium cations by lightly cross-linked polystyrenesulfonate gels appears to involve 'structure-enforced' bonding. The ΔV values are negative and the selective uptake of large R_4N^+ ions is determined by the relatively large entropy increase in the reaction. The volume decrease and entropy increase can be explained by assuming that the 'ice-like' structure of the external water is destroyed when large, quaternary ammonium ions enter the gel.

(4) The selective binding of the heavier halide ions, especially I^- ion, appears to be

a consequence of 'charge-transfer complex formation' with the aromatic groups of the exchanger.

(5) The Katchalsky theory of ion exchange selectivity in lightly cross-linked gels appears to be valid for the alkali-metal cations where field binding dominates. Multiply-charged cations show significant site-binding, however.

(6) The Rice-Harris theory of ion exchange selectivity which is based on the assumption of extensive ion-pair formation appears to be supported by the ΔV measurements. Solvent-separated and contact ion-pairs evidently occur.

References

1. Katchalsky, A. and Michaeli, I.: *J. Polymer Sci.* **23**, 683 (1957).
2. Feitelson, J.: *J. Phys. Chem.* **66**, 1295 (1962).
3. Rice, S. A. and Harris, F. E.: *Zeit. f. Physik. Chem.*, *N.F.* **8**, 207 (1956).
4. Miller, I. F., Bernstein, F., and Gregor, H. P.: *J. Chem. Phys.* **43**, 1783 (1965).
5. Hemmes, P.: *J. Phys. Chem.* **76**, 895 (1972).
6. Boyd, G. E.: 'Thermal Effects in Ion Exchange Reactions with Organic Exchangers: Enthalpy and Heat Capacity Changes', in *Ion Exchange in the Process Industries*, Proc. Internat. Conf., Society of Chemical Industry, London, 1970.
7. Boyd, G. E., Larson, Q. V., and Lindenbaum, S.: *J. Phys. Chem.* **72**, 2651 (1968).
8. Horne, R. A., Courant, R. A., Myers, B. R., and George, J. H. B.: *J. Phys. Chem.* **68**, 2578 (1964).
9. Hamann, S. D. and McCay, I. W.: *AIChE. Journal* **12**, 495 (1966).
10. Millero, F. J., 'The Partial Molal Volumes of Electrolytes in Aqueous Solutions', in R. A. Horne (ed.), *Water and Aqueous Solutions*, Wiley & Sons, Inc., New York, 1972, p. 519.
11. Linderstrøm-Lang, K. and Lang, H.: *Compt. rend. trav. lab. Carlsberg, Ser. Chim.* **21**, 313 (1938).
12. Strauss, U. P. and Leung, Y. P.: *J. Am. Chem. Soc.* **87**, 1467 (1965).
13. Conway, B. E., 'Solvation of Synthetic and Natural Polyelectrolytes', *J. Macromol. Sci., Reviews Macromol. Chem.* **C6** [2], 113 (1972).
14. Boyd, G. E., Lindenbaum, S., and Myers, G. E.: *J. Phys. Chem.* **65**, 577 (1961).
15. Gregor, H. P., Gutoff, F., and Bregman, J. I.: *J. Colloid Sci.* **6**, 245 (1951).
16. Boyd, G. E. and Bunzl, K.: *J. Am. Chem. Soc.* **96**, 2054 (1974).
17. Johansen, G.: *Compt. rend. Carlsberg, Ser. Chim.* **26**, 399 (1948).
18. Kratky, O., Leopold, H., and Stabinger, H.: *Z. angew. Phys.* **27**, 273 (1969).
19. Boyd, G. E.: *J. Chem. & Engr. Data*, in press (1975).
20. Redlich, O. and Meyer, D. M.: *Chem. Rev.* **64**, 221 (1964).
21. Conway, B. E., Verrall, R. E., and Desnoyers, J., *Trans. Faraday Soc.* **62**, 2738 (1966).
22. Skerjanc, J.: *J. Phys. Chem.* **77**, 2225 (1973).
23. Tondre, C. and Zana, R.: *Ibid.* **76**, 3451 (1972).
24. Ikegami, A.: *J. Polymer Sci.* **2A**, 907 (1976); *Biopolymers* **6**, 431 (1968).
25. Oosawa, I.: *Polyelectrolytes*, M. Dekker, Inc., New York 1971, p. 79.
26. Boyd, G. E. and Lindenbaum, S.: Unpublished researches (1956).
27. Lapanje, S. and Rice, S. A.: *J. Am. Chem. Soc.* **83**, 496 (1961).
28. Zundel, G.: *Hydration and Intermolecular Interaction – Infrared Investigations of Polyelectrolyte Membranes*, Academic Press, New York, 1969.
29. Kotin, L. and Nagasawa, M., *J. Am. Chem. Soc.* **83**, 1026 (1961).
30. Diamond, R. M. and Whitney, D. C.: *Ion Exchange*, Vol. I (ed. by J. A. Marinsky), Dekker, New York, 1966, p. 227.
31. Frank, H. and Wen, W. Y.: *Discussions Faraday Soc.* **24**, 136 (1957).
32. Narten, A. H. and Lindenbaum, S.: *J. Chem. Phys.* **51**, 1108 (1969).
33. Wen, W. Y. and Saito, S.: *J. Phys. Chem.* **68**, 2639 (1964).
34. Plane, R. A. and Taube, H.: *J. Phys. Chem.* **56**, 33 (1952).
35. Boyd, G. E. and Soldano, B. A.: *J. Am. Chem. Soc.* **75**, 6105 (1953).
36. Slough, W.: *Trans. Faraday Soc.* **55**, 1036 (1959).
37. Hamann, S. D., Pierce, P. J., and Strauss, W.: *J. Phys. Chem.* **68**, 375 (1964).
38. Fisher, F. H.: *J. Phys. Chem.* **66**, 1607 (1962).

SWELLING OF POLYELECTROLYTE GELS AND THERMODYNAMIC PARAMETERS OF SOLVATION

MARGUERITE RINAUDO

C.E.R.M.A.V. (C.N.R.S.), Domaine Universitaire, B.P. 53, 38041 Grenoble cedex (France)

Abstract. An overview of determination and interpretation of the swelling of ion-exchange gels is presented. First, the basic experimental techniques of determination are recalled and a series of selected results of existing literature presented, more especially concerning polysulfonic ion-exchange resins. Then, calculation of the thermodynamic functions from adequate measurements is illustrated and the influence of experimental variables discussed following a first pathway of interpretation of results. Lastly theoretical models of swelling of Gregor, Rice and Harris and Katchalsky *et al.* are recalled and discussed.

List of Symbols

V_e	external volume of a wet gel
V_m	external volume of a dry gel or matrix volume
W_e	weight of a wet gel
n_{H_2O}	number of water molecules in a wet gel
W_{H_2O}	weight of water in a wet gel
V_i	internal volume of a gel
V_0	external volume of a gel corresponding to the extrapolation for $W_{H_2O} = 0$
$\bar{V}_{H_2O}$	partial molar volume of water
V	bed volume of the gel
β	packing factor of the gel
H	height of the bed of gel
H_0	height of the bed of gel in the H^+ form
$H_{cs=0}$	height of the bed of gel in pure water
$X = p/p_0$	relative humidity over salt saturated solutions
$\Delta G, \Delta H, \Delta S$	thermodynamic parameters of swelling
Π	swelling pressure of the gel
a	module of elasticity of the gel
b	rest volume of a gel corresponding to $\Pi = 0$

This contribution is concerned with the swelling equilibrium of ion exchange gels. A number of experimental results and various theoretical treatments have been published by now on ion selectivities or to a somewhat lesser extent on swelling [2–5]; it appears useful to review briefly our knowledge on the latter.

In general a gel can be regarded as a tridimensional porous network of cross-linked polyelectrolytes their swelling being preponderantly dependent on the degree of cross-linking (Figure 1); in such a model one can also assume a macroscopically uniform distribution of the exchange sites throughout the gel phase.

We propose to discuss the experimental techniques more often used and the way in which the parameters characterising the swelling of the gel are deduced.

Following an analysis of results showing the influence of the nature of counter-ions and the fixed ionic sites, of the ionic strength of the medium and the degree of cross-linking, we present and discuss finally the different theoretical models proposed to interpret the behaviour of ionised gels.

Eric Sélégny (ed.), Charged Gels and Membranes I, 91–120. All rights reserved.
Copyright © 1976 by D. Reidel Publishing Company, Dordrecht-Holland.

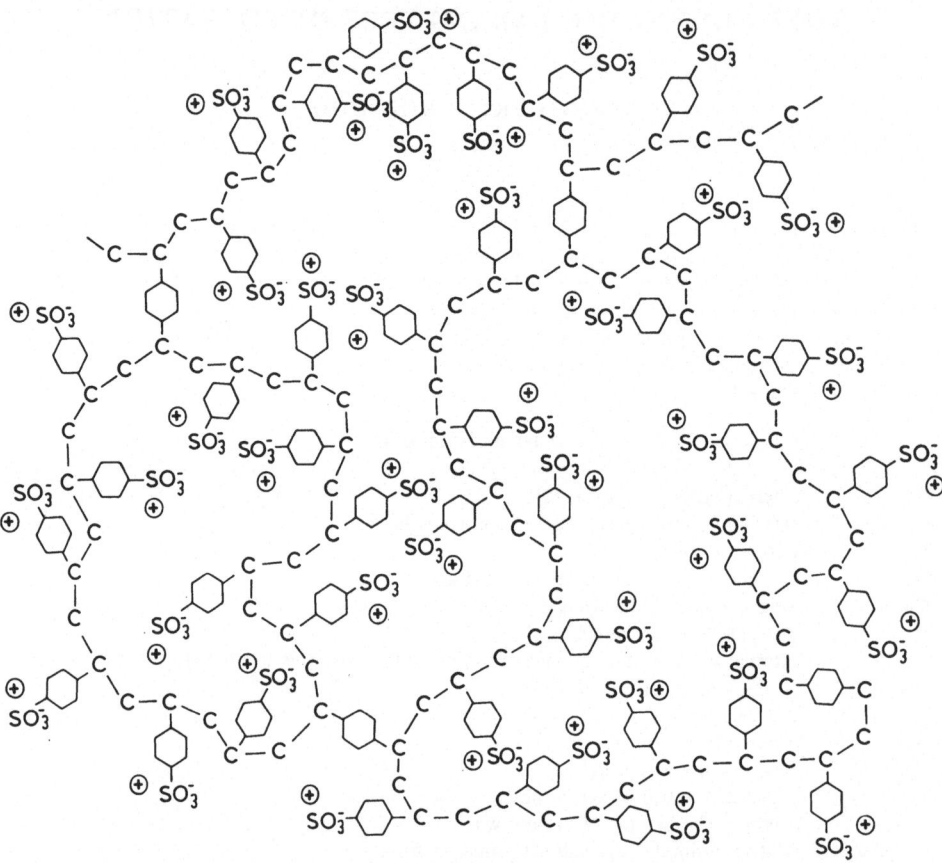

Fig. 1. Diagrammatical representation of the structure of polystyrene-sulfonic exchange resins. From Gregor *et al.* [1].

1. Experimental Determinations of the Degree of Swelling

The experimental results obtained on ion exchange gels and more especially on polysulfonic gels show that many parameters influence the degree of swelling.

Some results are given in Figures 2 to 4; the influence of the valence of counter-ions on swelling is apparent from Figure 2 at different degrees of cross-linking expressed as per cent of DVB content; the influence of the ionic strength of the supernatant solvent on the swelling is given in Figure 3. The influence of the nature of the counter-ion is represented diagrammatically in Figure 4 and shows the influence of the hydration of the cation; at first sight, the swelling increases for moderately crosslinked polysulfonic gels following the sequence $Cs^+ < Rb^+ < K^+ < Li^+$ but the order is reversed for fully neutralized polycarboxylic gels; in other cases, the sequence of swelling depends on the characteristics of the gel.

Several methods are convenient to test the degree of swelling; the different more useful ones will be presented and discussed successively.

To attain reproducibility, determinations must be conducted on samples of a gel

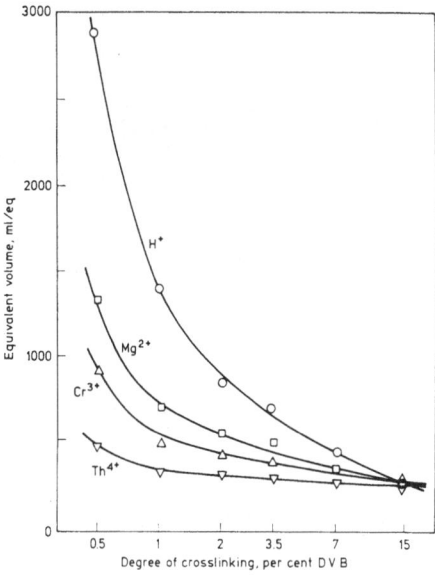

Fig. 2. Dependence of swelling on the degree of cross-linking and the valence of the counter-ion. From Calmon [6].

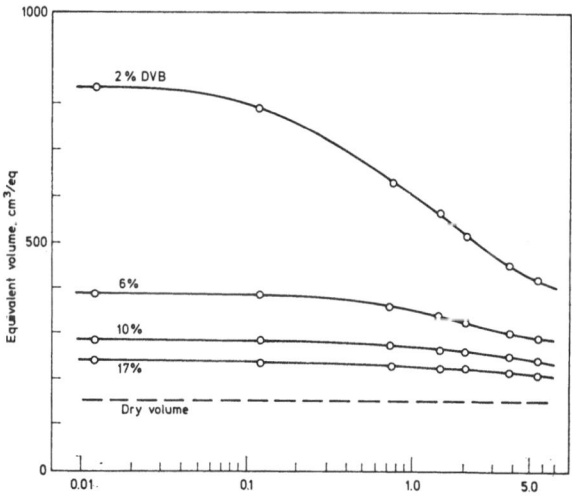

Fig. 3. Dependence of swelling on the concentration of the solution for a styrene type cation exchanger (NH_4^+ form) in NH_4Cl solutions. From Gregor et al. [7].

which has been carefully conditioned under given form and degree of relative humidity. Every experimental parameter is to be reported to the unit charge or unit weight of the gel in a given form; it is also interesting to express the quantity measured relative to one ionic site; generally, for sulfonic gels, all 'specific' quantities are reported to the unit weight in the H-form chosen as the standard state.

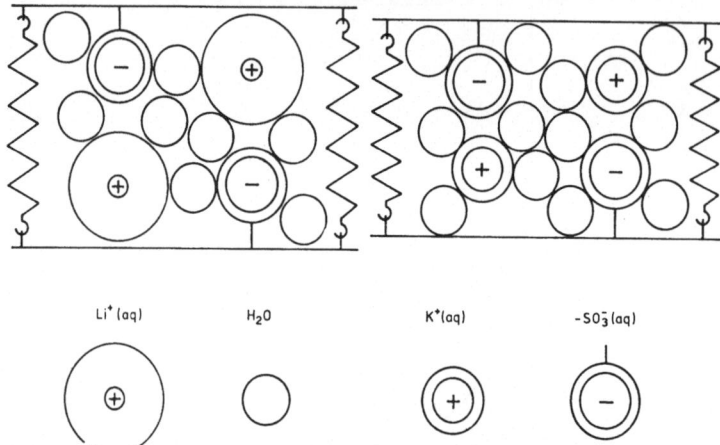

Fig. 4. Quantities of free water and hydration water in a swollen cation exchanger in Li+ or K+ form. From Buser *et al.* [8].

1.1. PICNOMETRIC DETERMINATION OF THE EXTERNAL VOLUME

The external volume V_e, a very important parameter to characterize a gel, is obtained by picnometry [9]. The method can be applied to dry gels, to gels equilibrated in a solution or to gels stabilized under different water vapor pressures.

The volume measured should be the total volume of the particle, so such an external picnometric liquid is to be used which does not enter the resin capillaries, nor desolvate the wet gel, nor dissolve the matrix.

Generally, with gels dried over P_2O_5 the inert picnometric liquid is octane; with wet gels the picnometric liquid used is octane or the equilibrium solution itself and both results should give the same result; when working with wet gels it is necessary to know the wet weight.

The picnometric method has been used extensively by Gregor *et al.* [9] and Lapanje and Dolar [10].

The principle of the determination is the following:
• knowing the weight of gel under a given condition, and the volume of the picnometer, the volume of the gel is obtained by the difference in weight of the picnometer full of the liquid only or containing a given quantity of gel.
• the specific gravity is obtained from the weight of the gel and its volume.

Usually V_m is related to the external volume of the dry gel and V_e to its wet volume. The external volume can be expressed as ml per equivalent or ml per gram of dry H-form.

1.2. WET WEIGHT OF THE GEL

The weight of a gel in a given salt form is determined from gravimetric measurements on the dry gel or after equilibration at different water vapor pressures. The difficulty is to determine the wet weight of a gel in a given solution. Gregor *et al.* [9] have compared the wet weights in water obtained by different techniques for Dowex 50 resins:

	Wet weight gr	Probable error
– suction drying	1.623	± 0.30
– blotting	1.693	± 0.25
– centrifugation	1.878	± 0.20
– extrapolation from wet weight to 100% relative		
humidity	1.870	± 0.50

It seems that centrifugation gave the most reproducible results but extrapolation to 100% RH could be more convenient because it does not need extraction of the water.

Discussion of the interrelation between V_e and W_e

From W_e, the weight (W_{H_2O}) or number of molecules (n_{H_2O}) of water bound by the gel are calculable. The following relation [11] gives more information about this water:

$$V_e = V_0 + W_{H_2O}/\varrho_{H_2O} \tag{1}$$

with V_0 the value of V_e extrapolated to $W_{H_2O}=0$; V_0 is found to be smaller than V_m. This contraction was related to the electrostriction of the first water layer estimated by Pepper to 0.6 ml per gram of polysulfonate exchanger or 6 water molecules per $-SO_3H$ [12].

The variation of $(V_e - V_m)$ as a function of the number of bound water molecules n_{H_2O} gives information on the nature of the resin-water interaction (Figure 5). In fact, the difference between free water and bound water has to be investigated.

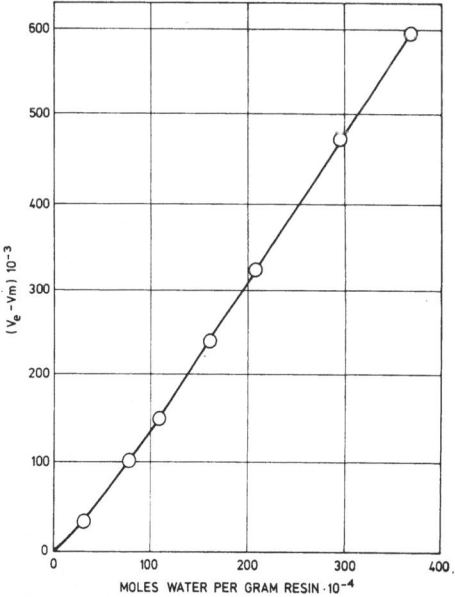

Fig. 5.   Increase in volume of Dowex-50 (K$^+$) as a function of sorbed water. From Gregor *et al.* [13].

1.3. BED VOLUME OF THE GEL

Determinations of the influence of the surrounding liquid are easily obtained by measuring the settled volume V of the bed in a calibrated tube or in a column sealed at its lower end to a porous glass. An advantage of these methods is that it is easy to equilibrate the gel in various liquids; they have been used by Braud and Sélégny [14] and Rinaudo [15]. There is an interrelation between the bed volume V and the external wet bed volume V_e:

$$V_e = (1 - \beta) V \qquad (2)$$

with β the packing factor. With relatively tight crosslinking and spherical resin particles, β is found to be constant [14]. In such a case the variations of the height of the bed in the column are directly proportional to those of V_e. Braud and Sélégny have expressed the relative modifications of the bed with the degree of neutralization by several counter-ions in water by H/H_0 (H_0 being the height of the bed in the H^+ form); they have also reported for these counter-ions the changes due to the ionic strength as $H/H_{cs=0}$ ($H_{cs=0}$ being the height in water) (Figure 6). With hydrophilic loose gels,

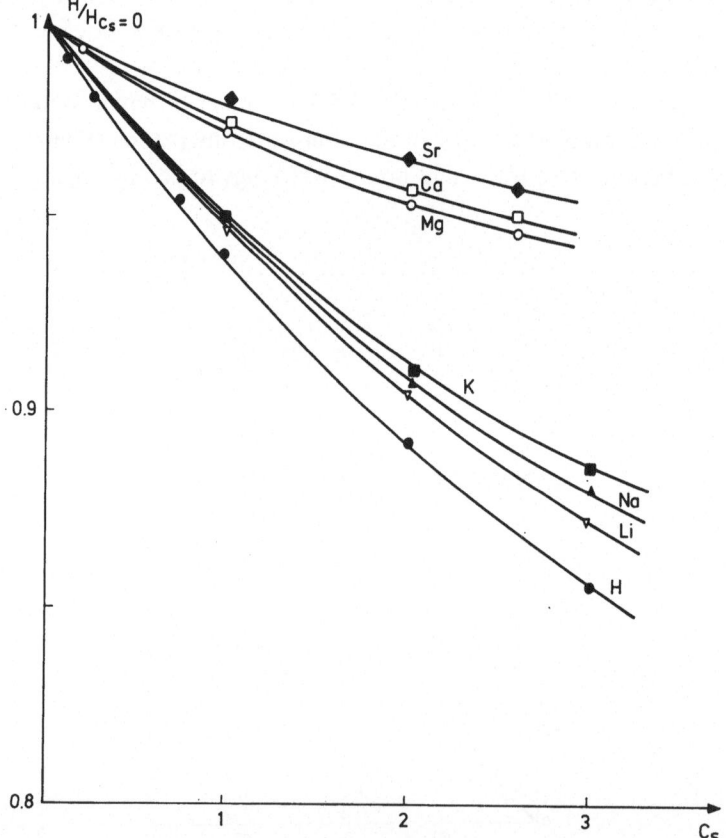

Fig. 6.   Variation of the relative thickness of the bed $H/H_{cs=0}$ with ionic strength (cs) (sulfonic exchanger ES 28). From Braud and Sélégny [14].

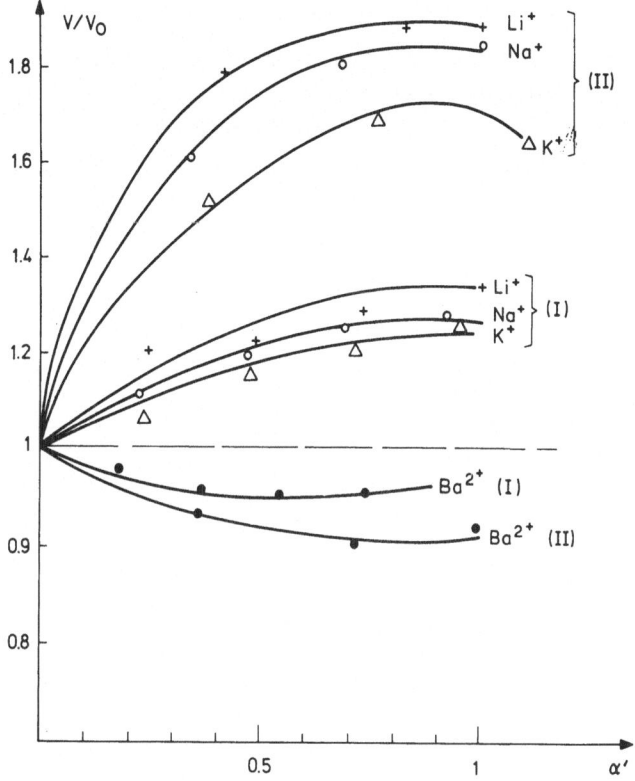

Fig. 7. Variation of the bed volume as a function of the degree of neutralization by different hydroxides on two alginic gels (Table VII) [15].

the validity of relation (2) may not be assumed and the variation of V is only a first approximation of those of V_e (Figure 7).

1.4. VAPOR SORPTION ISOTHERMS

The isopiestic determination of vapor-sorption isotherms has been described by Boyd and Soldano [11]. It must also be remembered that Gregor and coworkers have obtained very good concordance between values found when a humidistat or a McBain balance was used (Figure 8).

In Figure 9, isotherms for different degrees of crosslinking are given for the H-form of polystyrene resins [4]. The influence of cross-linkings is greater for $X > 0.6$.

These curves are important for the evaluation of the thermodynamic parameters and the swelling pressure in the Gregor model. The humidistat method only needs a series of desiccators containing different salt-saturated solutions. The equilibrium is usually obtained for a gel after a few days. The activity of water (X) in the vapor phase over a few salt-saturated solutions is illustrated in the Table I.

In Tables II and III some results of Gregor *et al.* are given for a polysulfonic resin Dowex 50 (10% DVB) under different salt forms and for different degrees of cross-linking.

98 MARGUERITE RINAUDO

1.5. Determination of the Enthalpy of Swelling

The ΔH of swelling can be obtained directly by microcalorimetric measurements or deduced from the isotherms. Gregor [13], Dickel [16], Dolar [10] and their co-workers have measured the ΔH by microcalorimetry on resin samples stabilized at various water vapor pressures in different salt forms by immersion in excess water.

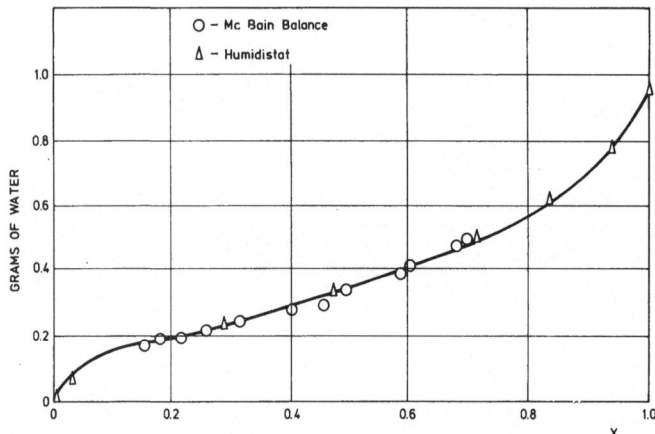

Fig. 8. Sorption measured with a McBain Balance (○) and humidistat (△). From Gregor *et al.* [13].

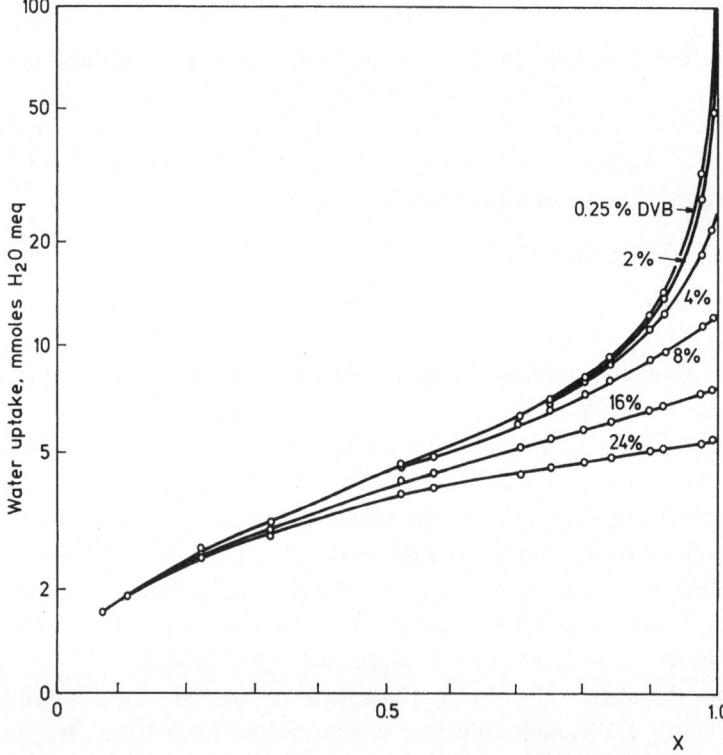

Fig. 9. Water-vapor sorption isotherms of ion-exchange resins (H⁺) [4].

TABLE I

Relative humidity over salt saturated solutions [16]

	values of X at 25 °C
NaCl	0.070
LiCl	0.111
$MgCl_2$	0.333
K_2CO_3	0.428
NaBr	0.577
KBr	0.807
KNO_3	0.925
$K_2Cr_2O_7$	0.980

TABLE II

Moles of water sorbed by equivalent weight as a function of relative humidity[a] [13]

ionic state \ X	0.025	0.29	0.47	0.71	0.83	0.94	1.00
H^+	1.08	2.84	4.00	5.85	7.24	9.15	10.18
K^+	0.60	1.61	2.27	3.35	4.32	6.14	7.63
Na^+	0.72	1.95	2.76	4.23	5.49	7.49	8.83
Li^+	0.93	2.22	3.17	5.02	6.49	8.63	10.05
NH_4^+	0.65	1.63	2.31	3.48	4.51	6.11	7.78
Ag^+	0.58	1.46	1.94	2.74	3.40	4.59	6.73
$(CH_3)_4N^+$	0.58	1.83	2.54	3.93	4.86	5.93	6.72
Mg^{++}	1.90	3.09	4.00	5.52	6.57	7.91	10.03
Ca^{++}	1.18	2.29	3.01	4.28	5.41	6.92	8.81
Ba^{++}	1.09	2.13	2.78	3.76	4.46	5.41	7.18

[a] Results on Dowex-50 sulfonic resin.

TABLE III

Weight of water sorbed[a] as a function of degree of cross-linking [13]

DVB % \ X	0.663	0.758	0.777	0.836	0.88	0.92	0.955	0.974
0.5	0.510	0.645	0.691	0.822	0.961	1.218	1.920	2.296
1	0.477	0.624	0.662	0.778	0.906	1.115	1.733	2.043
2	0.488	0.616	0.677	0.788	0.908	1.131	1.643	1.924
4	0.478	0.617	0.657	0.757	0.867	1.051	1.415	1.543
8	0.480	0.618	0.627	0.709	0.787	0.888	1.036	1.083
12	0.431	0.541	0.540	0.580	0.614	0.664	0.715	0.731
16	0.323	0.387	0.416	0.444	0.468	0.503	0.537	0.550
24	0.216	0.235	0.246	0.243	0.246	0.261	0.283	0.287

[a] Expressed in gram H_2O per gram of resin H^+.

In this case, one can write the reaction as:

$$\text{resin}(X = 0) + n\ H_2O\,(1) \rightarrow \text{resin} \cdot n\ H_2O\,(X).$$

Using Hess's law, the integral enthalpies of swelling are calculated for the process of swelling. In Table IV.1 are collected the integral ΔH values obtained by Gregor *et al.* [13] for a Dowex 50-K form resin (10% DVB) stabilized at different relative humidities, and in Table IV.2 the enthalpy of wetting of different commercial resins under different ionic forms measured by Matsuura [17].

TABLE IV

Experimental thermodynamic data

(1)

X	ΔH integral
0.040	2321
0.150	3754
0.280	4515
0.475	4804
0.725	5265
0.825	5391
0.947	5379
1	5265

(2)

resins	salt form	ΔH
amberlite IR-120 (SO₃H)	H	12400
	Na	9300
	K	5900
amberlite IR-112 (SO₃H)	H	12500
	Na	9200
	K	4800
amberlite IRC-50 (COOH)	H	3300
	Na	15300

(3)

	ΔH	ΔG	$T\Delta S$
R-H [10]			
0.5% DVB	10670	6830	3340
3% DVB	11030	6990	4040
8% DVB	11610	6530	5080
R-Na [19]			
1% DVB	7910	4810	3100
3% DVB	7860	4810	3050
8% DVB	8440	5000	3440
15% DVB	8700	4490	4210
R-K [20]			
3% DVB	4320	3930	
15% DVB	6180	3750	

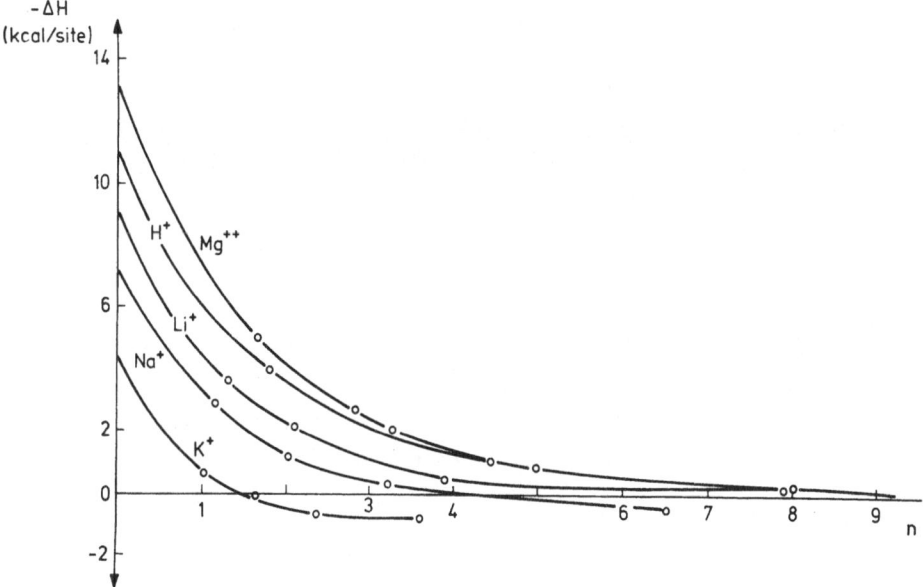

Fig. 10a.   Integral enthalpies of swelling per ionic site as function of the water content (n molecules H_2O by ionic site). From Dolar *et al.* [18].

Most extensive up-to-date results are from Dolar's laboratory [10, 18–20]. The enthalpies of hydration measured for different samples stabilized at different water contents are given in Figure 10a; by rotation of 180°, using Hess's law, the integrals are deduced (Figure 10b) and give the results of Table IV.3. All these results are in calories per equivalent of resin.

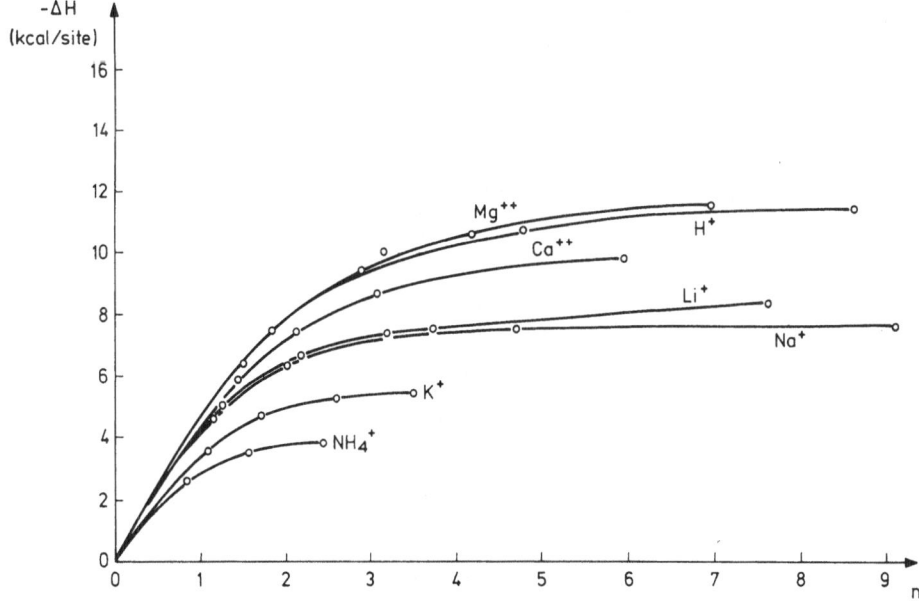

Fig. 10b.   Integral enthalpies of swelling for a polystyrene-sulfonic resin in different ionic states as a function of the final water content n. From Dolar *et al.* [18].

2. Interpretation of Experimental Results

2.1. THERMODYNAMIC FUNCTIONS

The free energy of swelling can be calculated directly from the sorption isotherms. Then, by combination with experimental ΔH values or from the dependence of ΔG on temperature, the entropy of swelling is obtained. As a result, all thermodynamic functions characterizing the swelling of a gel are easily deduced.

2.1.1. *Free Energy of Swelling*

The method described by Bull [21] and Dole and McLaren [22] is generally used to calculate the free enthalpy of swelling.

The free energy of the system can be expressed as the sum of two terms:

ΔG_1 for the addition of one mole of dry resin in an infinite excess of water vapor of activity X corresponding to:

$$\text{resin (dry)} + n \, H_2O(X) \rightarrow \text{resin} \cdot n \, H_2O(X)$$

$$\Delta G_1 = - RT \int_0^X n \, d \ln X$$

ΔG_2 for the transfer of n molecules of pure solvent $(X = 1)$ to an infinite quantity of resin equilibrated at vapor pressure (X)

$$n \, H_2O(X = 1) \rightarrow n \, H_2O(X)$$
$$\Delta G_2 = n \, RT \ln X$$

The total free enthalpy of swelling is given by

$$\Delta G = - RT \int_0^X n \, dX/X + n \, RT \ln X \qquad (3)$$

The graphical integration is made taking the area under the curve nRT/X as a function of X to obtain ΔG_1; the ΔG_2 term is calculated for every set of n and X data. Some examples (from Gregor) are given on Figure 11 with monovalent and divalent counter-ions; the free energy of the water sorption in the same sample is given in Table V.

2.1.2. *Enthalpy and Entropy of Swelling*

ΔH has also been obtained by Gregor *et al.* from the temperature dependence of ΔG (23). The experimental results are given in Table VI and Figure 12 for polystyrene-sulfonic resins. If the isotherms are drawn for two different temperatures, the ΔH^* (analogous to differential heat of dilution) is calculated for the same n, writing, according to Gregor:

$$\left[\frac{(\Delta H)}{n} \right]_{T, \, p} = \frac{RT_1 T_2}{T_2 - T_1} \ln \frac{x_1}{x_2} = \Delta H^*$$

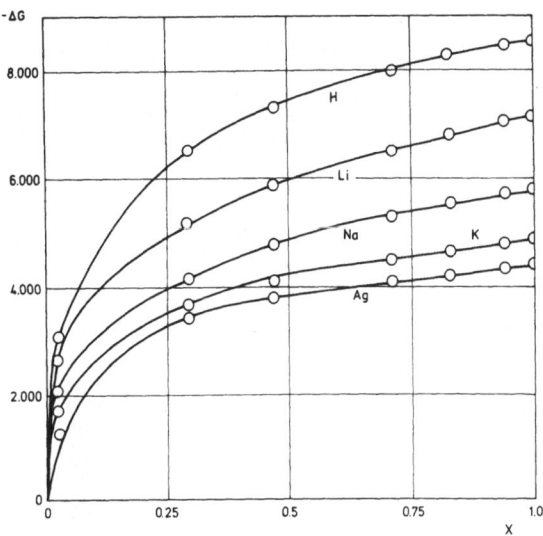

Fig. 11a–b. Calculated free energy changes of the water sorption process by Dowex-50. From Gregor *et al.* [13].

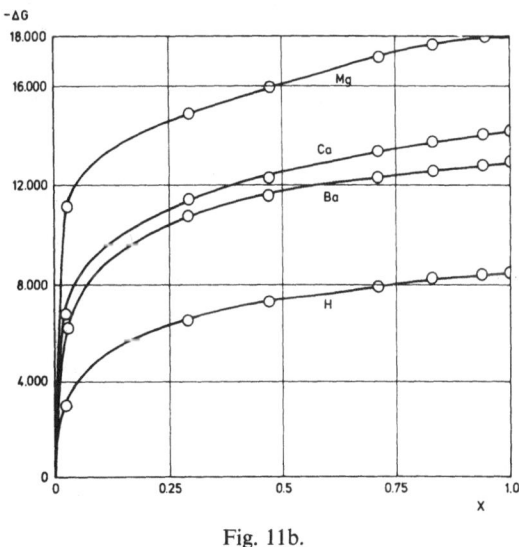

Fig. 11b.

For each composition n, ΔH^* is calculated (the same n is obtained for $X_1 < X_2$ corresponding to $T_2 > T_1$). The integration gives ΔH from which ΔS is obtained. Unfortunately, the variation of n with T is very small and the results are imprecise. The entropy ΔS is better obtained by the technique of direct calorimetric measurements combined with analysis of the isotherms mostly done by Dolar; a set of such results is given in Figure 13.

TABLE V

Decrease in free energy for water sorption process in various ionic states[a] [13]

X	Ca^{++}	Mg^{++}	Ag^+	NH_4^+	Li^+
0.025	3330	5553	1668	1881	2653
0.29	5680	7446	3451	3645	5188
0.47	6126	7989	3768	4073	5848
0.71	6667	8568	4055	4485	6478
0.83	6876	8804	4185	4697	6775
0.94	7034	8990	4319	4854	7017
1.00	7083	9040	4356	4897	7076

X	$(CH_2)_4N^+$	Na^+	Ba^{++}	H^+	K^+
0.025	1678	2076	3093	3061	1250
0.29	3822	4240	5362	6523	3680
0.47	4316	4764	5773	7289	4080
0.71	4816	5257	6131	7965	4495
0.83	5015	5505	6292	8258	4620
0.94	5158	5694	6399	8435	4780
1.00	5211	5753	6466	8484	4846

[a] Calculated values of Dowex-50 resin.

TABLE VI

Moles of water sorbed per equivalent as a function of relative humidity[a, b] [23]

DVB \ X	0.0218	0.050	0.0940	0.1345	0.287	0.443	0.667	0.759	0.924	0.957	1.000
0.4	0.69	0.99	1.25	1.39	2.05	2.97	5.50	7.06	13.70	18.50	–
2	0.77	1.11	1.51	1.70	2.55	3.58	5.53	7.10	12.93	16.49	55.17
4	1.05	1.28	–	1.79	2.50	3.38	4.90	6.06	11.39	14.38	28.43
8	1.17	1.33	1.69	1.90	2.72	3.64	5.30	6.14	9.69	11.02	14.05
10	1.43	1.61	–	2.12	2.91	3.90	5.50	6.40	9.08	9.86	11.79
13	1.54	1.69	–	2.09	2.66	3.41	4.93	5.60	7.97	8.81	10.38
17	1.30	1.48	–	1.81	2.28	3.04	4.30	4.98	6.60	6.99	7.61
23	0.87	1.14	–	1.86	2.51	2.98	3.87	4.26	5.62	6.13	7.01

[a] Temperature 25 °C

DVB% \ X	0.0135	0.0275	0.0580	0.1046	0.254	0.501	0.747	0.92
0.4	0.35	0.47	0.75	1.06	1.75	3.42	6.50	13.28
2	0.36	0.61	0.92	1.31	2.12	3.61	6.22	12.51
4	0.40	0.73	1.02	1.25	2.06	3.42	5.65	10.99
8	0.64	0.83	1.07	1.34	2.05	3.61	5.54	9.69
10	0.70	1.01	1.25	1.45	2.31	3.75	5.50	8.99
13	0.85	1.14	1.38	1.60	2.08	3.38	5.19	7.81
17	0.83	1.05	1.32	1.41	1.71	2.82	4.20	5.82
23	0.46	0.69	0.94	1.31	1.94	2.92	4.08	5.35

[b] Temperature 50 °C.

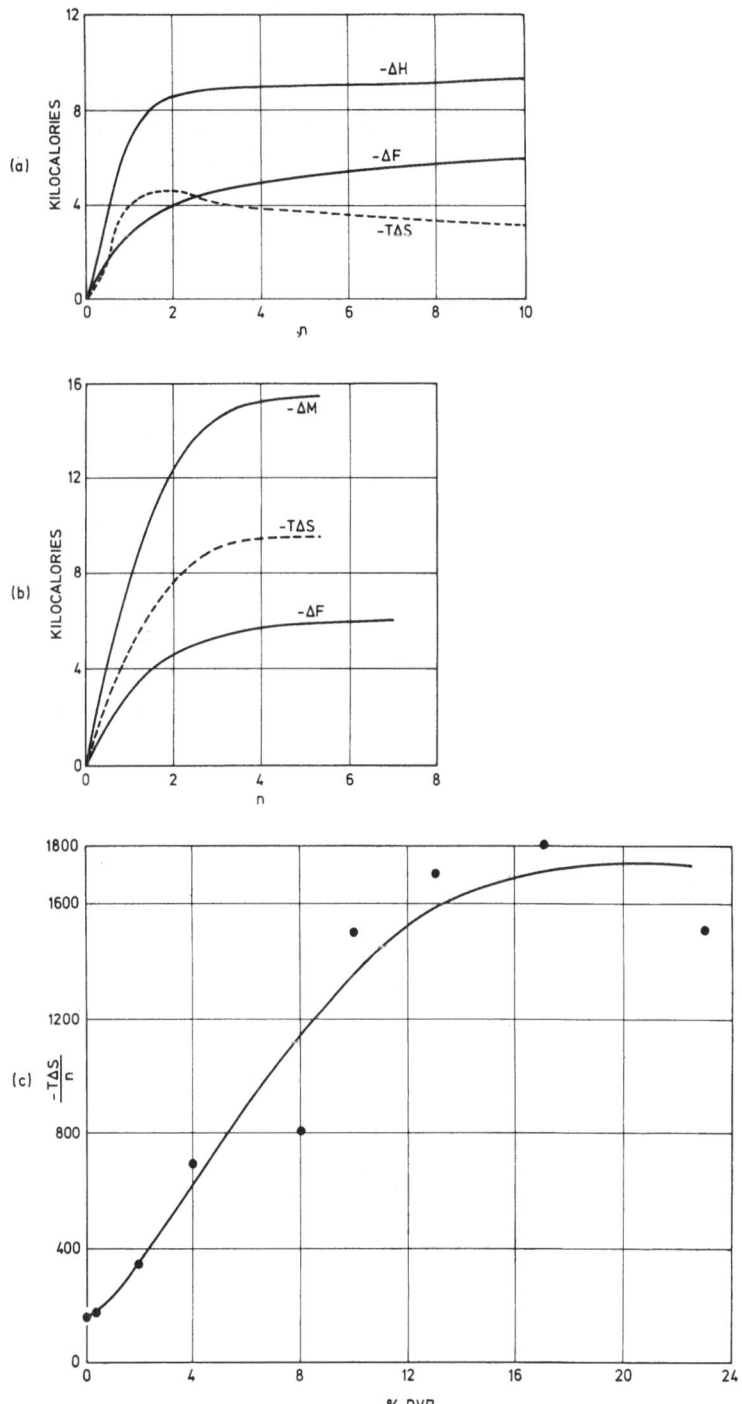

Fig. 12a–c.   Thermodynamic functions obtained from the temperature dependence of water sorption for polystyrene-sulfonic acid resins. From Sundheim *et al.* [23]. (a) DVB 0.4 %; (b) DVB 23%;
(c) variation in entropy function per mole of water sorbed at $X = 0.957$.

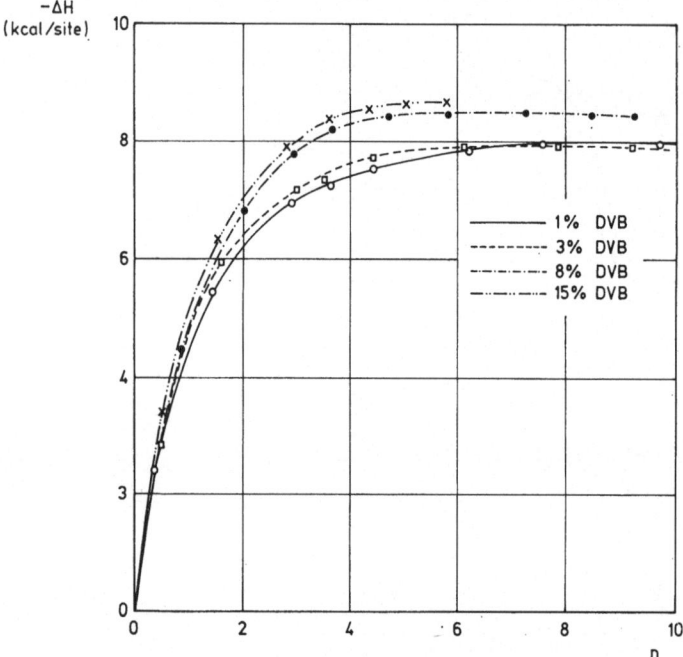

Fig. 13a–c. Thermodynamic functions for the swelling of a polystyrene-sulfonic acid resin. From Lapanje and Dolar [10]. (a) Integral enthalpies of swelling; (b) Integral free energy of swelling; and (c) Integral entropies of swelling.

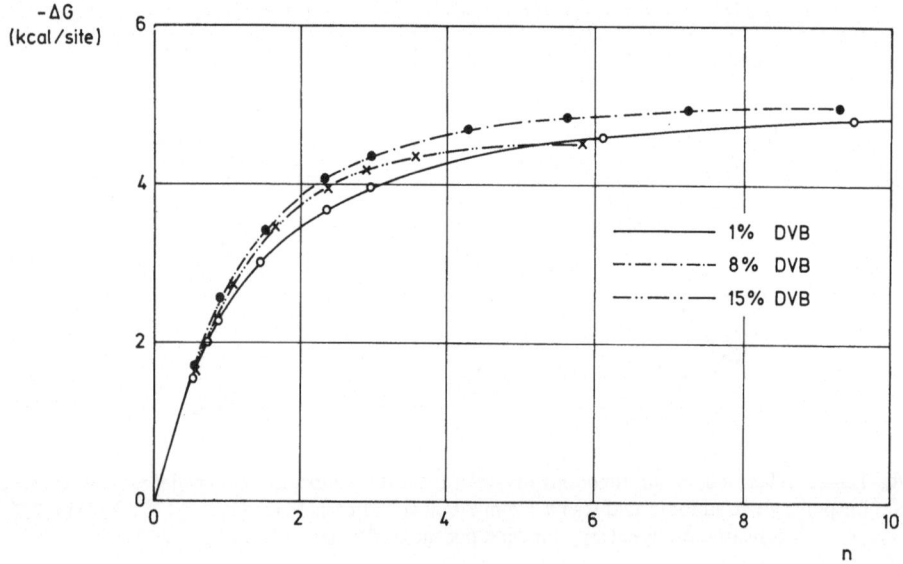

Fig. 13b.

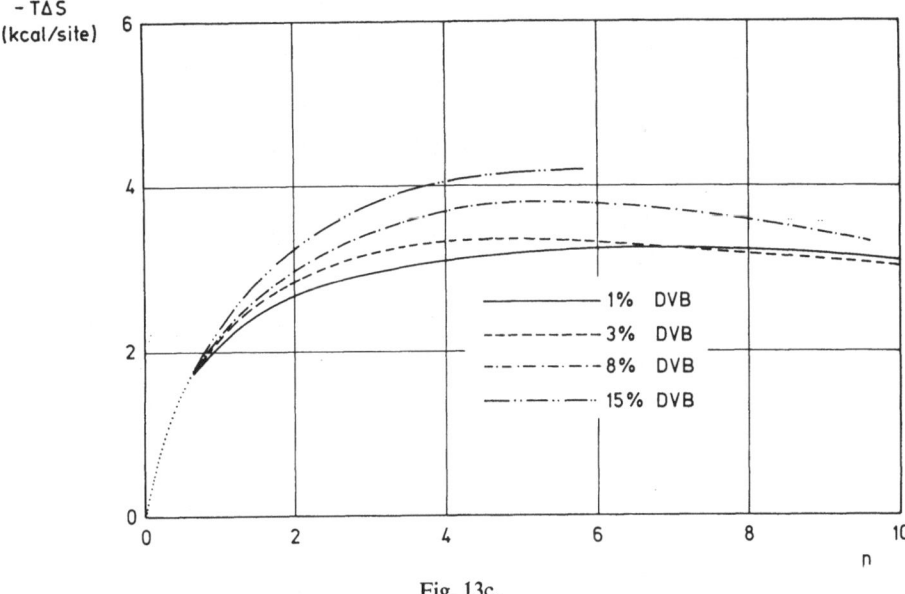

Fig. 13c.

2.2. SWELLING PRESSURE

From isotherms of sorption, illustrated in Figure 9, the swelling pressure π is deduced
and can be used for interpretation of results in the Gregor model. π is defined as the
pressure difference between the liquid inside the pores of a gel and the external solution. Boyd and Soldano [11] considered the exchanger as a polyelectrolyte which
owing to its crosslinking is subjected to internal forces equivalent to those arising from
an external hydrostatic pressure.

A diagram giving the principle of the determination is given in Figure 14. Boyd
suggests that for low degrees of crosslinking (0.5% DVB) the gel swells without tension and is under atmospheric pressure.

Now when two samples of the same degree of hydration but different crosslinkings
are compared, the equilibrium conditions correspond to respective water activities X
and X'; in the tighter gel the activity of water is increased just as under an externally
applied pressure. The isopiestic equilibrium is characterized by the equality of chemical
potentials:

$$\mu_{H_2O}(P', n') = \mu_{H_2O}(P, n)$$

with P' the total hydrostatic pressure on the internal gel solution in the more cross-
linked and P on the weakly crosslinked exchanger; the latter being assumed of one
atmosphere and $\bar{V}_{H_2O}$ constant over this range of pressure:

$$\mu_{H_2O}(P', n') = \mu_{H_2O}(1, n') + (P' - 1)\,\bar{V}_{H_2O}$$

then

$$\mu_{H_2O}(1, n) - \mu_{H_2O}(1, n') = \pi \bar{V}_{H_2O}$$

with π the swelling pressure equal to the pressure difference $(P' - P)$.

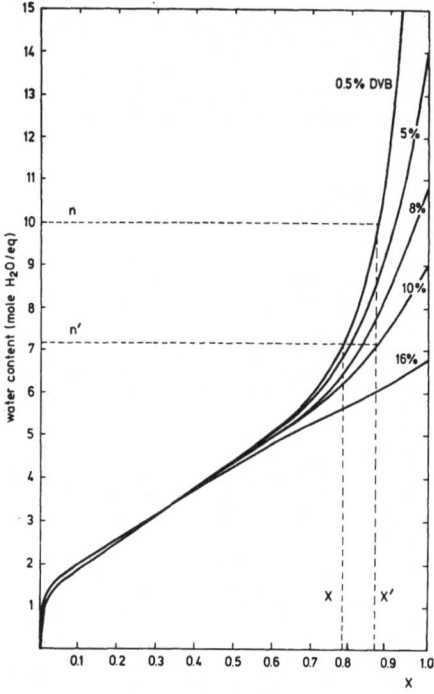

Fig. 14. Moisture absorption isotherms for a polystyrene-sulfonic acid resin of different DVB content for Π determination. From Boyd and Soldano [11].

On other hand, the following relations are valid:

$$\mu_{H_2O}(1, n) = \mu^0_{H_2O}(1) + RT \ln X'$$
$$\mu_{H_2O}(1, n') = \mu^0_{H_2O}(1) + RT \ln X$$

then

$$RT \ln X'/X = \pi \bar{V}_{H_2O} \qquad (4)$$

Thus from the isotherms of a loose and a more cross-linked gel compared for the same water content, the swelling pressure π is easy to calculate ($\bar{V}_{H_2O}$ being assumed equal to the molar volume of water). From Figure 9, one can find for a fully hydrated 16% DVB resin $\pi = 340$ atm., note that the same water content is obtained with a 0.24% DVB reference resin for $X = 0.78$. The variations of π as a function of DVB content calculated from the isotherms of Figure 9 are given in Figure 15.

3. Discussion of the Experimental Results

Based on the selected experimental data given above and a few complementary results the influence of the various parameters on swelling equilibria and thermodynamic functions of ion exchange gels can be summarized.

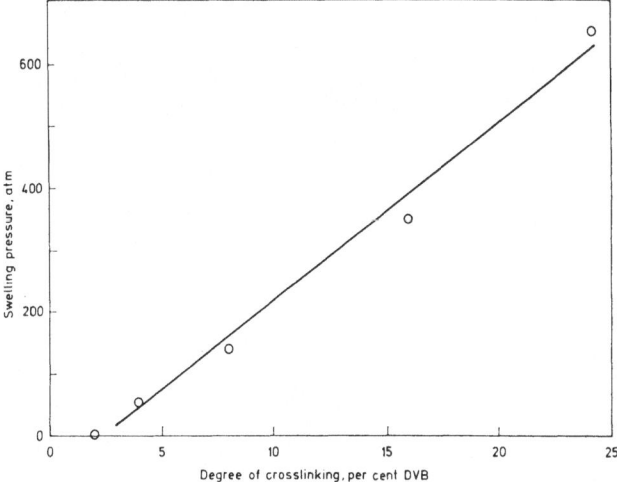

Fig. 15.   Influence of the degree of cross-linking on the swelling pressure [4].

3.1. INFLUENCE OF THE NATURE OF THE SITE

The swelling is dependent on the type of site as shown by the following results of Gregor [13]:

		$X = 0.47$	$X = 1$
– COOH resin (10% DVB)	H^+	0.70	3.82
	K^+	3.09	11.30
– SO₃H resin (10% DVB)	H^+	4.00	10.18
	K^+	2.27	7.63

The generally observed strong variation of the swelling of carboxylic gels versus the degree of neutralization is well illustrated in Figure 16 representing the modifications of the specific external volume and by the values of Table VII [15]. From Figure 7 showing the results obtained with alginic gels it is also seen that the bed volume, though directly connected to the degree of neutralization, depends very strongly on the nature of the counter-ion.

3.2. INFLUENCE OF THE COUNTER-IONS

The degree of swelling is greatly affected by the nature of the counter-ions (Figures 7 and 17). The sequence with monovalent counter-ions and polysulfonic gels is generally:

$$H^+ > Li^+ > Na^+ > K^+$$

On the contrary, with polycarboxylic gels, the sequence is

$$Li^+ > Na^+ > K^+ > H^+$$

with the possibility of inversions (Figure 16). With tight resins, the effect of divalent

TABLE VII

Swelling of two polyalginic gels[a]

Gel (I)

	$w_{H2O/g}$	V/V_0	$\overline{V}_{H2O}^{\text{ml g}^{-1}}$	V_{NaCl}/V_{H2O}
Li	6.77	1.35		
Na	6.69	1.29	4.40	0.86
K	6.42	1.27		
Ba	3.41	0.97		
H	1.67	–		

Gel (II)

	$W_{H2O/g}$	V/V_0	$\overline{V}_{H2O}^{\text{ml g}^{-1}}$	V_{NaCl}/V_{H2O}
Li	7.62	1.79		
Na	6.37	1.72	9.2	0.74
K	5.63	1.65		
Ba	1.42	0.930		
H	2.88	–		

V bed volume under various conditions; V_0 bed volume under H^+ form;
$\overline{V}$ bed volume per gram of gel.

[a] The gels were prepared by Dr Kuniak (Bratislava)

counter-ions is small as compared to hydrophilic carboxylic gels which shrink in presence of divalent ions (Figure 7).

The influence of the counter-ions is also reflected on the ΔH, ΔG and ΔS of solvation; some values are given in Tables IV.3, V and Figure 11; $|\Delta G|$ and $|\Delta H|$ increase when passing from the K to the H form of polysulfonic gels. The swelling can be related to the ionic affinity of the gel and a more bound counter-ion gives lower swelling. Information on the tightness of water binding to the matrix is given by investigation of the external volume V_e as a function of the number of bound water molecules.

3.3. INFLUENCE OF THE DEGREE OF CROSS-LINKING

The large dependence of the swelling on the degree of cross-linking is translated in the modifications of the number of bound water molecules, the external volume, the swelling pressure or of the thermodynamic functions with this variable. Results have been shown in different figures; the lower the degree of cross-linking, the lower values are the $|\Delta H|$. Dolar et al. [25] have compared the thermodynamic data of water sorption of a polymethylstyrenesulfonic acid and of ion exchangers; the values with the linear polymer are about the same as with the more cross-linked exchanger [10].

In fact the detailed interpretation of the swelling enthalpy ΔH is difficult because of the great number of influencing factors: extension of the matrix, dilution, dissociation of the ion pairs, hydration of site and counter-ions. It seems however that the two last terms are the prevailing ones.

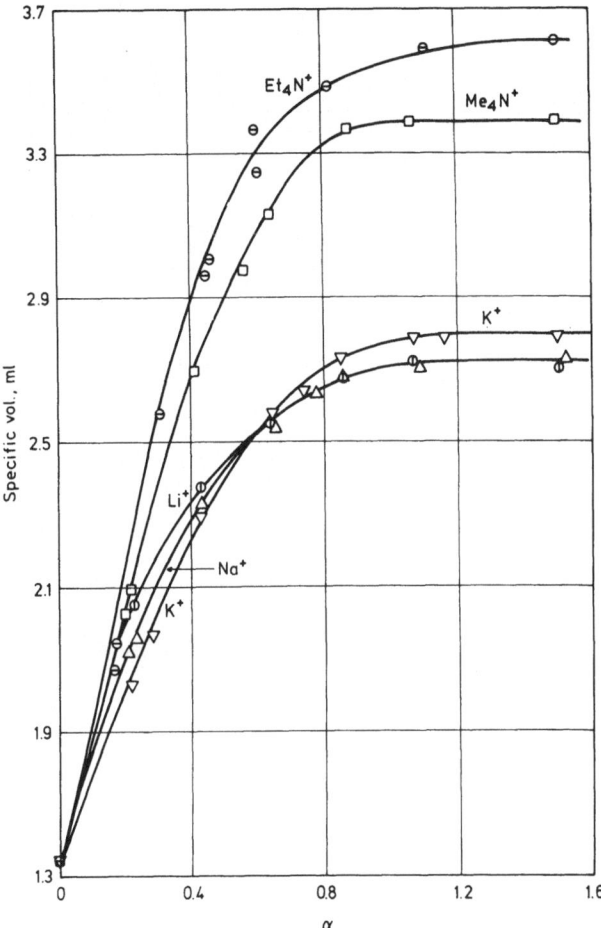

Fig. 16. Specific volume of a methacrylic acid resin as a function of the neutralization. From Gregor [24].

4. Theoretical Approach

Two types of models have been proposed to interpret the swelling properties of gels. The first one is the mechanical model of Gregor [26], the second types a molecular model treated by Rice and Harris [27] and Katchalsky and coworkers [28].

4.1. THE GREGOR MODEL (1951)

The diagram of the model is given in Figure 18. The matrix is represented by elastic springs which are stretched when the resin swells and exerts a pressure on the liquid inside the pores. This pressure affects swelling, sorption and ion exchange equilibria. The equilibrium of swelling corresponds to an equilibrium between two opposite forces: the thermodynamic osmotic pressure difference π between the more-concentrated solution inside the pore and the external solution on the one hand and the contraction of the elastic matrix on the other hand; this model does not include electrostatic interactions.

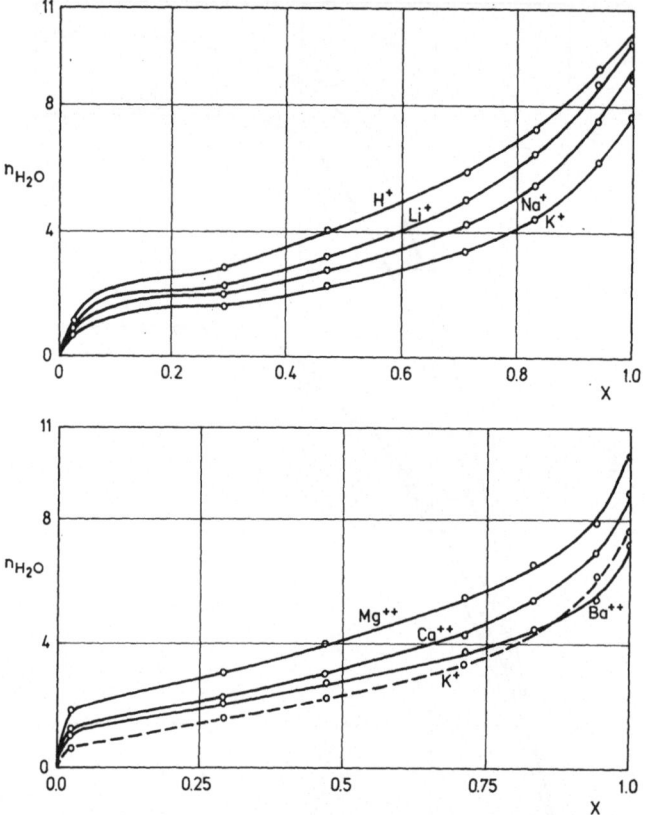

Fig. 17.   Moles of water sorbed per site on Dowex-50 (results used to obtain Figure 11). From Gregor [13].

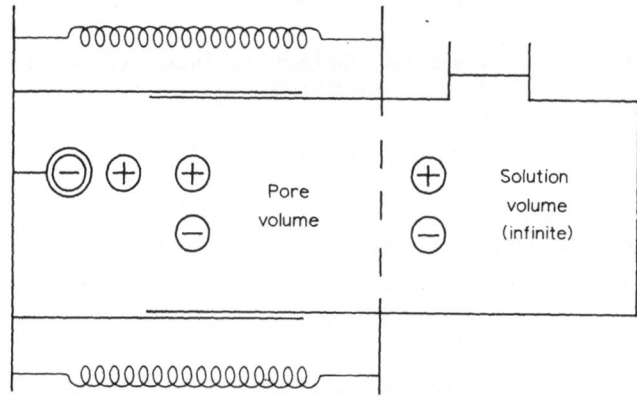

Fig. 18.   The Gregor model of ion exchangers. From Gregor [26].

As a first approximation, Gregor [26] and Boyd and Soldano [11] showed that a linear relation exists between π and the external volume:

$$V_e = a\pi + b \tag{5}$$

or

$$V_i = a\pi + b'$$

V_i=internal volume of the gel; a=module characterizing the elasticity of the resin; 'a' is larger when the gel is more cross-linked; b=volume of the gel when $\pi=0$; $b=V_m+a$ 'rest volume'; b is independent of the degree of cross-linking. The parameters a and b characterize the resin and are independent of the relative humidity and of ionic form (Figure 19).

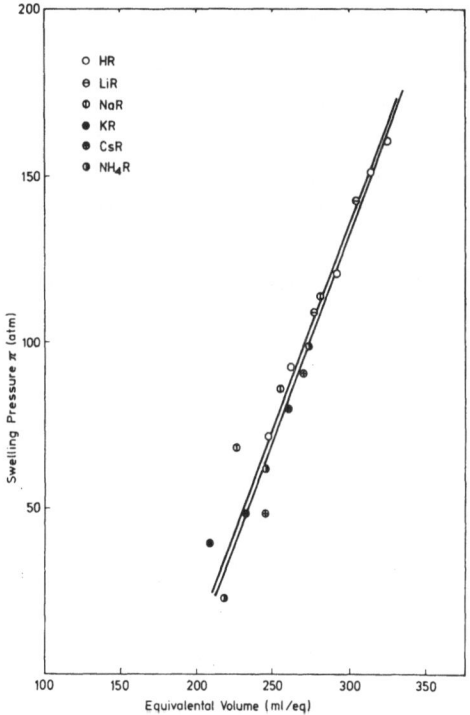

Fig. 19.   Equivalent volume as a function of the swelling pressure for a Dowex-50 cation exchanger. From Boyd and Soldano [11].

These relations describe in a general way the elastic properties of a cross-linked resin.

The swelling pressure affects the solvent uptake of the resin and thus the solvent activity in the resin; for a gel in equilibrium with pure solvent, the following relation is obtained [4]:

$$\pi = \frac{RT}{\bar{V}_{H_2O}} \ln \frac{a_{H_2O}^{ext}}{a_{H_2O}^{int}} = -\frac{RT}{\bar{V}_{H_2O}} \ln a_{H_2O}^{int} \tag{6}$$

$a_{H_2O}^{ext}$=solvent activity in the external phase; and $a_{H_2O}^{int}$=solvent activity in the internal phase.

The swelling pressure is evaluated by isopiestic measurements as shown previously; in this case, the equilibrium between the resin and the gas phase is expressed by:

$$RT \ln X = \pi \bar{V}_{H_2O} + RT \ln a_{H_2O}^{int} \tag{7}$$

The swelling pressure for a given resin is related to the internal concentration of counter-ions; the difference of swelling for different counter-ions must be related to a difference in their osmotic activity.

The swelling pressure is given by:

$$\pi = \frac{RTn^{int}}{V_e - V_m - c} \tag{8}$$

n^{int} = number of osmotically active ions; and c = effective volume of counter-ions.

In conclusion, swelling and vapor-sorption equilibria can be predicted when the osmotic coefficient (function of the internal molarity) and the constants a and b of the resin are known. The difficulty lies in the evaluation of the osmotic coefficient. Nevertheless, this theory explains the selectivity when the dimensions of solvated ions are introduced:

 - π and V_e increase when c increases at the same activity;
 - π decreases when the binding is larger thus reducing the activity of counter-ions

4.2. MOLECULAR MODELS

Two models proposed are respectively those of Rice and Harris [27] and Katchalsky *et al.* [28]. They can be considered as extensions of the polyelectrolyte theories and the gel is described as a network of rigid segments, each of them carrying one ionic site (Figure 20). The elasticity of the matrix is reflected by entropy changes; the con-

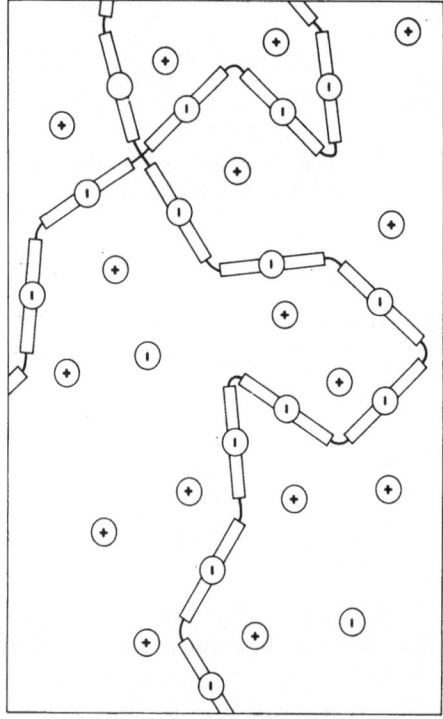

Fig. 20.   The Katchalsky molecular model [4].

formational entropy contribution is calculated from Flory's statistical theory of non-polar gels [29].

Electrostatic interactions are introduced in both models:

– The Katchalsky model is adapted to lightly cross-linked gels; it introduces interactions between neighbouring sites of the same chain.

– The Rice and Harris model is convenient for moderately and highly cross-linked gels; it takes care of interactions between sites the whole gel volume and introduces site binding.

In molecular models, the different forces which have to be considered are successively

  · free energy of mixing due to the dilution of the inner solution of the gel;
  · solvation of the matrix;
  · electrostatic repulsion between neighbouring ionic sites. The three contributions are counterbalanced by conformational entropy which increases when the matrix swells.

4.2.1. *The Rice and Harris Model*

This model gives an expression for the degree of swelling in equilibrium conditions for the minimum of the free energy. Dolar and Lapanje [20, 30] have applied this treatment with some modifications to swelling of polystyrenesulfonic resins. Nevertheless, this treatment introduces many empirical parameters and its application is hazardous.

The free energy of swelling ΔG is a sum of three contributions:

ΔG_{chem}: chemical contribution due to the formation of ion pairs;
ΔG_{conf}: due to the multiplicity of configuration of the non-charged chain and to free energy of mixing;
ΔG_{int}: taking account of all the electrostatic effects and free energy due to the distribution of free sites and ion pairs.

The minimalization of the free energy describes the swelling and the ion exchange as a function of the degree of cross-linking; as a consequence, the selectivity is related to the existence of ion pairs and the swelling is smaller when the number of ion pairs is higher.

The free energy of the system is then obtained; in the treatment, each contribution can be given for a resin which occupies a volume V and contains $\{n_\sigma\}$ ions of σ type associated to sites $(\sigma \neq 0)$.

The chemical contribution ΔG_{chem} associated with the formation of $\{n_\sigma\}$ ion pairs $(\sigma \neq 0)$ is given by the relation:

$$\Delta G_{chem}(\{n_\sigma\}) = \sum_{\sigma=0} n_\sigma kT \, |\ln K_\sigma - \ln C_\sigma| \qquad (9)$$

$K_\sigma =$ intrinsic equilibrium constant for ion pair formation; and $C_\sigma =$ concentration of ion σ in the external solution.

Dolar and Lapanje have shown the influence of the empiric parameter K; the dry resin and the separate solution are chosen as the standard states [20]. If $K = \alpha^2/V(1-\alpha)$ is the constant of dissociation and the concentration $C = \alpha/V$ with V the equivalent volume of the swollen resin, then for one equivalent of resin:

$$\Delta G_{\text{chem}} = - RT \alpha \ln\left[\alpha/(1-\alpha)\right] \tag{10}$$

The fraction of dissociated sites α cannot be experimentally obtained and calculation for different α gives:

α	0.7	0.85	0.9	0.95	0.97	0.99
ΔG_{chem}	350	910	1170	1660	2000	2700

The second contribution ΔG_{conf} contains two separate additive terms depending only upon V: an elastic contribution and a mixing contribution calculable from the Flory-Huggins solution:

$$\Delta G_{\text{conf}}(V) = \tfrac{3}{2}kT\nu\left[0.4\left(\frac{V}{V_0}\right)^{2/3} - 1 - 0.6\left(\frac{V_m}{V_0}\right)^{2/3}\right] \times$$

$$\ln\left\{1 - \left(\frac{V}{V_m}\right)^{2/3}\right\} - \frac{\nu kT}{2}\ln\left(\frac{V_0}{V}\right) + \frac{V - V_0}{V_s}kT \times$$

$$\left[\ln\frac{V - V_0}{V} + \chi_1\frac{V_0}{V}\right] \tag{11}$$

with ν the effective number of cross-linkings in the network; V_s = the volume of a solvent molecule; V_0 = the volume of the unswollen gel taken as the reference state; V_m the maximum volume to which the gel may swell (governed by the contour length of its constituent chains); and χ_1 interaction parameter.

Relation (11) describes the excess of the configurational free energy of the solvent-gel system over that of separate solvent and gel.

Lapanje and Dolar [30] determine V and V_0 the volume of dry resin by picnometry. The calculation of V_m implies an assumption for the 1% resin, $V = 0.9 \cdot V_m \cdot \chi_1$.

More recently, Dolar and coworkers [20] proposed to write the third term of Equation (11) in a new form:

$$n_{\text{H}_2\text{O}} RT \ln \frac{18.07 n_{\text{H}_2\text{O}}}{V} + \chi_1\left(\frac{V_{\text{matrix}}}{V}\right) \tag{12}$$

The free enthalpy per equivalent of swollen resin is obtained by multiplying the full expression (11) by N_A and introducing experimental V_0 and V values for the equivalent volume; $n_{\text{H}_2\text{O}}$, number of moles of water per equivalent of resin, is directly obtained from the experimental results. The parameter ν, effective number of cross-links divided by N_A can be equated in polystyrene resin to %DVB/100. In the expression (12), V_{matrix} is obtained as $V_{\text{matrix}} = V - 18.07 n_{\text{H}_2\text{O}}$.

The interaction χ_1 is equated to 1. The difficulty in applying this theory is now in the calculation of V_m. Dolar [20] proposed to fix an average of 2.5 chains per molecule of DVB from which the number of monomer units per chain n is deduced. If S_0 is the root-mean-square length of the chain in a relaxed network of volume V_0, S_m the contour length of the chain and l the C—C distance, then:

$$S_0^2 = nl^2 \qquad S_m = nl \tag{13}$$

According to Rice and Harris, the following relation exists:

$$\frac{\langle S^2 \rangle_V}{S_0^2} = \left(\frac{V}{V_0}\right)^{2/3} \quad \text{and} \quad \frac{S_m^2}{\langle S^2 \rangle_V} = \left(\frac{V_m}{V}\right)^{2/3} \tag{14}$$

S is the end to end distance of a chain in the swollen resin. From (13) and (14), the parameter V_m is extracted. ΔG_{conf} has been calculated for two polystyrenesulfonic resins and the following values are obtained:

resin	V_0ml/eq	V ml/eq	V_{matrix}ml/eq	$V_m$ml/eq	n_{H_2O}	ΔG_{conf}
3% DVB	145	528	127	710	22.18	430
15% DVB	151	220	127	653	5.17	810

The third contribution is the electrostatic one for which Dolar et al. [20] proposed:

$$\Delta G_{int} = -\frac{N_A q^2 \alpha x}{D_0 (1 + xa)} + \frac{N_A q^2 z\alpha\, e^{-xr_0}}{2 D_E r_0} +$$
$$+ RT[\alpha \ln \alpha + (1 - \alpha) \ln (1 - \alpha)] \tag{15}$$

given per equivalent of resin in which q is the electronic charge, D_E the average dielectric constant set equal to 40; D_E is the effective dielectric constant at the distance r_0 from an ion and calculated by the expression $D_E = 5(r_0 \quad 4)$; x is the Debye-Hückel parameter within the resin (considering only the counter-ions); 'a' is the distance of closest approach taken equal to 3.5 Å; the values of the average number of nearest neighbours z is chosen equal to 6 for a simple cubic lattice; r_0 is the average distance between two ionic groups and depends on the model chosen ($r_0 = (V/V_0)^{1/3} r_0^0$ with r_0^0 distance between two groups in the dry state; r_0^0 is calculated from V_0 with the hypothesis that one ionic site is in the centre of a cube of volume V_0.)

Calculations by Dolar [20] for the 3% DVB and 15% DVB resins give respectively r_0 values equal to 9.56 or 7.15 Å. ΔG_{int} is calculated in both these cases for different values α of the fraction of free counter-ions:

α	0.7	0.85	0.9	0.95	0.97	0.99
3% DVB			1410	1420	1410	1410
15% DVB	1370	1560	1610	1630		

The ΔG of swelling, sum of the three contributions, has the following values for different α:

resin			ΔG calculated		experimental
3% DVB	α 0.90	0.95	0.97	0.99	3930 cal
	3010	3510	3840	4540	
15% DVB	α 0.70	0.85	0.90	0.95	3750 cal
	2530	3280	3590	4100	

The difference existing between experimental and calculated values is not very large within the arbitrarily chosen range of α; it is necessary to select a higher α for 3% DVB than for 15% DVB which is in agreement with an increased formation of site binding as a function of the charge density.

The method of Rice and Harris lead to an expression for the free energy of the system for specified resin volumes and numbers of ion pairs of each kind.

The equilibrium state of swelling and exchange will be that of minimum free energy with respect to V and $\{n_\sigma\}$ for a set of values of introduced independent parameters. By differentiation of ΔG with respect to resin volume, the condition for mechanical equilibrium of a swollen resin is obtained. The model gives a calculation of the equilibrium of swelling (V) of the gel. Numerical values are given in the original paper of Rice and Harris.

4.2.2. The Katchalsky Model

This model combines the equations for the free energy of polyelectrolyte solutions and those for the free energy of a non-polar network. It introduces three factors:
  · the contractibility of the polymer network;
  · the ideal osmotic contribution of the macromolecules and free ions in solution; and
  · the electrostatic interactions between charged macromolecular chains and the free ions.

For the swelling in water, the application of the model is easy (Figure 20).

The model relates the swelling or the contraction to the change of ionisation α. If 1 ml of dry gel in a solvent occupies a volume V under a pressure P, upon changing the pressure or the degree of neutralization at constant temperature, the volume will change; the authors give an equation of state of the gel [28a].

The general equation reduces to simplified forms in most cases:

(a) in case of equilibrium swelling at zero degree of ionization ($P = 0$, $\alpha = 0$):

$$\frac{\lambda V^{2/3}}{Z} = \frac{1}{2Z} + V \ln \frac{V}{V-1} - 1 - \mu/V \tag{16}$$

in which λ is a correction factor for highly swollen gels

$$\lambda = 1 + \frac{0.6sV^{2/3}}{z - sV^{2/3}} \tag{17}$$

s = number of monomers per statistical element in Kuhn's theory; Z = number of monomer units between neighbouring cross-links; and μ solvent polymer interaction constant.

For moderate degrees of swelling, $\lambda \sim 1$ and Equation (16) reduces to the equation of Flory

(b) for equilibrium swelling at $\alpha > 0$, V is very large and the general equation becomes:

$$\frac{\lambda V^{2/3}}{Z} = \alpha \left\{ 1 + B \sqrt{\frac{\alpha V}{n}} \right\} \tag{18}$$

with n number of monomoles of cross-linked polymer; B is a complex expression given in the original paper [28a]. The equilibrium swelling of ionized gels is so high that λ increases considerably approaching infinity for fully extended chains.

Experimental values and those obtained from Equation (18) are given in Figure 21 (the value of s is chosen equal to 10). In this model, also, different arbitrary parameters are introduced which imply difficulties in drawing quantitative conclusions.

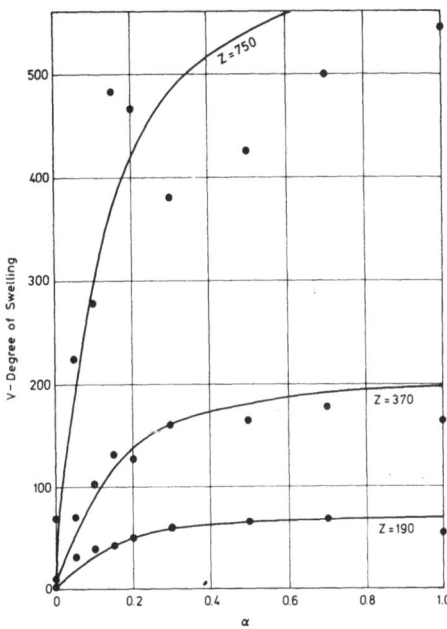

Fig. 21. Equilibrium swelling of polymethacrylic acid gels cross-linked with divinylbenzene. Calculated curves from Equation (18); $s = 10$; $T = 300\,°C$. From Katchalsky et al. [28a].

5. Summary

In a first part, we have exposed the different techniques useful for equilibrium swelling determinations of ion exchange gels in salt free solution.

Starting from some experimental results obtained by different authors, we have shown how it is possible to deduce the thermodynamic functions which characterize the swelling and examined the influence of the nature of the counter-ions and of the degree of cross-linking on the swelling.

Lastly we have recalled the theoretical models proposed respectively by Gregor, Rice and Harris, Katchalsky and coworkers and discussed briefly their applications. *Each model implies the introduction of empirical parameters making it difficult to predict correctly and quantitatively the theoretical values which can only be confronted with experimental results.*

References

1. Gregor, H. P. and Frederick M.: *Ann. N.Y. Acad. Sci.* **57**, 87–104 (1953).
2. Wheatan, R. M. and Seamster, A. H.: *Encyclopedia of Chemical Technology*, vol. 11, Interscience, 1966, p. 871.
3. Abrams, I. M. and Benezra, L.: *Encyclopedia of Polymer Science and Technology*, vol. 7, Inter-science, 1967, p. 692.
4. Helfferich, F.: *Ion Exchange*, Chapters 4 and 5, McGraw Hill, 1962.
5. Marinsky, J. A.: *Ion Exchange*, vol. I, M. Dekker, 1966.
6. Calmon, C.: *Anal. Chem.* **24**, 1456 (1952); **25**, 490 (1953).
7. Gregor, H. P., Gutoff, F., and Bregman J. I., *J. Colloid Sci.* **6**, 245 (1959).
8. Buser, W., Graf, P., and Grütter, W. F.: *Chimia* **9**, 73 (1955).
9. Gregor, H. P., Held, K. M., and Bellin, J.: *Analyt. Chem.* **23**, 620 (1951).
10. Lapanje, S. and Dolar, D.: *Z. Physik. Chem. Neue Folge* **18**, 11 (1958).
11. Boyd, G. E. and Soldano, B. A.: *Z. für Elektrochem.* **57**, 162 (1953).
12. Pepper, K. W. and Reichenberg, D.: *Z. Elektrochem.* **57**, 183 (1953).
13. Gregor, H. P., Sundheim, R. B., Held, K. M., and Waxman, M. H.: *J. Colloid Sci.* **7**, 511 (1952).
14. Braud, C. and Sélégny, E.: *Separation Sci.* **9**, 13 (1974); **9**, 21 (1974) – Braud, C.: Thesis of Speciality, Paris 1972.
15. Rinaudo, M.: unpublished.
16. Dickel, G., Degenhart, H., Haas, K., and Hartmann, J. W.: *Z. Physik. Chem. Neue Folge* **20**, 121 (1959).
17. Matsuura, T.: *Bull. Chem. Soc. Japan* **27**, 281 (1954).
18. Dolar, D. and Mocnik, M.: *Vestnik Slovensk Kem. Drustva* V, 35 (1958).
19. Lapanje, S. and Dolar, D.: *Z. Physik. Chem. Neue Folge* **21**, 376 (1959).
20. Dolar, D., Lapanje, S., and Paljk, S.: *Z. Physik. Chem. Neue Folge* **34**, 360 (1962).
21. Bull, H. B.: *J. Am. Chem. Soc.* **66**, 1499 (1944).
22. Dole, M. and McLaren, A. D.: *J. Am. Chem. Soc.* **69**, 651 (1947).
23. Sundheim, B. R., Waxman, M. H., and Gregor, H. P.: *J. Physik. Chem.*, **57**, 974 (1953).
24. Gregor, H. P., Hamilton, M. J., Becher, J. and Bernstein, F.: *J. Phys. Chem.* **59**, 874 (1955).
25. Dolar, D., Lapanje, S.. and Celik, L.: *Makromol. Chem* **41**, 77 (1960).
26. Gregor, H. P.: *J. Am. Chem. Soc.* **73**, 642 (1951).
27. Rice, S. A. and Harris, F. E : *Z. Physik. Chem. Neue Folge* **8**, 207 (1956).
28a. Katchalsky, A., Lifson, S. and Eisenberg, H.: *J. Polym. Sci.* **7**, 571 (1951).
28b. Katchalsky, A. and Michaeli, I.: *J. Polym. Sci.* **15**, 69 (1955).
29. Flory, P. J.: *Principles of Polymer Chemistry*, Cornell University Press, 1953.
30. Lapanje, S. and Dolar, D.: *Z. Physik. Chem. Neue Folge* **21**, 376 (1959).

III

IRREVERSIBLE THERMODYNAMICS

SOME USES FOR MEMBRANE TRANSPORT COEFFICIENTS

PATRICK MEARES

Dept. of Chemistry, University of Aberdeen, Old Aberdeen AB9 2UE, Scotland

Abstract. Although the fixed-charged theory of ionic membranes, expressed in terms of the Gibbs-Donnan distribution and the Nernst-Planck flux equation, has given a satisfactory account of many phenomena it is not quantitatively accurate in all cases. In particular, coupling processes are not individually taken into account.

Non-equilibrium thermodynamic formulations include all couplings but are restricted in application to situations close to equilibrium. They require the determination of many experimental data. A differential development of the non-equilibrium thermodynamic treatment permits application to macroscopic concentration intervals provided sufficient flux experiments are carried out to determine all coefficients as functions of concentration.

This has been done and the uses of such coefficients are discussed here. Resistance, coupling and friction coefficients are evaluated as functions of the concentration of NaBr in an almost homogeneous cation-exchange gel.

It is shown that a careful consideration of the magnitudes and trends in these coefficients reveals how the ions and water are distributed in the gel and how these distributions are affected by concentration. The extents of interaction between these components are evaluated and inferences drawn as to the detailed mechanisms whereby these interactions occur.

In the case of the ionic components there is a good correlation between resistance coefficients and tracer diffusion coefficients. A similar comparison in the case of water shows that this is transported primarily by convective flow within the swollen gel.

1. The Fixed Charge Theory

The first systematic treatments of transport in charged membranes were due to Teorell [1] and Meyer and Sievers [2]. They used the Nernst-Planck equation to relate the fluxes of the permeants with the forces on the system. According to this equation the flux ϕ_i of an ionic species i in any volume element is given by the product of its concentration c_i, mobility u_i and the force acting on it. The force is usually written as the negative gradient of the electro-chemical potential; we shall, however, adopt the simple expedient of defining positive fluxes as being in the direction opposite from that chosen to measure increases in potential. In this way we avoid having to write many minus signs.

The electrochemical potential can be expressed as a sum of contributions from the chemical activity $\gamma_i c_i$ (γ_i is the molar activity coefficient), pressure $p \times$ partial molar volume V_i and electric potential $\psi \times$ ionic charge $z_i F$. In an amorphous and rapidly relaxing system the absolute mobility u_i may be replaced by D_i/RT where D_i is the diffusion coefficient of i. Thus the Nernst-Planck flux equation can be written

$$\phi_i = c_i u_i (d\mu_i/dx)$$
$$= \frac{D_i c_i}{RT} \left(RT \frac{d \ln c_i}{dx} + RT \frac{d \ln \gamma_i}{dx} + V_i \frac{dp}{dx} + z_i F \frac{d\psi}{dx} \right) \tag{1}$$

where isothermal conditions and uni-dimensional gradients have been assumed.

The early theories were concerned mainly with explaining the potential differences observed across relatively impermeable membranes. The great significance of the fixed charge model was that it correctly assigned the major role in determining these potentials to concentration differences, rather than to mobility differences, between the oppositely charged ions in the membrane.

On the basis of Equation (1), the flow of i must be regarded as unaffected by the flows of other permeants in the membrane. It is accepted, however, that their presence may influence D_i or the force on i by affecting γ_i or ψ. Thus the flow of i is sensitive to the presence but not to the net motions of other components.

By combining equations analogous to (1) for each of the ions present and integrating the equations across the membrane a comprehensive theory of ionic and electrical phenomena in membranes was built up. This gave a good account particularly of the potential differences [3].

2. High Flux, Gel Membranes

The introduction in the later 1940's of homogeneous polyelectrolyte gel membranes, developed for early electro-dialysis cells, and, simultaneously, the newly arrived availability of a wide range of radiotracers permitted precise measurements of ionic fluxes to be made. These experiments revealed that the existing theory did not account quantitatively for the ion fluxes.

The discrepancies were traced to three main sources. The ionic distributions across the solution/membrane interfaces could not be described by the ideal form of the Gibbs-Donnan equation, especially at low concentrations in the solution phases [4]. The ionic mobilities or diffusion coefficients in the membrane were often strong functions of the ionic concentrations in the membrane [5]. Each ionic flux occurred against a background of non-random motions on the part of the molecules of the other flowing components. Consequently non-random momentum transfer took place between the particles of the different flowing components [6, 7].

Although the flux Equation (1) is independent of the Gibbs-Donnan distribution, this was introduced in the original fixed charge theory to obtain the boundary conditions for the integration of Equation (1) across the membrane. Integration was then carried out under the assumption of constant D_i, independent of x and c_i. There is no term in Equation (1) which takes into account the effect on ϕ_i of the fluxes of or forces on components other than i.

The contradictions between the three statements in the last paragraph and the observations referred to in the preceding one are obvious. In this discussion attention will be given to the second and third of these problems. In principle, at least, concentration dependent mobilities could be measured and taken into account in the integration of Equation (1). It is not, however, immediately obvious how they may be measured because tracer diffusion coefficients are not identical with D_i owing to the frequently overlooked phenomenon of tracer/non-tracer interaction [8].

3. The Effects of Mass Flow

Interactions between flows may be of two kinds, long range and short range. The coulombic forces between ions are long range. Flows of oppositely and of like charged ions may interact quite strongly without requiring direct contact between the particles. Such interactions may be significant even at low concentrations as the theory of the equivalent conductances of strong electrolytes well shows [9]. The flows of non-electrolytes interact with these of other non-electrolytes or with those of ions only through short range intermolecular forces i.e. through what are usually regarded as molecular collisions. Direct transfer of momentum between particles then occurs.

Additionally, a semantic complication arises when dealing with the flows of strongly hydrated ions and water. An undetermined number of water molecules may be regarded as bound to the ion for periods long compared with the intervals between molecular collisions. It the unhydrated ions are defined as a component of the system then their flux must generate a flux of bound water even in the absence of an osmotic gradient or of coupling between the flows of hydrated ions and non-hydration water. On this definition of components the strong interaction between the ion and hydration water will appear as a contribution averaged over all the water molecules in the system. Nevertheless, this precise definition of components is preferable to any ambiguous definitions involving specifying hydration numbers.

In highly swollen membranes the flux of water may be considerable as a consequence of osmotic and, more especially, electro-osmotic effects. Attempts have been made to take into account the effects of this convection stream on the ion fluxes [6, 7, 10]. Corrections have been made, while retaining the fixed-charge membrane model and the Nernst-Planck flux equations, by transferring the frame of reference to which the flux equations were held to apply from the fixed membrane to the flowing stream of fluid within it. Equation (1) then becomes

$$\phi_i = \frac{D_i c_i}{RT}\left(RT\frac{\mathrm{d}\ln c_i}{\mathrm{d}x} + RT\frac{\mathrm{d}\ln \gamma_i}{\mathrm{d}x} + V_i\frac{\mathrm{d}p}{\mathrm{d}x} + z_i F\frac{\mathrm{d}\psi}{\mathrm{d}x}\right) + \alpha v c_i \qquad (2)$$

where v is, according to the various authors' preferences, the velocity of the local centre of volume or of the local centre of mass in the positive flux direction. α is an empirical factor usually taken as unity. It is a measure of the extent of coupling between water and ion flows.

Experimental results showed that when α was taken as unity there was over compensation for the effects of the convection stream on the flux of counterions because there was a strong attraction between these and the stationary matrix which carried the fixed charges. Values of α between 0.4 and 0.6 were found [11]. Setting α universally at 0.5 seems to offer a reasonable empirical compromise.

In the case of the co-ions no simple pattern of behaviour for α was apparent. This was attributed to the interactions of the co-ions with the counterion flow being of comparable importance to their interaction with the solvent flow. Interactions between ionic flows were not taken separately into account in the extended Nernst-Planck equation (2).

4. Non-Equilibrium Thermodynamics

While the physical nature and causes of these problems with the original fixed-charge theory were being identified a formalistic solution of them was being developed from another direction. The basic concepts of linear non-equilibrium thermodynamics, [12] which stem particularly from the work of Onsager, provide for flux equations in which the flow of each molecular component is expressed by a sum of terms, one for each of the forces acting on the system, including those on each of the other independent molecular components. A familiar version of these non-equilibrium thermodynamic flux equations is

$$\phi_1 = l_{11} \frac{d\mu_1}{dx} + l_{12} \frac{d\mu_2}{dx} \cdots\cdots l_{1k} \frac{d\mu_k}{dx}$$

$$\phi_k = l_{k1} \frac{d\mu_1}{dx} + l_{k2} \frac{d\mu_2}{dx} \cdots\cdots l_{kk} \frac{d\mu_k}{dx}$$

$$(3)$$

where there are k independent components. In these equations, the l_{ik} are permeabilities which are independent of the forces but dependent on the state of the system. Thus they are functions of the local intensive variables but not of their gradients. This so-called linear hypothesis holds locally close to equilibrium and the Onsager reciprocal relations $l_{ik} = l_{ki}$ are then valid.

The straight coefficients l_{ii} have an obvious counterpart in the simple theory. It is easily seen that, in the absence of coupling between flows,

$$l_{ii} = c_i D_i / RT \tag{4}$$

would hold. The physical meaning of the l_{ik} cross coefficients is less obvious. It may be imagined that the direct consequence of the force on k is to generate a flux of k which can then couple with and modify the flux of i. In this somewhat indirect way l_{ik} is a measure of the extent of coupling between the flows of i and k. Confirmation of the logic of this interpretation may be obtained by noting the thermodynamic requirement

$$l_{ii} l_{kk} \geqslant l_{ik}^2 \tag{5}$$

Thus if k had zero mobility and hence zero flux irrespective of the force on it, i.e. $l_{kk} = 0$, then l_{ik} must also be zero, i.e. the force on k would not affect the flux of i.

Although Equation (3) enables flux equations to be set down which provide for mutual effects among the fluxes of all mobile components it does not tell us how to determine the coefficients nor how to interpret or predict them.

5. Discontinuous Membrane Systems

Alternative sets of linear equations, consistent with the requirements of non-equilibrium thermodynamics but written in terms of fluxes and forces which could be more readily measured and controlled, were developed during the 1950's by Staverman [13], Kirkwood [14], Katchalsky and Kedem [15].

These equations were formulated so as to recognise that, in practice, whereas differences between the electrochemical potentials in the solutions on either side of the membrane are measurable, their local gradients within the membrane are not. Thus the potential differences were chosen as forces.

The procedure was adopted of writing linear relations between fluxes and potential differences. This raised doubts however about the ultimate practical value of the whole approach, as opposed to doubts about its fundamental correctness. After integration, the Nernst-Planck flux equations do not lead to linear equations for a fixed charge membrane. It seemed possible therefore that the linear non-equilibrium thermodynamic representation might hold only so close to equilibrium that it would not yield any interesting results. This objection was important because a great deal of effort would be needed to measure the permeability coefficients and one wished to be assured of a significant outcome.

A good deal of attention has now been given to this problem and the value of the non-equilibrium thermodynamic approach to membrane phenomena has been demonstrated. In the remainder of this article it will be indicated briefly how this has been achieved.

6. Coping with Concentration Dependence

The difficulty regarding the range of applicability of linear equations is at it most acute when these are applied in global form to a discontinuous system, i.e. one in which potential differences rather than gradients are chosen as forces. As Equation (4) shows, the l_{ii} are proportional to 'concentration'. However, concentration has no clear meaning in the case of a membrane across which a macroscopic concentration difference exists and in which the concentration profile is unknown. On the other hand, the linear relations may hold locally over practically useful ranges of the forces provided these can be expressed as local gradients of the potential.

A way to cope with this problem was first indicated by Kedem and Michaeli [16]. It was developed into a practical scheme by Krämer and the author [17]. Since that time the author and his colleagues have carried out the experimental work necessary to apply the scheme to systems consisting of a single salt and water permeating a cation exchange membrane of the homogeneous gel type. The scheme has been described in detail elsewhere [18] and only a bare outline is given here for the sake of intelligibility.

In the system under consideration there are three independent mobile components. Usually the cations, anions and water are chosen and the membrane matrix is regarded as fixed. Thus there are three independent fluxes and three forces. Six phenomenological coefficients: three straight and three symmetrical coupling coefficients, inter-relate the fluxes and forces in a set of linear equations. In place of the fluxes of anions and water two other fluxes, which are more easily measured, are chosen. They are the electric current i and the defined volume flux ϕ_v given by

$$\phi_v = \phi_1 V_s + \phi_w V_w \tag{6}$$

where V_s and V_w are partial molar volumes of salt and water respectively. ϕ_1 and ϕ_w are the molar fluxes of cations and water.

Forces to conjugate with these fluxes can be found from the expression for the entropy production. The linear flux equations can then be formulated. They are

$$\phi_1 = L_\pi [\pi(1 + \bar{C}_s V_s)/v_1 \bar{C}_s] + L_{\pi p}(p - \pi) + L_{\pi E} E \tag{7}$$
$$\phi_v = L_{p\pi} [\pi(1 + \bar{C}_s V_s)/v_1 \bar{C}_s] + L_p(p - \pi) + L_{pE} E \tag{8}$$
$$i = L_{E\pi} [\pi(1 + \bar{C}_s V_s)/v_1 \bar{C}_s] + L_{Ep}(p - \pi) + L_E E \tag{9}$$

Here π is the osmotic pressure difference and p the hydrostatic pressure difference between the solutions on opposite sides of the membrane and E the potential difference between the solutions measured with electrodes reversible to the anions. The coefficients $L_{\alpha\beta}$ are average coefficients which are dependent on the composition profile in the membrane. $\bar{C}_s$ is a logarithmic mean concentration correctly defined in terms of salt activities. For ideal solutions only it is given by

$$\bar{C}_s = \Delta \bar{C}_s / \Delta \ln C_s \tag{10}$$

Unfortunately the values of $L_{\alpha\beta}$ which satisfy Equations (7), (8) and (9) are not the values specifically applicable to the concentration $\bar{C}_s$. It is these specific values of the coefficients which we seek as functions of the concentration.

Three coefficients present little difficulty. If p and π are set at zero the membrane is in equilibrium with a single solution of known concentration. The membrane composition is then uniform and determinable. L_E, $L_{\pi E}$ and L_{pE} can be obtained from measurements of the electric current, cation transport number and electro-osmotic flow under unit potential difference. Measurements made at a series of concentrations give the true values of the coefficients, which will be written $\mathcal{L}_E$, $\mathcal{L}_{\pi E}$ and $\mathcal{L}_{pE}$, as functions of precisely known concentrations.

7. Differential Discontinuous Systems

By setting p and i at zero, the salt and volume flows, ϕ_1 and ϕ_v, can be measured as functions of π only. The membrane potential E can also be measured but this does not give an independent equation. Through ϕ_1 and ϕ_v, measured under these free diffusion conditions, two relations between the remaining three $L_{\alpha\beta}$ coefficients may be obtained.

A third relation is needed and this may be obtained by measuring ϕ_v as a function of pressure p under the open circuit conditions $i=0$. In such an experiment π cannot be separately controlled because the low pressure side of the membrane is bathed by the liquid emerging from the membrane. The concentration of this is ϕ_s/ϕ_v. This concentration, as well as ϕ_v, has to be measured. π can then be obtained from the known concentration C_0 on the high pressure side of the membrane and the emergent concentration on the low pressure side [19].

In each of the three experiments just described there is a concentration difference across the membrane and hence an ambiguity as to the reference concentration for the values of $L_{\alpha\beta}$. This difficulty can be resolved in the following way. A series of measurements is made, all at the same ingoing concentration C_0, and at a series of emergent concentrations C_e, or pressures p which determine the emergent concentration. Curves are plotted of the fluxes versus the emergent concentration.

It may be shown that for an intrinsically homogeneous membrane the slopes of these curves at any chosen concentration C are independent of the value of C_0 which was fixed for the series of measurements. These slopes are of the form $(\partial\phi/\partial C)_{C_0}$ or $(\partial\phi/\partial\pi)_{C_0}$ and they refer to the specific value of C at which the slope is measured. Several such slopes can be taken from a single flux versus concentration curve.

The slopes are substituted into a set of linear equations, like Equations (7), (8) and (9) which relate vanishingly small fluxes and forces at a constant and known concentration of the bathing solution. On solving this set of equations the remaining coefficients $\mathcal{L}_\pi$, $\mathcal{L}_p$ and $\mathcal{L}_{p\pi}$ are obtained as functions of concentration.

8. Experiments on Zeo-Karb 315

The above procedure has now been carried out for several salts permeating the homogeneous gel membrane Zeo-Karb 315 which contains about 75% water by volume and carries the cation-exchanging fixed-charges $-CH_2SO_3^-$ at a mean concentration of about 0.55 molar in the swollen membrane.

A thorough examination of the data has confirmed the underlying principles of the thermodynamic analysis [19]. The coefficients are found to comply with the restrictions of thermodynamics and the reciprocal relations $\mathcal{L}_{\pi E}=\mathcal{L}_{E\pi}$ and $\mathcal{L}_{\pi p}=\mathcal{L}_{p\pi}$ are closely obeyed. ($\mathcal{L}_{pE}$ and $\mathcal{L}_{Ep}$ has not yet been tested.) The differential discontinuous coefficients $\mathcal{L}_{\alpha\beta}$ are found to be dependent on concentration but independent of the gradients dE/dx and dc/dx over the ranges of observation. A small apparent pressure dependence was found in the coefficients but it has not yet been established whether this is a dependence on the gradient dp/dx or whether it arises because the pressure applied to the upper surface of the highly swollen gel resting on a porous support causes some exudation of fluid. This would mean that the composition of the membrane was pressure dependent and it was the composition variations which altered its permeability properties.

While we have been carrying out this work other groups [20, 21] have also measured sets of coefficients although they have always introduced some extra and not fully tested assumptions in order to avoid making measurements under a difference of hydrostatic pressure.

In order to encourage more workers to undertake similar programmes of measurements so as to characterise a wider range of membrane and permeants. We shall now discuss some of the uses to which these non-equilibrium thermodynamic coefficients can be put.

9. Other Sets of Phenomenological Coefficients

The $\mathscr{L}_{\alpha\beta}$ coefficients describe only the relations between certain readily observed fluxes and forces. The l_{ik} coefficients of Equation (3) are a more fundamental set. These relate the molar flux densities of the mobile components to the chemical potential gradients. The connection between the two sets is given precisely by the non-equilibrium thermodynamic treatment. It is

$$(l_{ik}) = \delta\Gamma^{-1}(\mathscr{L}_{\alpha\beta})\Gamma^{-1T} \tag{11}$$

where Γ is given by

$$\Gamma = \begin{pmatrix} 1 & 0 & 0 \\ V_s/v_1 & 0 & V_w \\ z_1 F & z_2 F & 0 \end{pmatrix} \tag{12}$$

is the matrix transforming the molar fluxes to the practical fluxes and δ is the thickness of the membrane on which the $\mathscr{L}_{\alpha\beta}$ were determined. The term δ introduces an extra dimension of length into the coefficients. In this operation we transfer from coefficients of the discontinuous system defined under a microscopic difference of potentials to the coefficients of the continuous formulation of Equation (3).

It was pointed out earlier that the mechanism whereby the flux of i is influenced by the force on k is indirect. The coefficients l_{ik} do not therefore characterise simple molecular events. A further step is taken by inverting the matrix of flux Equations (3) to obtain a reciprocal set of relations between forces and fluxes:

$$\partial\mu_1/\partial x = r_{11}\phi_1 + r_{12}\phi_2 \cdots\cdots r_{1k}\phi k$$

$$\tag{13}$$

$$\partial\mu_k/\partial x = r_{k1}\phi_1 + r_{k2}\phi_2 \cdots\cdots r_{kk}\phi_k$$

Here also the reciprocal relations

$$r_{ik} = r_{ki} \tag{14}$$

hold.

Since the l_{ik} are known the evaluation of r_{ik} from them is straightforward because

$$r_{ik} = A_{ik}/|l| \tag{15}$$

where A_{ik} is the co-factor of l_{ik} and l is the determinant of the coefficients in the matrix of l_{ik}. Unfortunately the experimental errors in the l_{ik} probably lead to somewhat greater uncertainties in the r_{ik}.

10. Resistance and Coupling Coefficients

The physical meaning of the r_{ik} coefficients is relatively clear. This may be seen by setting all the fluxes except one, ϕ_i say, in Equations (13) equal to zero and then putting ϕ_i at unity. There are then two kinds of equations to consider

$$\partial \mu_k / \partial x = r_{ki}; k \neq i \qquad (16)$$

and

$$\partial \mu_i / \partial x = r_{ii} \qquad (17)$$

The coefficient r_{ki} measures the force per mole which has to be applied to component k in the direction opposite from the flux ϕ_i in order to hold k stationary ($\phi_k = 0$). Thus the drag force on k per mole exerted by unit flux density of i is $-r_{ki}$.

To the extent that this drag effect can be compared with macroscopic friction between two bodies in relative motion, $-r_{ki}$ is a positive force and r_{ki} is negative numerically. r_{ki} thus characterises the interaction between components i and k and is independent of the motions of the other components, but it is dependent on their presence because they help to determine the physical state of the system.

r_{ii} measures the force per mole that has to be applied to i in order to drag it at unit flux while all other components are held stationary. r_{ii} is therefore the sum of drag forces exerted by i on all the stationary components with respect to which it is moving. These include all the mobile components other than i and also the stationary matrix of the membrane.

These resistance coefficients are useful in two ways; they can be used to obtain molecular friction coefficients, which will be discussed a little later, and they can be used to measure the degree of coupling between flows. Since individual couplings are not taken into account in the Nernst-Planck formulations it is useful to know when they are likely to be important.

The degree of coupling q_{ik} was introduced by Caplan [22]. It is defined by

$$q_{ik} = - r_{ik} / \sqrt{r_{ii} r_{kk}} \qquad (18)$$

The coefficient q_{ik} represents the extent to which a flow of i is produced by a flow of k in the absence of an external force on i and when all flows other than those of i and k are zero. Ordinarily, q_{ik} would be expected to vary between zero in the absence of coupling and unity when the particles of k drag those of i with them at the same speed. In some circumstances, where a repulsive interaction occurs between i and k e.g. if i and k are different isotopes of the same ionic species, q_{ik} can take negative values.

11. Interpretation of Coupling Coefficients

The coupling coefficients q_{12}, q_{13} and q_{23}, where $1 =$ cations, $2 =$ anions and $3 =$ water, for NaBr in Zeo-Karb 315 are plotted in Figure 1 against the concentration of the NaBr solution with which the resin is in equilibrium.

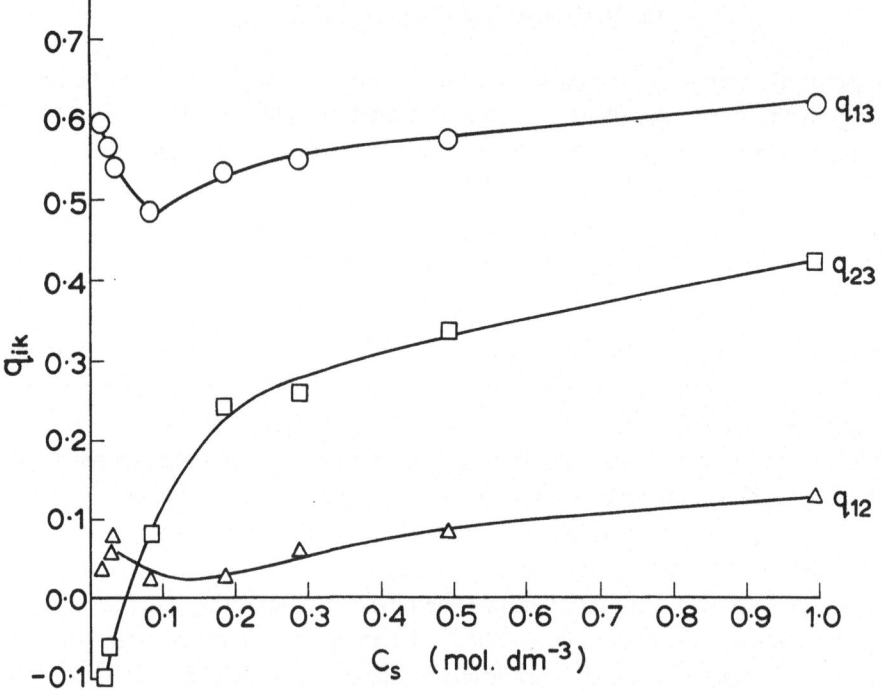

Fig. 1.   Coupling coefficients q_{ik} versus concentration of NaBr in the external solution C_s. 1=Na⁺,
2=Br⁻, 3=H₂O.

When interpreting any data obtained from a theory which treats the membrane as intrinsically homogeneous, it has to be realised that the coefficients refer to a mean situation in which all the components are identically and uniformly distributed. This is never exactly fulfilled in nature. In a cross-linked gel the distances between cross-links, chemical or physical, must be considerably larger than the dimensions of the transported particles. Otherwise the gel would be impermeable. Thus there is a built-in non-uniformity at the molecular level. Furthermore, in an amorphous gel, the cross-links themselves cannot be exactly uniformly distributed throughout the network.

One may counter that, if the gel network is sufficiently elastic, the heat motions of its segments will have a sufficient range to permit the situations in every volume element to be identical on a time average basis. With a charged gel this is unlikely to be true. The osmotic pressure of the counter-ions and the electrostatic repulsion of the fixed charges along the chains swell the network until the elastic tensions in the chains just counter balance these stresses. These tensions restrict the lateral motions of the segments between crosslinks and severely limit the freedom of the cross-links themselves.

Even if the local time average compositions were identical, this would not eliminate the difficulty because the time scale of the averaging must be determined by the frequency of the network segmental motions. These will be of the same order as or, more probably, a good deal less frequent than those of the diffusing molecules whose fluxes are under discussion. There seems no escape from built in inhomogeneities at the micro-level, i.e. over distances comparable with the length of a single molecular jump, in a charged gel when viewed from a diffusing molecule. Different types of transported particles would experience different force fields in their interactions with the surroundings. They will tend therefore to adopt different average pathways through a charged gel membrane. This concept is of overriding importance in the mechanistic interpretation of any class of phenomenological transport coefficients.

We may now attempt an interpretation of the q_{ik} coefficients in Figure 1.

q_{13} measures the extent of coupling of the flows of water and the counterionic species and so is influenced by the strength of the interaction between the counterions and water and by the extent of involvement of the stream of one component with that of the other. q_{13} is substantial at all concentrations; it shows a slight decrease with decreasing concentration and appears to pass through a shallow minimum at low concentrations. Although it is not possible to guarantee the reliability of every small trend in these coefficients because so many calculation steps intervene between the experimental data and the evaluation of the q_{ik}, q_{13} is certainly the most accurately known.

At high external concentrations there are in the membrane comparable number of cations balancing the fixed charges and present, with equal numbers of equivalents of mobile anions, as sorbed electrolyte. Under these conditions the balancing counterions mostly lie close to the fixed charges while the sorbed electrolyte is distributed relatively uniformly through the whole system, and is in good contact with the water stream. Close to the fixed charges the counterions are exposed only to a relatively slowly moving water stream because of the retarding frictional interaction of the network chains on the water. The counterions themselves must be retarded by direct interaction with the fixed charges and with the network chains.

The observed value of q_{13} at high concentrations is 0.60 and this represents an average over the whole distribution of cations. The decrease in q_{13} with decreasing concentration reflects the increasing preponderance of the balancing counterions over the sorbed ions. The decrease is, to some extent, offset at lower ionic strengths because the ionic atmospheres around the fixed charges expand bringing the counterions into more complete contact with the water. At very low concentrations there are scarcely any sorbed ions present and further dilution is effective mainly in expanding these ionic atmospheres. These two opposing effects of dilution may well explain the existence of the minimum in q_{13}.

Although we can see that there are subtle trends in q_{13} which reflect changes in the mean distributions of the counterions with changes in concentration, it would, nevertheless, not be far wrong to say that q_{13} is never far from 0.5. This statement is equivalent to the observation that the empirical parameter α in the extended Nernst-Planck equation can be satisfactorily represented by 0.5.

The behaviour of q_{23} is quite different from that of q_{13}. It is always smaller than q_{13} and this may be due in part to the lower degree of hydration of anions. q_{23} falls to very small values in dilute solutions and, in NaBr, it goes slightly negative at the lowest concentrations. Although these values may be zero within the uncertainty of the data, it should be borne in mind that at such low solution concentrations the concentration of sorbed co-ions in the membrane falls to extremely low values.

The treatment of the sorption of co-ions in a charged membrane, introduced by Glueckauf [23], is helpful in understanding the behaviour of q_{23}. In view of the micro-inhomogeneity of the membrane the local fixed charge density varies from point to point. As a result, the local mean concentration of sorbed co-ions varies from point to point also. In fact, the local variations in the co-ion concentration are far more pronounced than the variations in the fixed charge concentration, owing to the form of the Donnan exclusion law.

The values of q_{13} and q_{23} can be traced back mainly to measurements of the electro-osmotic flow. The force driving this flow is proportional to the local space charge. The regions of low fixed charge density are the ones most preferred by the sorbed electro-lyte. Here the local water velocity may be low and the co-ions may therefore be ex-posed to a water stream which is moving with a lower velocity than the mean in the membrane. This consideration, together with the difference in hydration between cations and anions, explains why at every concentration q_{23} is lower than q_{13}.

In very dilute solutions there is only a small number of co-ions present and they are confined almost entirely to regions of such low fixed charge density that they constitute essentially micro-cavities in the network. It could arise that the water molecules in the immediate vicinity of these anions may tend to be dragged by them in the direction opposite from the net electro-osmotic flow. This would explain not only the very small positive values of q_{23} in very dilute solutions but also the small negative values as well.

The interionic coupling coefficient q_{12} is always small and positive. It appears to decrease with decreasing concentration but this small coefficient is known only with low accuracy and one is reluctant to speculate too freely on the significance of minor trends. Positive values of q_{12} would be expected between mutually attracting particles. Evidently the dependence on the concentration of co-ions is relatively minor. This suggests that the distance of the nearest Na^+ ions to any Br^- does not vary greatly with the concentration of the Br^- ions. This may be a consequence of two factors; the relatively constant ionic strength within the resin as a whole and the influence of the almost constant characteristic dimensions of the network on the distribution of the particles within it.

The foregoing remarks have indicated how the coupling coefficients may be inter-preted so as to deduce something of the distributions and motions of the mobile particles in a charged gel. They have been based on the supposition that the interac-tions between particular pairs of particles at close distances of approach do not depend on minor variations in the ambient conditions. The same kind of analysis may be extended to other types of coefficients.

12. Molar Friction Coefficients

The idea of describing steady transport processes in the membrane as being governed by balances between the thermodynamic forces on the system and frictional interactions among the components has been explored by several authors [8, 24]. They have been concerned with the resemblance between the r_{ik} and macroscopic friction. A particularly clear and useful version of these concepts was introduced by Spiegler [25] and it will be adopted here.

The fundamental principle of this frictional model is that, for each species, the gross thermodynamic force acting on one mole of the species is balanced by the sum of the interacting forces between that amount of the species and all of the other species present in its environment. Further, the interactions are analogous to frictional forces in that each can be expressed as the product of a friction coefficient f_{ik} and the difference between the mean net velocities of the molecules of i and k.

These mean velocities v_i and v_k can be expressed in terms of the flux densities and molar concentrations of the components. Thus

$$v_i = \phi_i/c_i \tag{19}$$

Continuing, as before, to take fluxes and velocities positive in the direction opposite from that of potentials increasing, gives

$$\frac{\partial \mu_i}{\partial x} = \sum_{k \neq i}^{m} f_{ik}\left(\frac{\phi_i}{c_i} - \frac{\phi_k}{c_k}\right) \tag{20}$$

Here the sum contains terms in which k represents in turn each of the permeating species other than i and also a term representing the interaction between i and the membrane m. By adopting the membrane as the stationary velocity reference, this term in the sum becomes simply $f_{im}\phi_i/c_i$.

f_{ik} is the force per mole of i owing to its interaction with the amount of k normally in its environment at unit relative velocity of i and k. The concentrations of i and k are not normally equal and the balance of forces requires that

$$c_i f_{ik} = c_k f_{ki} \tag{21}$$

Equation (20) can be rearranged to

$$(\partial \mu_i/\partial x) = \phi_i \sum_{k \neq i}^{m} f_{ik}/c_i - \sum_{k \neq i} (\phi_k f_{ik}/c_k) \tag{22}$$

and when this is compared with Equation (13) it is seen that

$$r_{ik} = -f_{ik}/c_k \tag{23}$$
$${\scriptstyle k \neq i}$$

while

$$r_{ii} = \sum_{k \neq i}^{m} f_{ik}/c_i \tag{24}$$

The set of r_{ik} coefficients has already been determined but before the f_{ik} coefficients can be evaluated from them the concentrations c_i of the cations, anions and water in the membrane have to be determined as functions of the external solution concentration. This is a straightforward matter of analysis and the methods used are well documented. They will not be discussed here.

Once the concentrations have been determined, two coefficients, f_{ik} and f_{ki}, can be evaluated from each r_{ik} cross coefficient. By making use of these coefficients and Equation (24) one f_{im} can be obtained from each r_{ii}.

13. Interpretation of Friction Coefficients

When attempting to interpret trends in the f_{ik} coefficients one must, once again, bear in mind that by representing the velocity of a component by the quotient ϕ_i/c_i, it is implicitly assumed that the distributions of the particles and of their fluxes are the same. Thus the local flux density is assumed to be in direct proportion to the local concentration. In a system in which interionic forces are important this is very unlikely to hold. Furthermore, taking the relative velocities of the components as the differences between pairs of these quotients implies that the distributions of concentration and flux are the same for all types of particles. This is essentially the assumption whose breakdown formed the basis for much of the discussion of the q_{ik} coefficients.

The friction coefficients for Na^+, Br^- and water in Zeo-Karb 315 are plotted in Figures 2, 3 and 4 as functions of the concentration of Br^- co-ions in the membrane. Figure 5 gives plots of $c_i f_{im}$ times the concentration of the membrane matrix in equivalents per unit volume. Figure 5 thus indicates the relative frictional interaction of each mobile component i with the fixed membrane. The quantities plotted may be thought of as f_{mi} where the membrane matrix has been assigned a defined molecular weight given by its equivalent weight.

On examining Figure 2 in detail, it is seen that friction with the co-ions f_{12} is always unimportant in its effect on the overall behaviour of the counter-ions, even at the highest co-ion concentrations. Over most of the range examined f_{12} is quite insignificant.

When comparing interactions of the counter-ions with water and with the membrane it must be remembered that water occupies about 75% of the total volume of the swollen gel. Thus, if the components were uniformly distributed and the force fields between water and Na^+ were the same as those between membrane matrix and Na^+, one would expect to find $f_{13} \simeq 3f_{1m}$. Probably the force of interaction of the ions with water is greater than with the organic matrix and this might lead to $f_{13} \geqslant 3f_{1m}$.

At high concentrations, where the counter-ions are most nearly uniformly distributed, it is found that $f_{13} = 2.4f_{1m}$. Evidently the electrostatic attraction of the counter-ions to the matrix causes them to have an enhanced interaction with the matrix at all concentrations.

At lower concentrations of sorbed electrolyte the electrostatic attraction of the matrix is greater per counter-ion and hence their molar frictional interaction with the matrix increases. This is clearly seen in the behaviour of f_{1m}. The change in concentration between the most and the least concentrated external solutions examined here would account for about two-fold increase in f_{1m}. The observed increase is nearer

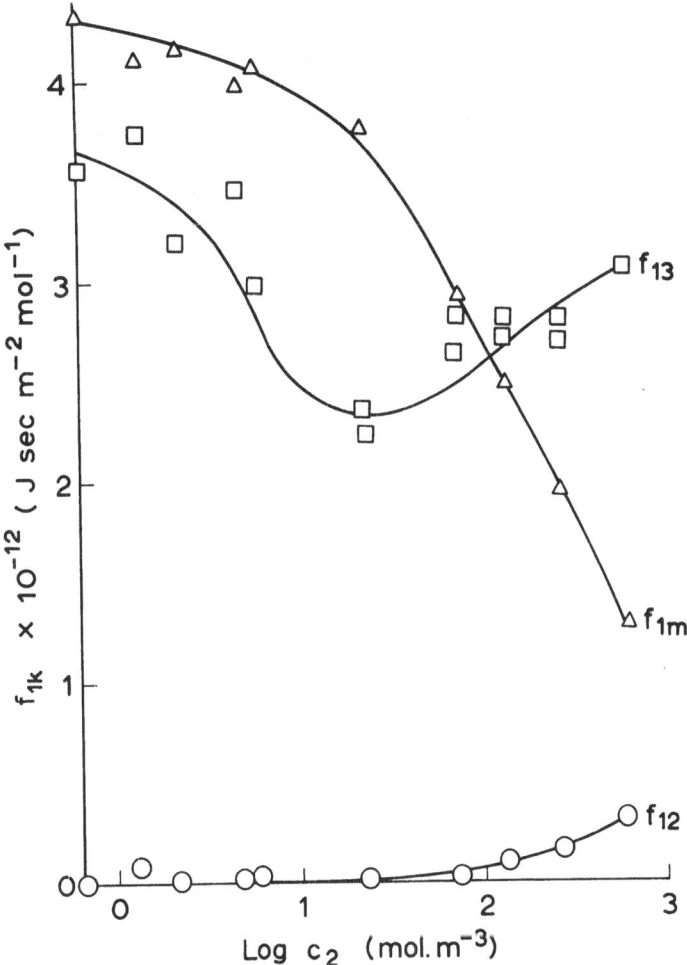

Fig. 2. Molar friction coefficients f_{1k} for Na^+ ions in Zeo-Karb 315 versus logarithmic concentration of sorbed Br^- ions, log c_2.

3.4-fold. The extra effect may be attributed to the closer association between the counter-ions and fixed charges when there are only few co-ions present to lower the electrostatic potential peaks between the matrix chains.

The friction coefficient per mole of counter-ions with the water f_{13} is greater than f_{1m} in concentrated solutions and less in dilute solutions. A simple but crude explanation of this is that in dilute solutions the counter-ions make more contact with the matrix and must therefore make less contact with the water. This must however be an over simplification because, in the most dilute solutions f_{13} is a little greater than, instead of less than, its value in the most concentrated solutions. Between these extreme concentrations it passes through a definite minimum. The decrease in f_{13} with decreasing concentration at the more concentrated end of the range may be accounted for by the decrease in the number of sorbed cations. The sorbed cations are in closer contact with the water than the charge-balancing cations. The subsequent increase in

f_{13} at the low concentration end of the scale, where the proportion of sorbed to fixed-charged-balancing cations is trivial, requires a different explanation. Readjustment of the counter-ionic atmospheres around the matrix chains at low concentrations can increase the electro-osmotic flow and hence the apparent interaction between cations and water [26]. Furthermore, the water closest to the chains may be electro-stricted or structured because of the presence of the hydrophobic organic groups. If this notion,

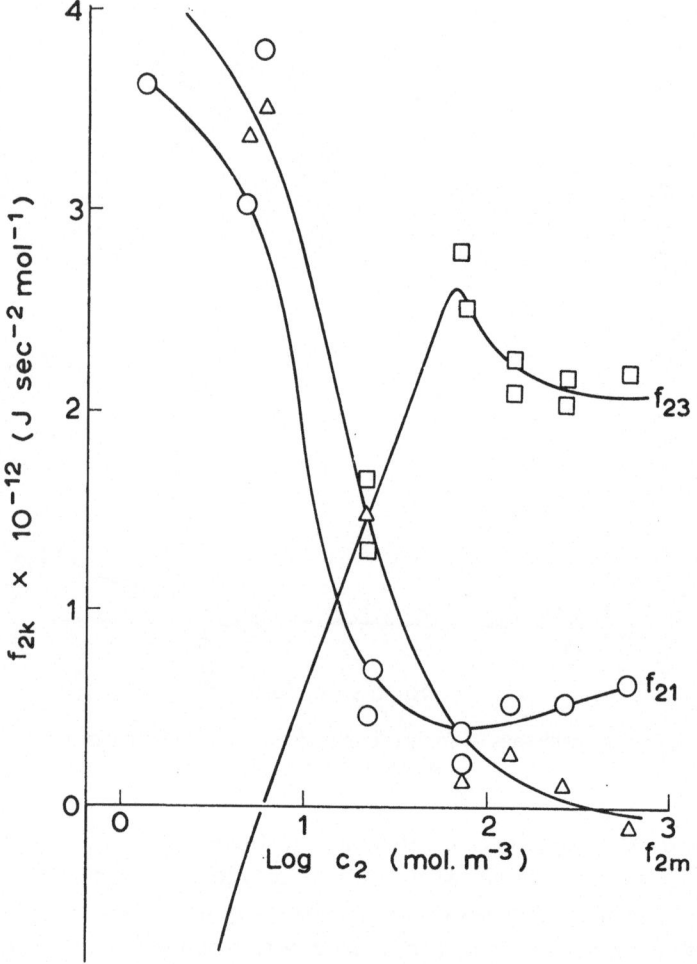

Fig. 3. Molar friction coefficients f_{2k} for Br^- ions in Zeo-Karb 315 versus logarithmic concentration of sorbed Br^- ions, log c_2.

which is often advanced, were correct the cations in dilute solution would be surrounded by water of somewhat enhanced viscosity. This would explain why f_{13} was larger in very dilute solutions than in concentrated ones.

Turning now to the friction experienced by the co-ions, the curves in Figure 3 are, at first sight, perplexing. At the higher concentrations, i.e. above about 0.15 molar external solutions, the observations appear understandable, The electrostatically repulsive term f_{2m} is very small and varies from a slightly negative to a slightly positive value. The electrostatically attractive term f_{21} is positive and accounts for about 20% of the co-ion friction. The remaining friction is due to interaction with the water. f_{23} is therefore the most important term and is similar in size to f_{13}. The moderate increase in f_{23} as the concentration of co-ions is decreased probably reflects the tendency of the remaining co-ions to prefer the regions of lowest fixed charge density which are more highly swollen than the average.

At lower concentrations there appears to be a rather sudden change in the situation. The co-ion concentration is here very low and experimental errors in c_2 have a big effect on the observed values of f_{2k}. However, the problems in determining c_2 at low concentration are such that c_2 is more likely to be too large than too small. Hence the f_{2k} coefficients would be expected to turn out numerically too small at low concentrations.

An increase in f_{21} is not hard to understand because in dilute solutions each mole of co-ions has so many counter-ions in its environment with which to interact. The sharp decrease in f_{23} to large negative values is the perplexing observation. The related behaviour of q_{23} has already been discussed. Although more interaction with the cations must leave fewer contacts for the anions to make with the water, this effect can hardly account for the whole change in f_{23}. The author believes that the steep fall in f_{23} is in part an artefact which arises from applying a uniform distribution treatment to a component, the co-ions, which at low concentrations was very non-uniformly distributed throughout the gel. Because f_{2m} is calculated by difference from r_{22}, it is bound to show a steep increase in order to compensate for the behaviour of f_{23}.

The frictional interactions on the water are shown in Figure 4. At high concentrations the cation concentration in the membrane is slightly less than twice the anion concentration. It may be seen that f_{31} is 2.6 times f_{32}. The difference between these factors is mainly attributable to the greater tendency to hydration of the cations than of the anions. It was suggested also, in connection with q_{13} and q_{23}, that the co-ions may experience, in electro-osmosis, a water stream which is moving more slowly than the average. At lower concentrations the influence of the co-ions on the water, per mole of the latter, naturally decreases to insignificant values, as may be seen from Figure 4.

The minimum in f_{31} is a reflection of the minimum already discussed in f_{13}. It is noteworthy that f_{31} is always much greater than f_{3m}, i.e. the frictional drag on water forced through a membrane at zero electric current, due to the stationary counter-ions is more important than that due to the matrix material of the membrane. The implications of this observation in relation to attempts to produce fixed-charge reverse osmosis membranes should not be overlooked.

It may be noted that f_{3m} decreases significantly with decreasing concentration. This might appear to conflict with the suggestion of structured water around the matrix material. It must be remembered however that f_{3m} is an average value taken over all the water molecules. In dilute solutions, when the counter-ionic atmospheres around the matrix are expanded, the velocity profile in the electro-osmotic stream changes so

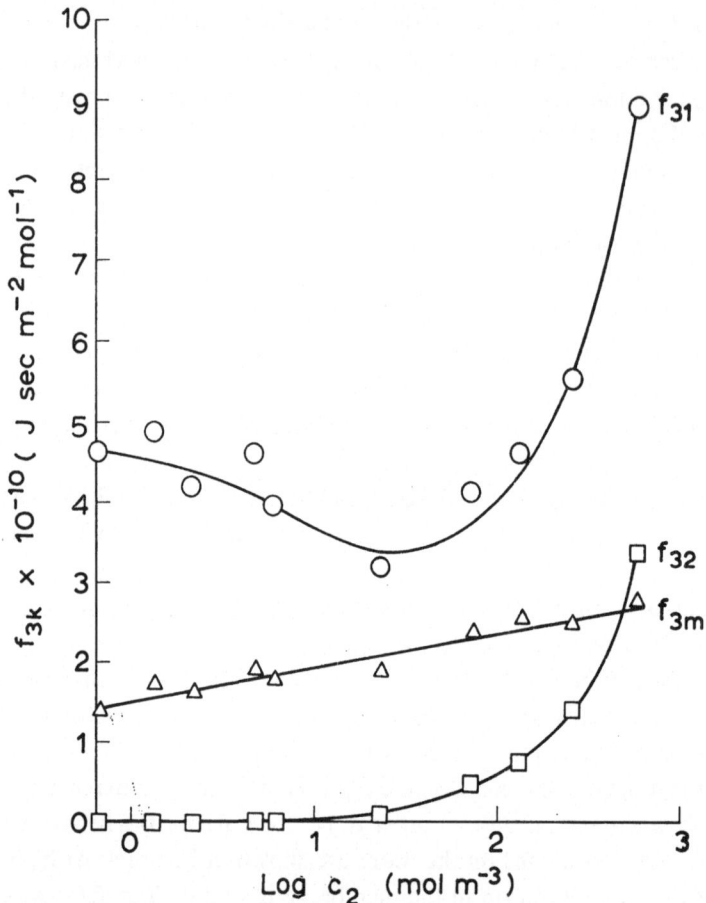

Fig. 4. Molar friction coefficients f_{3k} for water in Zeo-Karb 315 versus logarithmic concentration of sorbed Br^- ions, $\log c_2$.

as to be less steep close to the matrix chains [26]. Thus, in terms of the uniform space charge distribution model, f_{3m} would appear to decrease.

All the coefficients f_{3k} are about two orders of magnitude lower than the other f_{ik} coefficients. The reason for this is that each water molecule is in contact mainly with other water molecules. Their interactions do not appear in the force balances in the equations used here because experiments with tracer water molecules have not been considered.

Finally, the friction coefficients per equivalent of fixed-charge matrix can be examined (Figure 5). As anticipated, f_{m2} is always negligible because there are relatively few co-ions and they are repelled by the matrix. f_{m3} decreases with decreasing concentration in parallel with f_{3m} because the concentrations of water and of matrix scarcely change over the whole range of co-ion concentrations examined.

f_{m1} is always greater than f_{m3} despite the much lower concentration of cations than water in the membrane. Furthermore, f_{m1} increases as the concentration of cations decreases. These facts confirm that there is a strong interaction between the counter-

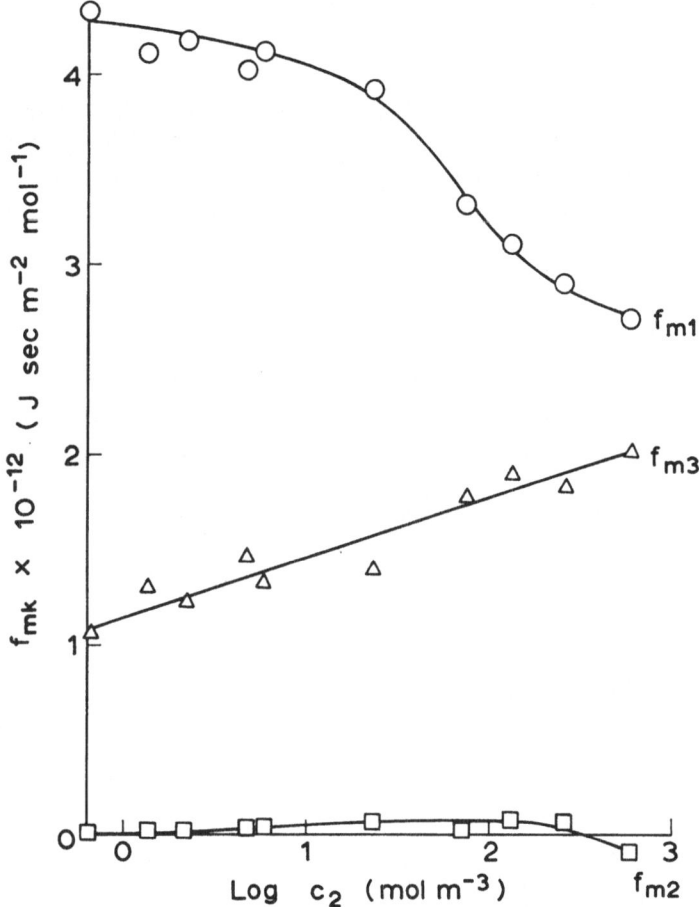

Fig. 5. Friction coefficients per equivalent of membrane matrix f_{mk} of Zeo-Karb 315 versus logarithmic concentration of sorbed Br⁻ ions, $\log c_2$.

ions and the matrix. At low concentrations there appears to be some non-specific binding of counter-ions, just as in polyelectrolyte solutions, in view of the fact that f_{m1} increases despite the decrease in the total concentration of cations.

14. The Mobilities of Ions and Water

It was pointed out earlier that, in the widely used Nernst-Planck formulation of membrane transport, integration across the membrane is usually carried out under the assumption that the ionic mobilities u_i inside the membrane are independent of the concentrations of the solutions outside. The non-equilibrium thermodynamic data permits an examination of the implications of this assumption.

It is generally supposed that in a gel membrane, in which diffusion occurs by the random molecular heat motion, the Nernst-Einstein relation between absolute mobility and diffusion coefficient holds for the ions in the form

$$u_i RT = D_i \qquad (25)$$

Equation (4) was derived on this basis for the case when all cross coefficients $l_{ik} = 0$. In this case RTl_{ii}/c_i would be equal to the diffusion coefficient D_i. RTl_{ii}/c_i can be evaluated from the transport and sorption data and its variation with concentration examined.

It has to be understood that, in a membrane where flux coupling does occur and the cross coefficients are not all zero, l_{ii}/c_i defines a mobility under unit force on i in which i is transported against a moving background. This complexity arises because fluxes of species other than i are generated by coupling with the flux of i which has been caused by the force applied to i.

The reciprocal quantity $1/r_{ii}c_i$ also defines a mobility. Inspection of Equation (13) shows that $RT/r_{ii}c_i$ would correspond with a thermodynamic diffusion coefficient determined under restrictions whereby all fluxes other than ϕ_i were zero.

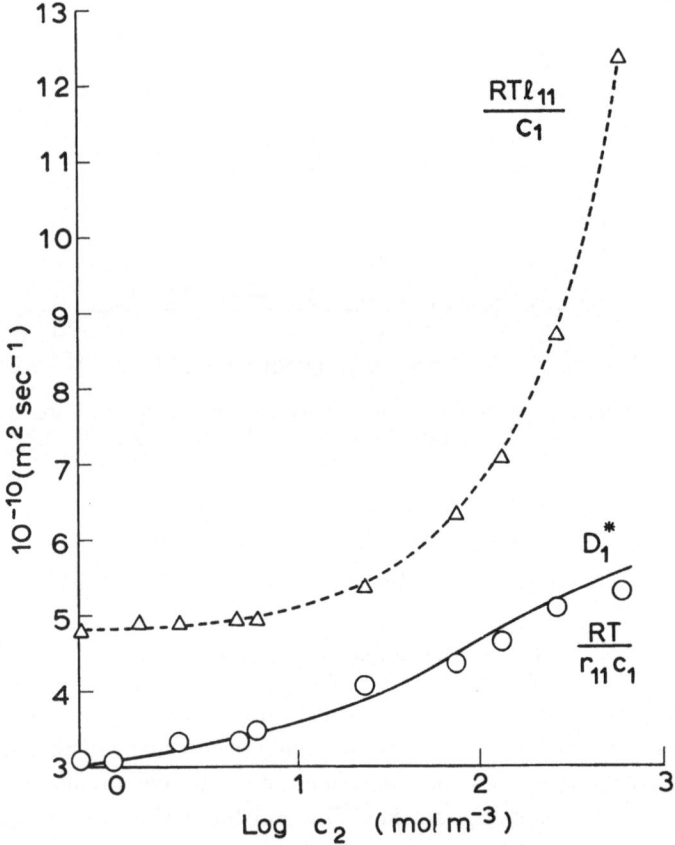

Fig. 6.   Mobility indexes of Na^+ in Zeo-Karb 315 versus logarithmic concentration of sorbed Br^- ions, log c_2. The full line is the self-diffusion coefficient D_1^* and the circles are $RT/r_{11}c_1$. The triangles and dashed curve are RTl_{11}/c_1.

So-called self-diffusion coefficients D_i^* are determined by measuring the flux of an isotope of one component in the membrane under conditions of uniform chemical composition and zero external forces [27].

The relations between these three measures of mobility can be seen for Na^+ in the membrane in Figure 6 and for Br^- in Figure 7.

It can be seen that in the case of counter-ions and co-ions RTl_{ii}/c_i is far from independent of concentration. In each case this index of mobility increases by more than 2.5-fold in going from the most dilute to the most concentrated solution examined. It also bears little resemblance to the self-diffusion coefficient which is shown as a full line in each figure. (The data have been published previously. [27])

$RT/r_{ii}c_i$ also increases with increasing concentration for both ions, but by a much smaller factor than does RTl_{ii}/c_i.

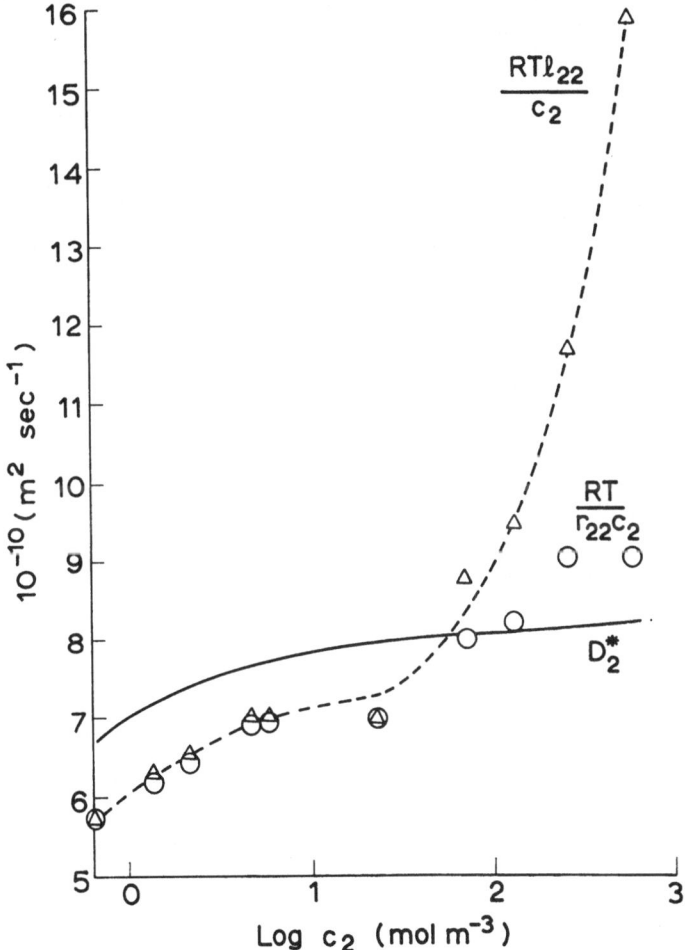

Fig. 7. Mobility indexes of Br^- in Zeo-Karb 315 versus logarithmic concentration of sorbed Br^- ions, $\log c_2$. The full line is the self-diffusion coefficient D_2^* and the circles are $ET/r_{22}c_2$. The triangles and dashed curve are RTl_{22}/c_2.

In Figure 6 it is seen that $RT/r_{11}c_1$ agrees excellently with D_1^*. In the case of the co-ions (Figure 7) the agreement between D_2^* and $RT/r_{22}c_2$ is less good but the experimental difficulties in obtaining accurate values of r_{22} may account for much of the scatter in the points.

It may be concluded that the measurement of self-diffusion coefficients can lead to good estimates of r_{ii} but not of l_{ii}. This suggests also that $1/r_{ii}c_i$ is probably the more appropriate mobility to use in the extended form of the Nernst-Planck treatment of Equation (2).

The variation of $1/r_{ii}c_i$ with concentration is certainly not negligible in the case examined here. When experiments over wide concentration intervals are under consideration, this variation should be taken into account when integrating Equation (2) across the membrane.

Figure 8 shows two quantities RTl_{33}/c_3 and $RT/r_{33}c_3$ as functions of concentration. As in the case of the ions, the former is usually larger than the latter because the dragged or coupled flows ease the passage of the driven flow. Both quantities increase somewhat with decreasing concentration. A very small part of this increase may be accounted for by the increased swelling of the membrane in more dilute solutions [26]. The major effect which permits $RT/r_{33}c_3$ to increase by a factor of 2.5-fold is the absence, when in dilute solutions, of sorbed electrolyte. This is present when in concentrated solutions and obstructs the flow of water under the defined conditions of all fluxes other than water held at zero. This restriction is not contained in the definition

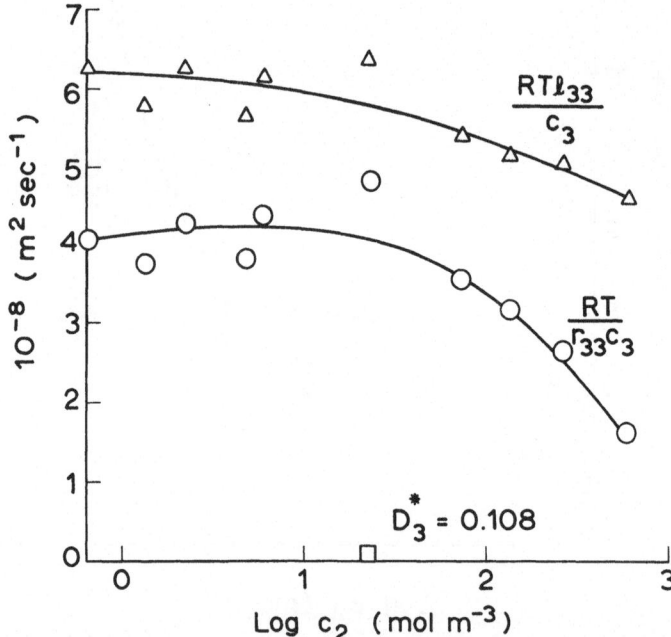

Fig. 8.   Mobility indexes of water in Zeo-Karb 315 versus logarithmic concentration of sorbed Br^- ions, log c_2. The square point shows the self-diffusion coefficient on the same scale.

of $RT l_{33}/c_3$ and, in consequence, this quantity does not vary so markedly with concentration.

Also indicated on Figure 8 is the self-diffusion coefficient of water in the membrane measured with D_2O [28]. The value is between 2% and 6% of $RT/r_{33}c_3$. This observation emphasises the significance of the comment made earlier that water molecules make most of their collisions with other water molecules. D_3^* refers to a true diffusion process of D_2O against a background of H_2O and other molecules. $RT/r_{33}c_3$, on the other hand, characterises a flow process which is mainly convective in character and is limited by viscous forces. The values make it clear that diffusional transport makes only a small additional contribution to the total transport of water. In this case the measurement of the self-diffusion coefficient is no guide to the value of r_{33} or l_{33}.

15. Conclusions

This discussion of the various non-equilibrium thermodynamic coefficients which can be determined as functions of concentration in a charged gel membrane shows that very detailed information may be deduced from them about the physical situation at molecular level and about the molecular mechanisms of the transport processes in such a gel. Such information could not readily have been obtained from an unsystematic series of flux and potential measurements, however extensive. The non-equilibrium thermodynamic discipline of designing experimental programmes and analysing the data from them is recommended to all who are interested in the transport properties of essentially homogeneous gel membranes.

The author extends his cordial thanks to his many collaborators whose efforts have provided most of the data and many of the concepts on which this paper has been founded.

References

1. Teorell, T.: *Proc. Soc. Expt. Biol. Med.* **33**, 282 (1935).
2. Meyer, K. H. and Sievers, J. F.: *Helv. Chim. Acta* **19**, 649, 655, 987 (1936); **20**, 634 (1937).
3. Teorell, T.: *Prog. in Biophysics* **3**, 305 (1953).
4. Mackie, J. S. and Meares, P.: *Proc. Roy. Soc.* A **232**, 485 (1955).
5. Meares, P.: in J. Crank and G. S. Park (eds.), *Diffusion in Polymers*, Academic Press, London 1968, Chapter 10.
6. Mackie, J. S. and Meares, P.: *Proc. Roy. Soc.* A **232**, 298, 510 (1955).
7. Schlögl, R. and Schödel, U.: *Z. Phys. Chem. N.F.* **5**, 372 (1955).
8. Laity, R. W.: *J. Phys. Chem.* **63**, 80 (1959). *J. Chem. Phys.* **30**, 682 (1959).
9. Fuoss, R. M. and Onsager, L.: *J. Phys. Chem.* **61**, 668 (1957).
10. Koefoed-Johnsen, V. and Ussing, H. H.: *Acta Physiol. Scand.* **28**, 60 (1953).
11. McHardy, W. J., Meares, P., Sutton, A. H., and Thain, J. F.: *J. Coll. Interf. Sci.* **29**, 116 (1969).
12. Katchalsky, A. and Curran, P. E.: *Non-Equilibrium Thermodynamics in Biophysics*, Harvard Univ. Press, Cambridge Mass., 1965.
13. Staverman, A. J.: *Trans. Faraday Soc.* **48**, 176 (1952).
14. Kirkwood, J. G.: in H. T. Clarke (ed.), *Ion Transport Across Membranes*, Academic Press, New York, 1954, p. 119.
15. Kedem, O. and Katachalsky, A.: *Trans. Faraday Soc.* **59**, 1918 (1963).
16. Michaeli, I. and Kedem, O.: *Trans. Faraday Soc.* **57**, 1185 (1961).
17. Krämer, H. and Meares, P.: *Biophys. J.* **9**, 1006 (1969).

18. Foley, T. and Meares, P.: *Experientia Suppl.* **18**, 313 (1971).
19. Foley, T., Klinowski, J., and Meares, P.: *Proc. Roy. Soc. A* **336**, 327 (1974).
20. Gardner, C. R. and Paterson, R.: *J. Chem. Soc. A* 2254 (1971); *J. C. S. Faraday I* **68**, 2030 (1972).
21. Demarty, M. and Sélégny, E.: *Compt. Rend. Acad. Sci. (Paris) C* **276**, 1549 (1973).
22. Caplan, S. R.: *J. Phys. Chem.* **69**, 3801 (1965); *J. Theoret. Biol.* **10**, 209, 346 (1966).
23. Glueckauf, E.: *Proc. Roy. Soc. A* **268**, 350 (1962).
24. Klemm, A.: *Z. Naturforsch.* **8a**, 397 (1953).
25. Spiegler, K. S.: *Trans. Faraday Soc.* **54**, 1408 (1958).
26. Mackay, D. and Meares, P.: *Trans. Faraday Soc.* **55**, 1221 (1959).
27. McHardy, W. J., Meares, P., and Thain, J. F.: *J. Electrochem. Soc.* **116**, 920 (1969).
28. Mackay, D.: *J. Phys. Chem.* **64**, 1718 (1960).

INTERACTION IN ION EXCHANGERS

Ionic Interactions in a Polyelectrolyte Gel System with Counter-Ions of Different Valence Types and Varying Dielectric Constants

THORVALD S. BRUN

Department of Chemistry, University of Bergen, Norway

Abstract. Transport properties and some thermodynamic properties have been measured in synthetic polyelectrolyte gels of the phenolsulphonic acid-formaldehyde type with potassium, magnesium and hexamethylputrescinium ions as counter-ions and chloride ions as co-ions.

With the univalent counter-ions a complete set of transport properties was measured at vanishing concentration of non-exchange salt and with water as the pore liquid, and the results were correlated with properties of free solutions by the friction coefficient formalism of irreversible thermodynamics.

At finite concentrations of non-exchange salt, the diffusion coefficient and adsorption equilibria were studied for all three valence types with water as the pore liquid and at different temperatures. A mean friction coefficient was compared with the corresponding quantity in free solutions.

The electric conductivity of the gels was measured for all three valence types as functions of concentration and temperature at different values of the dielectric constant of the pore liquid. The results were used for the discussion of the interactions between counter-ions and fixed ions, and between counter-ions and co-ions.

Finally some statistical and Monte Carlo calculations on a simple lattice type model are given, the subsystem being a spherical cavity with pore liquid and mobile ions, preserving some important characteristics of the gels.

1. Introduction

There are three main approaches to the study of polyelectrolyte gels and membranes:

(a) The study of thermodynamic properties and transport properties.

(b) The direct study of microstructure and microdynamics by spectroscopic and other methods.

(c) The theoretical study of models by statistical methods and numerical methods.

Our group in Bergen has mainly worked with the thermodynamics and transport properties; we have also tried to enter the numerical simulation approach. Some of the results are presented and discussed in terms of ionic interactions.

2. Incomplete Dissociation in Polyelectrolyte Gels with Univalent Counter-Ions and Vanishing Concentration of Non-Exchange Salt

It is always important to see whether the properties of these complicated systems may be understood in terms of the properties of ordinary electrolytes or not. The friction coefficient formalism of the thermodynamics of irreversible processes is suitable for this purpose. We use Laity's [1] formulation as a starting point, and introduce a parameter α defined to be the fraction of free counter-ions. $(1 - \alpha)$ is then the fraction of fixed-ion – counter-ion pairs. This is a simple two-state model.

We now define unsymmetrical mean friction parameters [2] expressing the interaction between one particle and its surroundings, and express these friction parameters by operationally defined quantities. The quantity α thus enters explicitly only. The equations are [3]:

$$R_{+s} = \tau \cdot V_s \cdot z_+ \cdot p/(\sigma \cdot M_s) \tag{1}$$

$$R_{+(-u)} = (F^2/\kappa + \tau^2 \cdot V_s/\sigma) \cdot z_+ \cdot A \cdot \alpha - \tau \cdot V_s \cdot z_+ \cdot p/(\sigma \cdot M_s) \tag{2}$$

$$R_{s+} = \tau \cdot V_s \cdot A \cdot \alpha/\sigma \tag{3}$$

$$R_{s(-u)} = (V_s/\sigma) \cdot (p/M_s - \tau \cdot A \cdot \alpha) \tag{4}$$

The meaning of the symbols is:

M_s – molecular weight of solvent.

α – 'degree of dissociation' parameter.

κ – specific electric conductivity at zero pressure gradient.

τ – moles solvent transported per Faraday at zero pressure difference.

σ – hydrodynamic permeability at zero electric current.

V_s – partial molal volume of the solvent.

p – grams H_2O per cm^3 gel volume.

A – equivalents sulfonic acid groups per cm^3 gel volume.

It is seen that the counter-ion to solvent friction coefficient R_{+s} does not contain the parameter α whereas the mean friction coefficient for counter-ions with free fixed ions and pairs $R_{+(-u)}$ contains the parameter. The valency of the resin component z_- which is an arbitrary number of sulfonic acid groups does not occur anywhere.

We now inspect the experimental results for three polyelectrolyte gels representing different stages of evaporation of a reaction mixture of phenol sulfonic acid and formaldehyde.

It is seen that the R_{+s} friction coefficient increases with the stoichiometric fixed ion concentration. The corresponding friction coefficient for a free KCl solution is in-

TABLE I

Friction parameters in units dyn cm^{-1} s $mole^{-1} \times 10^{-15}$ and related dimensionless quantities

		Membrane type		
		A	B	C
1	R_{+s}	2.1	0.80	1.25
2	$R_{+(-u)}$	$6.5\alpha - 2.1$	$3.9\alpha - 0.8$	$4.7\alpha - 1.3$
3	$R_{+(-u)}$	4.4	3.1	3.4
	$a = 1$			
4	$R_{+(-u)} = R_{+-}(KCl)$	0.19	0.14	0.16
	α	0.36	0.25	0.31
5	$R_{+(-u)} = 0$			
	α	0.32	0.21	0.28
6	$t = \tau \cdot M_s \cdot A/p$	0.52	0.50	0.48
7	$p = gH_2O \ cm^{-3} \ gel^{-1}$	0.92	0.80	0.76
8	$A = eq \ SO_3H \cdot 10^3 \ cm^{-3}$			
	gel^{-1}	1.24	0.68	0.83

dependent of concentration in a large range of concentration and is equal to the middle of the three values measured for our gels (ref. [3]). The low value of R_{+s} in the soft membrane type B may be explained by assuming a certain fraction of charge-free pores making the hydrodynamic permeability too large. We now make different choices of the parameter α for purpose of comparison, Table I.

(i) $\alpha = 1$ corresponds to complete dissociation. The cation to anion and pairs friction coefficient is then larger than the corresponding quantity in free solution by a power of ten. This makes another choice of α natural.

(ii) If the friction parameter $R_{+(-u)}$ in the gel is made equal to the corresponding quantity in aqueous solution, there results a value of α of 25–35% according to the membrane chosen, and the lowest value is obtained in the membrane with the lowest fixed ion concentration.

(iii) The lowest reasonable value of $R_{+(-u)}$ is zero. This corresponds to an ideal solution without co-volume effects. The choice gives vales of α of about the same magnitude as before, the difference being 10–15% only.

(iv) The quantity t in the 6th row does not contain α and is identical with the anion transfer number of the gel if it is treated as an ordinary electrolyte solution and using the water component as frame of reference.

It is seen that this quantity is constant for the three gel types. This is quite reasonable as the gels correspond essentially to different stages of evaporation of a reaction mixture and transfer numbers show little sensitivity to concentration changes.

The conclusion is that the study of transport properties in these gels with univalent counter-ions and vanishing concentration of non-exchange salt indicates that a description based on a simple two-component model with incomplete dissociation is sufficient.

It can also be mentioned that if the existence of a simple ideal Donnan equilibrium is postulated for an external solution of the same concentration as used by us we find an effective fixed ion concentration of the same magnitude as the one derived from the transport properties. However, it is well known that the simple Donnan model is incapable of explaining the salt adsorption in these systems.

3. Interactions with Ions of Different Valence Types and Non-Exchange Salt Equilibria

The determination of the six independent transport properties at finite concentrations of non-exchange salt has been outside our reach. What we have done to a certain extent are measurements of the diffusion coefficient of non-exchange salt under nonstationary conditions. We then also get values for the amount of non-exchange salt absorbed. The valence types of the counter-ions have been varied by using KCl, $MgCl_2$ and the bolaform electrolyte hexamethylputrescinium di-chloride [4, 5, 6].

The diffusion coefficient contains contributions from several interactions. As our main purpose is to be able to carry out a comparison with the properties of free solutions, the diffusion coefficient is now expressed by the friction coefficients. It is possible to define a function F of the friction coefficients which is related to operationally defined quantities only. This function is suitable for comparison of our limited amount of information with the properties of aqueous solutions. The calculation consists in a straightforward application of non-equilibrium thermodynamics, and the use of Scatchard's [7] assumption: for every section in the membrane diffusion field there exists an external equilibrium solution. By use of the Gibbs-Duhem equation, this allows connection with the experimental diffusion coefficient. The experimental situation is zero electric current, zero gradient in chemical potential of the polyelectrolyte, and zero volume flux. This set of conditions is compatible with Scatchard's assumption. Typical values of the ratios between mole fractions of water, high polymer electrolyte (considered as uni-valent), and non-exchange salt are

$$x_s : x_e : x_i = 1000:20:1$$

The formulae obtained are now simplified disregarding higher than second order terms. The remaining result for the experimental diffusion coefficient D then reads

$$(z_+ + 1) \cdot RT/(\alpha \cdot D) = r_{cs} \cdot x_s' \cdot z_+ + (r_{c+} \cdot z_+ \cdot x_+ + r_{c-} z_+ x_-) +$$
$$r_{+s} \cdot x_s' \cdot x_c/(z_+ x_+) \overset{\text{def}}{=} z_+ \cdot F \qquad (5)$$

Here c, s, $+$, $-$ refer to co-ions, solvent molecules, gegen-ions, and polyelectrolyte unit respectively. X denotes particle fractions and r friction coefficients. z_+ is the valence of the gegen-ions, and the co-ions are assumed to be uni-valent (chlorine ions throughout). The quantity α is an empirical parameter from an equation by Glueckauf and Watts [8] describing Donnan-equilibria:

$$\log C_i(\text{membrane}) = \log A + \alpha \log C_i(\text{equilibrium solution}). \qquad (6)$$

The experimental data consist of exchange capacities, water content, Donnan equilibria, diffusion coefficients, external geometry and swelling data. The latter were obtained by weighing. Some typical experimental data are given in Table II.

TABLE II

Membrane data and non-exchange salt equilibrium parameters (ref. [3])

Membrane type	I	II	III
Millimol SO₃H pr gram dry K-form	1.83	1.70	2.08
Per cent dry matter in K-form	32.6	41.3	38.9
Parameter α, Equation (6), K-systems	1.47	1.30	1.50
Parameter α, Equation (6), Mg-systems	1.00	–	1.10
Parameter α, Equation (6), HMP-systems	0.96	–	0.98

It is seen that with $MgCl_2$ and $HMPCl_2$ there is practically no Donnan exclusion; with KCl we get a $\frac{3}{2}$ power dependence on the external concentration in accordance with the empirical findings of Glueckauf and Watts [8]. An ideal Donnan behaviour gives a second power dependence.

In Figure 1 the mean friction coefficient F is represented as a function of the concentration of the external equilibrium solution. The interval of variation of the corresponding quantity F_0 in free solution up to the stoichiometric fixed ion concentration is also shown.

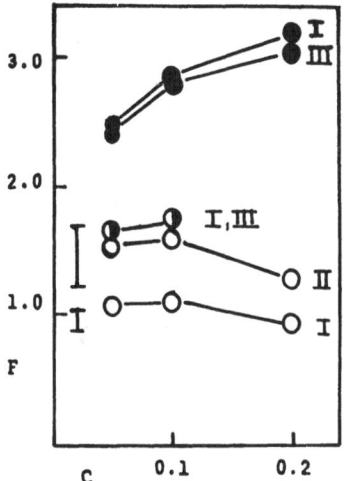

Fig. 1.   Friction coefficient function F in units $RT\,10^5$ against concentration C of external equilibrium solution. Systems K $\bigcirc$, Mg $\pmb{\bullet}$, HMP $\pmb{\bullet}$.

We see that in the KCl system the values are similar to those in free solution. In the dense membrane type II we have $F > F_0$. For $MgCl_2$ we also have $F \simeq F_0$ for types I and III. For the HMP system the value of F_0 is only known at infinite dilution; F is approximately equal to the double of this value.

The temperature coefficient is also larger than the normal value (a 2 kcal increase in the apparent activation energy).

Exchange equilibria show a strong preference of the tetramethylputrescinium ion over the potassium ion.

The conclusion is that for magnesium ions there is a strong interaction with the fixed ions, but the non-exchange magnesium chloride moves rather non effected by this; the coupling between the nonexchange salt and the gel is but weak. For KCl the interaction is weaker as shown by the Donnan-like exclusion, but the diffusion of the non-exchange KCl is uninhibited, except in the very dense membrane type II where sieve effects may be present. In the hexamethylputrescinium systems the interaction with the gel fixed ions is strong, but there is a certain coupling between the diffusion of nonexchange HMP and the polyelectrolyte gel.

A 'loop' like movement may be possible with bolaforms in general.

4. Systems with Lowered Dielectric Constant and Temperature

It is of interest to increase ionic interactions by lowering the dielectric constant of the medium and by lowering the temperature. The increased experimental difficulties make the choice of the electric conductance a natural one, as this is the transport quantity most easily determined [9].

The results of some measurements are shown in Figure 2. The temperature coefficient of the electric conductance corresponds to apparent activation energies of 7–9 kcal mole^{-1}. The potassium systems have the lowest values. The specific conductivity is much lower for Mg and HMP than for the potassium system. The ratio κ_{Mg}/κ_{HMP}

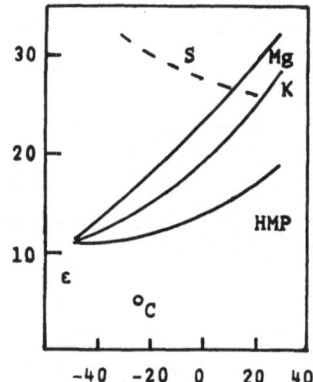

Fig. 2.   Dielectric constant versus temperature. Curve S: pure solvent (ethanol).

is constant and equal to 2.4. The interaction between cations and the high polymer anion increases in the series K, Mg, HMP as the ratio between the conductivity in the membrane and that of th external solution decreases strongly in that order. The high overlapping values of the activation energy indicate interactions of the same type in the various systems. In the Mg and HMP systems the chloride ions must be all-important and charged ion pairs like $MgCl^+$ and $HMPCl^+$ could well be the important kinetic units for conduction. If so interaction must be the interactions of these positive ion pairs with the fixed SO_3-groups.

The average distance between free fixed charges must be the largest in the Mg-systems. The dielectric constant is the largest for the Mg-systems, smallest for the

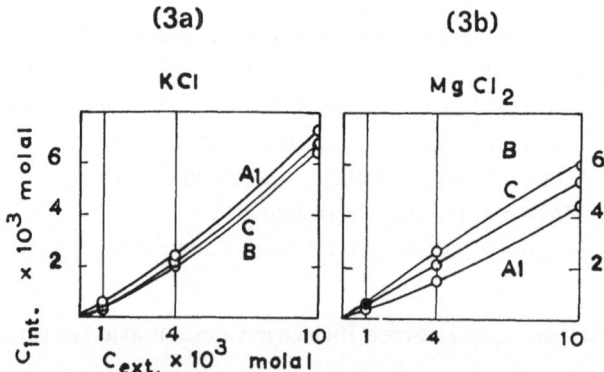

Figs. 3a–b.   (a) KCl systems. (b) MgCl₂ systems Concentration of non-exchange salt as a function of the external equilibrium concentration in three membrane types.

bolaform gegen-ion systems. The increase of the dielectric constant with temperature may indicate increased electric assymmetry and an increase in degrees of freedom.

We now present some results showing the conductance of polyelectrolyte gels in potassium form and in magnesium form at low dielectric constant and varying external concentration [10].

The equilibrium values (Figures 3a, b) show a weak Donnan exclusion with the free systems, none at all with the $MgCl_2$ systems.

The concentration of non-exchange salt must be calculated on the basis of the total gel volume; the values then are 40–70% of the external equilibrium solution. If it is calculated on the basis of the volume of the pore liquid only, the internal concentration is twice as large as that of the external solution. It is not quite clear whether this must be interpreted as a salting – in effect or as a change in effective volume when interparticle distances and interpore distances are of comparable magnitude in electrochemical systems.

We now consider the conductance curves (Figure 4). In the $MgCl_2$ systems the residual conductance is very small. Practically the whole of conductance must be ascribed to the non-exchange salt, presumably in the form of $MgCl^+$ ions.

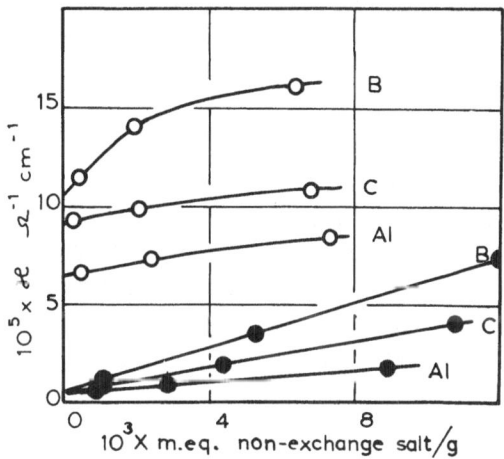

Fig. 4. Specific electric conductivity at 25 °C and 8800 c s^{-1} in function of the concentration of non-exchange salt. Open circles: potassium systems. Filled circles: magnesium systems.

For KCl the residual conductance is the leading term. Curvature due to Donnan exclusion is strongest in the softest membrane, type *B*.

The lowest temperature coefficient is shown by the $MgCl_2$-system with the softest membrane. This system has the largest distance between the fixed ions.

The less the distance between SO_3-groups, the larger is the contribution from the gegen-ions.

The contribution of the counter-ions is highest with univalent counter-ions. This is the explanation of the high apparent activation energy for transport of these ions.

It is thus clear that the non-exchange salt adsorption equilibria and the equilibrium state of the gel is the most difficult problem in these systems. The transport properties do not seem to offer new problems at this macroscopic level of description.

We have therefore tried to do some model calculations on the thermodynamic properties.

5. Model Calculations

It is important to keep the model so simple that its properties can be calculated without uncertain assumptions. The value of treating an almost oversimplified model exactly is to see whether or not it can simulate real systems. Our model is a collection of sub-systems each of which is a hollow sphere with twelve fixed ions on the inner wall, and a diameter equal to a typical pore diameter in polyelectrolyte gels i.e. 30 Å. The primitive electrolyte model is used with the cations as hard spheres in a structureless dielectric. This subsystem preserves a few of the most important characteristics of the real system: electric symmetry properties, fixed ion concentration and distance parameter. The model is in principle a lattice model in the sense that it consists of a large number of independent and distinguishable subsystems, each being a hollow sphere as mentioned. Work on this model has been done by E. Alver and D. Vaula in Bergen [11]. The main purpose is to calculate the limiting law for the non-exchange salt adsorption. A simple point of departure is to allow four possible states only for the subsystems:

0. subsystem alone (reference state)
1. subsystem + 1 gegen-ion
2. subsystem + 1 co-ion
3. subsystem + 1 gegen-ion and a co-ion.

Monte Carlo calculations have been done for these states and a Güntelberg charging process carried through. Further, a statistical calculation by means of the usual methods has been carried through giving expressions for the salt adsorption isotherms. Table III shows the calculated sorption parameters

$$w^2 \stackrel{\text{def}}{=} \frac{q_1 q_2}{q_0^2}; \quad b \stackrel{\text{def}}{=} \frac{q_1 q_2}{q_0 q_3}$$

TABLE III
Calculated parameters for lattice model

Free energies in charging process, $(\Delta A_j)_2$

j	1	2	3
$(A_j)_2/kT$	0.877 ± 0.033	-1.693 ± 0.029	-1.437 ± 0.023

Characteristic parameters

Volume of cavity	$v = 14140$ Å³
Free volume	$V_1^* = 12490$ Å³
Sorption parameter	$b = 0.516$
Sorption parameter	$w = 5257$ Å³

and the free energies $(\Delta A_j)_2$ in the charging process. The quantities q are the canonical ensemble partition functions for the respective four states. A simple linear dependence between mean potential and charging parameter was found, suggesting relatively weak interaction.

The sorption isotherm is found to deviate both from ideal Donnan behaviour and from the Glueckauf-Watts empirical equation. It agrees with the general qualitative observation that for many ion exchangers the concentration of the adsorbed salt is of the same order as that of the external solution, and has a first degree dependence on the external concentration as a leading term at the lowest concentration. With the present choice of numerical values, the coefficient of this term was found to be 0.4.

An important feature of this and more general models is whether states corresponding to spreading of the charges are included or not. We have not exhausted the model and it seems worthwhile to do more work on it.

References

1. Laity, R. W.: *J. Phys. Chem.* **63**, 80–83 (1959) and **30**, 682 (1959).
2. Spiegler, K. S.: *Trans. Farad. Soc.* **54**, 1408–28 (1958).
3. Brun, T. S. and Oftedal, T. A.: *Årbok Univ. Bergen* (1965), *Mat.-Naturv.* Serie No. 6.
4. Bauge, K. and Brun, T. S.: *Årbok Univ. Bergen* (1959), *Natruv.* Serie No. 10.
5. Brun, T. S. and Meisingseth, E. A.: *Årbok Univ. Bergen* (1960), *Mat.-Naturv.* Serie No. 13.
6. Brun, T. S. and Songstad, J.: *in preparation.*
7. Scatchard, J.: *Am. Chem. Soc.* **75**, 2883 (1953).
8. Glueckauf, E. and Watts, R. E.: *Proc. Roy. Soc. (London)* **A 268** 339 (1962).
9. Bauge, K. and Brun, T. S.: *Årbok Univ. Bergen* (1961), *Mat.-Naturv.* Serie No. 9.
10. Brun, T. S. and Oftedal, T. A.: *Årbok Univ. Bergen* (1963), *Mat.-Naturv.* Serie No. 12.
11. Alver, E. and Vaula, D.: *Acta Chem. Scand.* **26**, 3 (1972).

INTERPRETATION OF MEMBRANE PHENOMENA, USING IRREVERSIBLE THERMODYNAMICS

Comparison of Observed Transport Properties with Those Predicted from a Salt Model Calculation

RUSSELL PATERSON, RONALD G. CAMERON, and IAN S. BURKE

Department of Chemistry, University of Glasgow, Glasgow G12 8QQ, Scotland, U.K.

Abstract. The discipline of irreversible thermodynamics provides a clear mathematical description of the processes of diffusion and transport within membrane systems. From a sufficient number of transport experiments on a membrane, it is possible to obtain the frictional, $(\bar{R}-)$, or mobility, $(\bar{l}-)$ coefficients which describe the coupling interactions between membrane components. These transport coefficients cannot be estimated theoretically.

In this paper an accessible model is developed, based upon the transport properties and irreversible thermodynamic parameters of an aqueous salt solution. The frame of reference for flows is changed from solvent to fixed anion (chloride, for sulphonated membranes). The resultant flows, frictional and mobility coefficients of the model are compared to similar data obtained from membrane studies.

The model is deliberately simple, but is shown to predict observed membrane characteristics with semi-quantitative accuracy. It is therefore concluded that specific polymer effects (apart from tortuosity), exert secondary rather than primary influences on membrane performance for most univalent forms and that the solution analogy is both useful and instructive.

The discipline of irreversible thermodynamics provides a precise mathematical description of the processes of transport and diffusion in membrane systems. Its application to membrane processes is a natural development of the basic theory of Onsager and has been developed by Staverman, Kedem, Katchalsky, Caplan, Meares and others in an extensive and expanding literature [1].

The present discussion will be confined to ion exchange polymer membranes under isothermal conditions and excludes the possibility of chemical reaction within the membrane. The simplest practical membrane system consists of a monofunctional exchanger in equilibrium with an aqueous binary electrolyte [1, 2]. The membrane contains four chemical species, counter-ion, 1; co-ion, 2; and water, 3. The fourth component is the membrane matrix, which carries fixed ionogenic groups. For the purposes of a thermodynamic description the matrix of polymer is considered as species 4 and the number of moles of 4 is taken to equal that of the fixed ionic groups on the matrix. Species 4 may therefore be considered to consist of the fixed charge and adjacent polymer segments, which together constitute the repeat units of the matrix. It is therefore possible that kinetic coupling interactions for 4 will include not only the contribution of fixed charge, but also specific polymer effects if such exist. This is an important qualification to the description of species 4 as simply an ion, although it appears from earlier studies [2, 3, 4] that such effects are small, at least in certain polystyrene sulphonic acid membranes.

For slow processes, under conditions not far from equilibrium, the flows of membrane species, $\bar{J}_i$, may be related by linear phenomenological equations to the thermodynamic forces in the system. These forces, $\bar{\mathbf{X}}_i$, are defined by the negative gradients of electrochemical potential within the membrane. The dissipation function Φ is defined as the sum of products of flows and conjugate forces Equation (1)

$$\Phi = \sum_{i=1}^{4} \mathbf{J}_i \bar{\mathbf{X}}_i \geqslant 0 \tag{1}$$

From the Gibbs-Duhem equation it is easily shown that $\sum_{i=1}^{4} \bar{C}_i \bar{\mathbf{X}}_i = 0$ where $\bar{C}_i$ is the concentration of i in the membrane, usually in moles cm^{-3} of total membrane volume. The four forces in Equation (1) are therefore not independent. One may be eliminated. For practical purposes the force on the matrix, $\mathbf{X}_4$, is usually chosen, the dissipation function may then be defined in terms of the experimentally measured flows, $\bar{\mathbf{J}}_i^4$, relative to the stationary membrane, Equation (2).

$$\Phi = \mathbf{J}_1^4 \bar{\mathbf{X}}_1 + \mathbf{J}_2^4 \bar{\mathbf{X}}_2 + \mathbf{J}_3^4 \bar{\mathbf{X}}_3 \geqslant 0 \tag{2}$$

It is pertinent to note that in studies of electrolyte solutions which are formally similar the *natural* frame of reference is stationary solvent. In a meaningful comparison of the transport properties of electrolytes and membranes it will be necessary to use transport parameters such as $\bar{R}_{ik}$ (below), which are independent of frame of reference.

In Equation (2) flows and forces are independent and, subject to the usual restrictions, linear phenomenological equations may be written in terms of mobility, (l_{ik}), or frictional, $(\bar{R}_{ik})$ coefficients, Equations (3) and (4).

$$\mathbf{J}_i^4 = \sum_{k=1}^{3} l_{ik} \bar{\mathbf{X}}_k \qquad i = 1, 2, 3 \tag{3}$$

or

$$\bar{\mathbf{X}}_i = \sum_{k=1}^{3} \bar{R}_{ik} \mathbf{J}_k^4 \qquad i = 1, 2, 3 \tag{4}$$

The Onsager reciprocal relations, (O.R.R.), require $l_{ik} = l_{ki}$ and $\bar{R}_{ik} = \bar{R}_{ki}$, reducing the number of coefficients to describe all isothermal vectorial transport properties to six.

Although it is more convenient to express measured transport properties such as conductivity in terms of mobility coefficients, (Appendix), frictional coefficients are to be preferred for interpretation. Unlike mobility coefficients they are independent of the frame of reference for flows and frictional coefficients between mobile species and the matrix-4 may be obtained from the summations Equation (5). [5]

$$\sum_{i=1}^{4} \bar{C}_i \bar{R}_{ik} = 0 \qquad k = 1, 2, 3, 4 \tag{5}$$

Useful information may be obtained about the relative importance of frictional interactions using the Spiegler [6], frictional coefficient, $(-\bar{C}_i \bar{R}_{ki})$, commonly given the symbol X_{ki} [6] or f_{ki} [7] which is the frictional coefficient between one mole of k and those i in unit volume of the environment. It has the advantage of being analogous to the coefficient of kinetic friction in mechanical terms and represent the frictional force between i and k, when the relative velocity $(\mathbf{V}_i - \mathbf{V}_k)$ is unity, Equation (6)

$$\overline{\mathbf{X}}_i = \sum_{k \neq i} f_{ik} (\mathbf{V}_i - \mathbf{V}_k) \qquad i = 1, 2, 3 \tag{6}$$

From these analyses the relative importance of frictional contributions to measured transport properties may be assessed and the effect of salt uptake, water content, and capacity may be determined. In this way a more fundamental evaluation of the source of functional properties of membranes may be obtained and within the scope of a specific series of experiments on a given membrane, or series of related membranes, more detailed predictions of, for example, concentration profiles or composite membrane properties may be obtained.

Irreversible thermodynamics is a macroscopic discipline. Transport coefficients, obtained experimentally, cannot be calculated from molecular theory and the physical parameters of the system. The only exceptions appear to be very dilute electrolyte solutions in the range of the Onsager limiting law for electrical conductance [8].

The interpretation, and even the prediction of the membrane properties might be advanced if suitable and accessible analogue systems might be found.

The analogy between transport in an aqueous electrolyte and in a charged membrane might be considered. It has proved useful in earlier interpretation of thermodynamic problems such as selectivity [9]. The first choice for a model would fall most naturally on the polyelectrolyte salt solution analogous to the cross-linked polyelectrolyte gel, which constitutes the membrane. Imbibed electrolyte in the membrane would require the model to be a ternary electrolyte and transport could be compared at equal molalities in membrane and model. There are however insufficient data on the transport properties of polyelectrolyte solutions and their ternary mixtures with simple salts to allow meaningful calculations at this time. The attraction of such models may stimulate such work and provided the model polyelectrolyte does not undergo conformational or other changes, which are not allowed in the cross-linked membrane, the model should be excellent.

For the time being, therefore, model systems must remain more modest and this discussion will adopt a limiting or extreme position by assuming that the properties of an ion exchange polymer membrane are similar to simple aqueous electrolyte. Much of the discussion will concern the observed properties of AMF C60 polystyrene sulphonic acid membranes in the sodium form containing sodium chloride as invading salt. As a first step towards an analogue system it is proposed that the sodium polystyrene sulphonic acid membrane may be modelled by an equimolal solution of sodium chloride. Although the theories developed are applied to several membrane systems and to various ionic forms with apparent success, the salt model calculation presented below was developed primarily from observations on the C60 systems, although it is to be hoped that the results and conclusions drawn will have more general application to the field of membrane transport and gel permeation.

The AMF C60 membrane (American Machine and Foundry Co.) is prepared from

TABLE I

Physical characteristics of AMF C60 membranes [3]

C60N: Dry weight of leached membrane disc in Na form = 0.2345 g, Ion-exchange capacity = 1.57 mequiv. g^{-1} dry membrane

ext. soln. NaCl/M	wet wt./g	% water w.r.t. dry wt.	diameter/ cm	thickness/ cm	volume/ cm³	$\bar{C}_1$	$\bar{C}_2$	$\bar{C}_3$
						mequiv. cm⁻³		
0.1	0.3579	52.6	3.698	0.0335	0.360	0.980	0.0024	19.03
0.5	0.3449	46.7	3.655	0.0314	0.329	1.104	0.0360	18.48
1.0	0.3320	40.8	3.611	0.0308	0.315	1.209	0.0946	16.86
2.0	0.3176	33.7	3.551	0.0299	0.296	1.419	0.2312	14.89

C60E: Dry weight of leached membrane disc in Na form = 0.2249 g. Ion-exchange capacity = 1.70 mequiv. g^{-1} dry membrane

ext. soln NaCl/M	wet wt./g	% water w.r.t. dry wt.	diameter/ cm	thickness/ cm	volume/ cm³	$\bar{C}_1$	$\bar{C}_2$	$\bar{C}_3$
						mequiv. cm⁻³		
0.1	0.3998	77.7	3.847	0.0333	0.387	0.960	0.0052	25.07
0.5	0.3807	68.6	3.754	0.0321	0.355	1.112	0.0717	24.11
1.0	0.3629	59.8	3.697	0.0317	0.340	1.270	0.1832	21.94
2.0	0.3470	50.5	3.616	0.0308	0.316	1.621	0.4533	19.95

low-density polyethylene, containing 35% styrene and 2% divinylbenzene. The sulphonating agent is oleum. Using the method of Arnold and Koch [10], an expanded form C60E was prepared by immersing the membrane in water at 95 °C for thirty minutes. The physical properties of the normal, C60N, and expanded, C60E, mem-

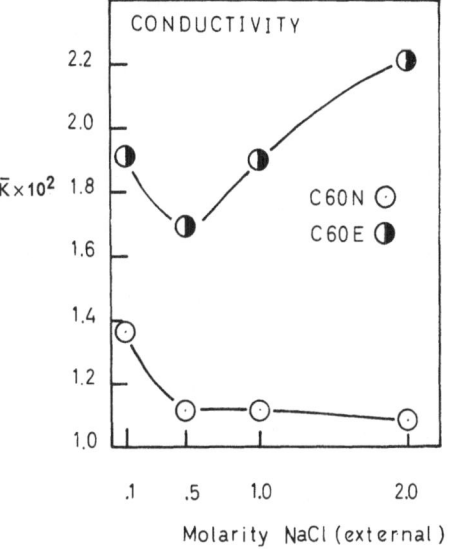

Fig. 1.   Specific conductivities ($\bigcirc$, C60N; $\bullet$, C60E) as functions of external concentration of salt [3].

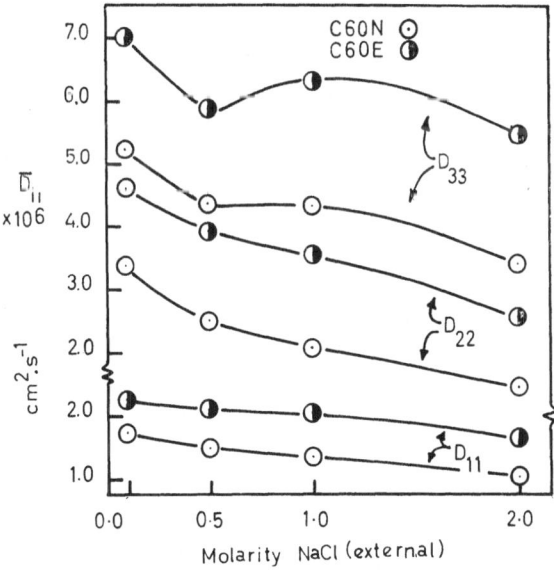

ISOTOPE – DIFFUSION

Fig. 2.   Variation of self-diffusion coefficients for water, $\bar{D}_{33}$; co-ion, $\bar{D}_{22}$; and counter-ion, $\bar{D}_{11}$ as functions of external salt concentration. $\bigcirc$ and $\bullet$, denote C60N and C60E membranes respectively [3].

branes in solutions of sodium chloride in the range 0.1–2.0 M are given in Table I [3]. The more expanded membrane takes up more salt, $\bar{C}_2$, at each concentration and salt uptake ranges from 0.2%–22% (of the total capacity) for the system. The results of electrical and isotopic diffusional studies are shown in Figures (1–4). The flows of

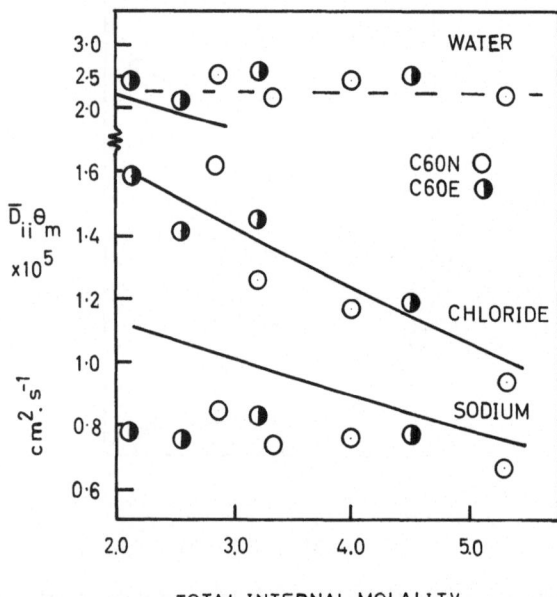

Fig. 3.   Tortuosity corrected diffusion coefficients $\bar{D}_{ii}\theta_m$, compared with those of water, chloride and sodium ions in equimolal solutions of sodium chloride; solid lines. ○ and ◑ denote C60N and C60E membranes respectively.

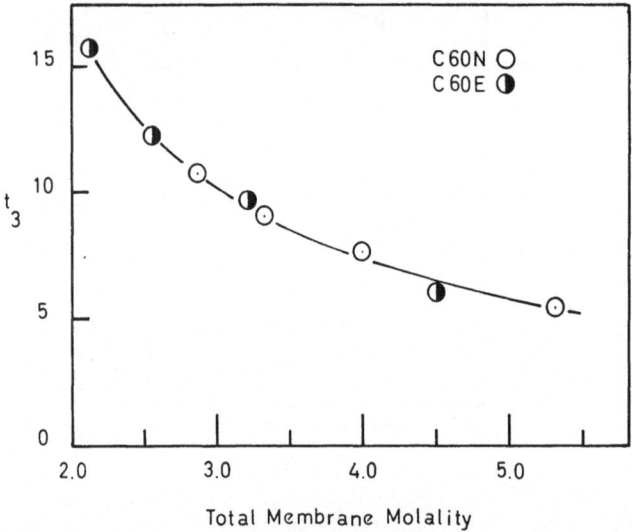

Fig. 4.   Electro-osmotic transference number, t_3, as a function of molality, m, of the membrane electrolyte.

ions and water are larger in the more open structure of the expanded membrane and in general decrease in both as the membrane shrinks in more concentrated solution, in part at least, due to the increased obstruction of the polymeric matrix as diffusional pathways become more tortuous and the fractional pore volume decreases. In electrical conductivity, Figure 1, there is the added effect of increased salt up take, which should increase conductivity. These two opposing effects almost cancel for C60N above 0.5 M NaCl (ext), but increased salt uptake in the C60E membrane more than balances the effect of increased tortuosity and, at higher concentrations, specific conductivity increases. Isotopic diffusional coefficients for counter-ion, co-ion and water (using tritiated water) show similar diversity between C60N and C60E and there is little to suggest the common source of these membranes at this stage, Figure 2. When scaled by the tortuosity factor, θ_m, (the value calculated by Meares [11] for path length tortuosity) the two sets of data for C60N and C60E effectively coincide close and are to the corresponding values for sodium [12], chloride [12] and tritiated water [13] in equimodal solutions of aqueous sodium choride, Figure 3. (Since θ_m is used outwith the terms of reference of Meares' derivation [11] it is to be noted that using Prager's estimate, θ_p, [14] (which is obtained from a more generalised treatment), alternative values, $\overline{D}_{ii}\theta_p$, are some 15% lower). In either case these purely geometric scaling factors bring C60N and C60E data into correspondence and to good agreement with equimolal sodium chloride. Values for sodium counter-ion, $\overline{D}_{ii}\theta$, are lower than for sodium chloride and this point can be considered when isotope-isotope effects are discussed in later sections.

Electro-osmotic transference numbers, t_3, defined by Equation (7), measure the number of moles of water transferred across the membrane by one Faraday of electricity.

$$t_3 = \overline{J}_3^4 F/I \qquad I = \text{amp cm}^{-2} \tag{7}$$

Experimental values range from 15.77 for C60E (0.1 M) to 5.48 for C60N (2.0 M). At corresponding external salt concentrations the transference numbers for C60E are greater than for C60N, but plotted against molality of electrolyte in the membrane, both sets combine in a single curve, Figure 4. Transference numbers are independent of tortuosity, since they are defined by the ratio of flows, but are, in this system, dependent on total membrane molality and not the proportion of free salt. Spiegler [6] has shown that a large number of exchangers obey the empirical equation $t_3 = \beta \overline{C}_3/\overline{C}_1$, where $\beta \approx 0.5$. Figure 5, shows that the ratio t_3/t_1 is a linear function of the concentration ratio $\overline{C}_3/\overline{C}_1$ and has slope 0.576. Once more the properties of normal and expanded membranes are shown to obey a common law and it is of interest to examine the significance of this relationship.

The definitions of t_3 and t_1 as $\overline{J}_3^4 F/I$ and $Z_1 \overline{J}_1^4 F/I$ respectively may be expanded in terms of concentrations $\overline{C}_i$ and velocities relative to the fixed membrane, V_i^4. Since $\overline{J}_1 = \overline{C}_i V_i^4$

$$t_3/t_1 = \overline{C}_3 V_3^4/Z_1 \overline{C}_1 V_1^4$$

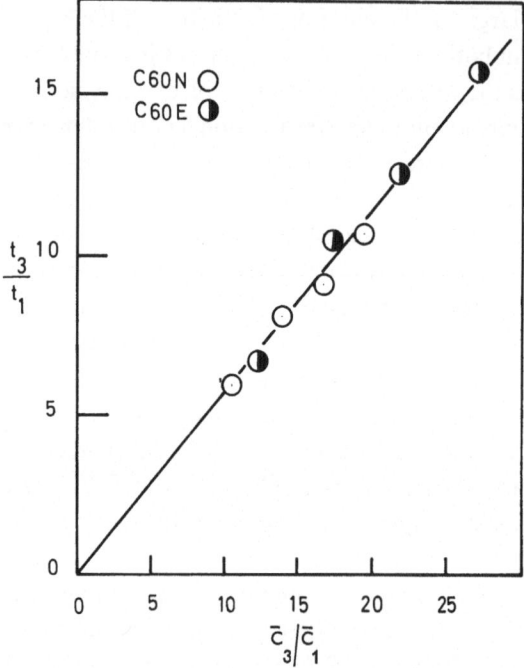

Fig. 5. t_3/t_1 against $\overline{C}_3/\overline{C}_1$: ○, C60N; ◑, C60E. [3]

where Z_1 is the signed valency of the counter-ion 1. The ratio V_3^4/V_1^4 is therefore a constant (0.576) for both membranes under all conditions studied and in agreement with the results quoted by Spiegler, where, in general counter-ion transport numbers might be expected to be close to unity. To examine the analogy with aqueous solution the velocities V_i^4 may be converted to a solvent-fixed frame of reference V_i^3. Then

$$V_3^4/V_1^4 = -V_4^3/(V_1^3 - V_4^3) = 0.576$$

and so the ratio of the velocities of sulphonate-matrix to sodium, $-V_4^3/V_1^3 = 1.36$. This shows that these two ions have similar velocities relative to water solvent. The sulphonate-matrix ion has therefore much in common with a simple order destroying anion such as nitrate or chloride which have somewhat larger mobilities than sodium in aqueous solutions in the same concentration range as internal electrolyte concentrations in the membranes. In concentrated sodium choride the mobility ratio $V_{Cl}^3/V_{Na}^3 = 1.70$. The fact that increasing concentrations of salt in the membranes, up to 22%, has no effect upon the mobility ratio is fair indication that, from a kinetic standpoint, sulphonate and chloride are broadly similar.

1. Calculation of Membrane Properties for C60E (1.0) using Ternary Solution Analogues

Irreversible thermodynamic analyses for C60N and C60E membranes have been obtained in 0.1 and 1.0 M sodium chloride solutions. The methods used for the evaluation of frictional and mobility coefficients are described in detail in the original papers [2, 4]. The C60E membrane in 1.0 M sodium chloride, (C60E (1.0)), has a sulphonate-matrix concentration of 2.75 molal and total ionic molality 3.21. The C60N (0.1) membrane is almost salt free (0.2%) and has molality 2.87. The two membranes have therefore almost equal molalities of sulphonate and C60E (1.0) may be considered to be equivalent to C60 (0.1) with added sodium chloride. It is therefore useful to consider C60E (1.0) as a ternary electrolyte of sodium sulphonate and sodium chloride and apply the empirical methods developed by Miller [15], which have been used successfully to calculate the transport properties of a ternary electrolyte solution from a knowledge of the solvent-fixed mobility coefficients of the binary electrolyte components. To apply this method to membrane calculations, solvent-fixed coefficients $\bar{L}_{ik}$, are obtained from the experimental values of l_{ik} for C60N (0.1). These coefficients refer to the bulk membrane and include tortuosity effects and are therefore not compatible with the solvent fixed mobility coefficients for aqueous sodium chloride. To convert membrane $\bar{L}_{ik}$ to solvent-fixed coefficients, which are representative of the aqueous or pore volume of the membrane the scaling factor θ/v must be used, where v is the fractional pore volume of the exchanger. (The scaling factor is dealt with more explicitly on the section devoted to salt model calculations, below.)

$\bar{L}_{ik}\theta/v$ are therefore equivalent to mobility coefficients for a non-tortuous membrane. Once obtained for C60N (0.1), these may be combined with corresponding mobility coefficients for aqueous sodium chloride at 3.21 m [16] in the LN calculation [15, 4]. The resulting coefficients may then be corrected for tortuosity

TABLE II

Membrane-fixed mobility coefficients for C60E in 1.0 molar sodium chloride (25 °C).
($\bar{C}_2/\bar{C}_{\text{Total}} = 0.144$)

Method	l_{11}	l_{12}	l_{13}	l_{22}	l_{23}	l_{33}
			$\times 10^{12}$ (mole^2 J^{-1} cm^{-1} s^{-1})			
Experimental						
(a)	2.09	0.210	32.0	0.372	12.3	1010
(b)	2.09	0.199	30.0	0.361	10.0	1076
Ternary Model						
(m)	2.24	0.33	25.3	0.31	7.1	695
(p)	2.41	0.32	26.7	0.40	7.7	735
SMC						
(s)	2.40	0.29	28.6	0.29	5.1	570
(m)	2.48	0.30	29.6	0.30	5.3	591
(p)	3.33	0.41	39.6	0.41	7.1	792

Assumptions made in the evaluation of parameters from experimental data [4]

(a) $\dfrac{l_{23}}{C_2 C_3} = \dfrac{2l_{33}}{\bar{C}_3{}^2} - \dfrac{l_{13}}{\bar{C}_1 \bar{C}_3}$, and

(b) $R_{22}{}^{*} = 0$

and pore volume corresponding to those of the C60E (1.0) membrane and converted to a membrane or 4-fixed frame of reference for comparison. The results of these calculations are shown in Table II, where calculated l_{ik} coefficients may be compared with experimentally derived data. The ternary-model calculations have been made using both Meares and Prager tortuosity corrections in the manner described above. Both sets of calculated l_{ik} results are in good agreement, and show the same trends and magnitudes as the experimentally derived data with average discrepancy of $\approx 20\%$ between observed and calculated, l_{ik}, coefficients. Calculated and observed conductivities transport and transference numbers are given in Table III.

TABLE III

Observed and predicted transport measurements for C60E (1.0)

		Specific conductivity $\kappa \times 10^2$ $\Omega^{-1}\,cm^{-1}$	t_1	t_3
Observed		1.90	0.92	9.70
Ternary Model	(m)	1.80	1.00	9.53
	(p)	1.94	0.98	9.13
SMC[a]	(s)	1.96 ⎫		
	(m)	2.03 ⎬	1.00	11.16
	(p)	2.71 ⎭		

[a] Salt model calculation

Again very reasonable agreement is obtained, and it would appear that this empirical method of Miller may be used with success to predict membrane performance. The method appears however to over-estimate coupling between sodium and chloride, l_{12}, resulting in low values of t_2, the co-ion transport number, and to consistently under-estimate the direct mobility for water l_{33}.

2. Salt Model Calculations

The results of the ternary-model calculation are satisfactory and indicate that fair agreement between may be obtained using this method notwithstanding the empirical nature of the LN approximation and the uncertainties in the calculation of tortuosity. In every case the predicted data, Table III, are more accurate than obtained by Nernst-Planck calculation [4].

The method can be regarded as illustrative rather than practical or for general use.

To provide an accessible model, the method of calculation is based upon the properties of a single electrolyte for which a complete irreversible thermodynamic analysis is available at concentrations which include the total ionic molality of the exchanger. From the experimental evidence cited above on the C60 membrane systems, there is a similarity between the function of sulphonate-matrix fixed charge in the membrane and free chloride ion in solution. This similarity suggests that chloride and sulphonate have broadly similar kinetic behaviour. Since the co-ion in these experimental studies was also chloride, a model of the exchanger may be conceived, in which chloride of two 'sorts' is present; chloride, species 2, co-ion in the membrane and chloride, species 4, which is fixed relative to the membrane matrix. The relationship of these two forms is taken to be that between chemically identical, but physically distinguishable isotopes. This postulate allows a precise definition of the frictional and mobility coefficients of the ternary solution and, in particular, the absolute magnitudes of frictional coefficients R_{2i} and R_{4i}, $(i = 1, 2, 3, 4)$, the frictional coefficients between co-ion and fixed charge and the other components of the system. Implicit in this model is the assumption that the polymer matrix has no chemical influence upon the transport process, other than by its presence, to obstruct diffusional pathways in the membrane phase and render them tortuous. Major deviations between the predictions of the model and observed transport parameters can therefore be considered as indications of specific polymer effects and the possibility with certain ions that sulphonate and chloride may have quite different complexing or ion pairing characteristics would obviously undermine the model. It might also be expected that dilute, homogeneous membranes or membranes with significant voids would have local ionic distributions which were grossly different from those in simple aqueous solution of equal molality, and consequently the processes of transport would be polymer dependent and specific processes such as polymer chain diffusion or pseudo-mosaic effects would be significant.

If, however, the aqueous and ion-permeable regions of the membrane constitute an essentially homogeneous phase, it is conceivable that the distribution of charges might approximate to those in an aqueous electrolyte solution and in this range a salt model would predict membrane properties, which were of the correct magnitude.

The choice of a salt model based upon a ternary isotopic solution allows precise evaluation of model coefficients. The theory of isotopic diffusion and the identification of isotope-isotope coefficients has been developed by Laity [17] and by Kedem and Essig [18]. The shortened theoretical treatment given below owes much to these papers and is developed primarily to express frictional interaction in the isotopic ternary solution in terms of those of the parent binary electrolyte and the isotopic diffusion coefficient for co-ion.

Since the analysis requires a change of frame of reference from solvent, (the normal frame for discussion of solutions), to ion-4-fixed in the membrane model, a development is presented using frictional coefficients, R_{ik}, which are independent of frame of reference.

The phenomenological equations for a binary electrolyte [1, 2] are given in matrix form by Equation (8), in which 1 represents the counter-ion and 2, co-ion.

$$\begin{vmatrix} \mathbf{X}_1 \\ \mathbf{X}_2 \end{vmatrix} = \begin{vmatrix} R_{11} & R_{12} \\ R_{21} & R_{22} \end{vmatrix} \begin{vmatrix} \mathbf{J}_1^3 \\ \mathbf{J}_2^3 \end{vmatrix} \tag{8}$$

The flows of counter-ion, $\mathbf{J}_1^3$ and co-ion $\mathbf{J}_2^3$ are given on a solvent fixed frame of reference. Thermodynamic forces on counterion and co-ion, defined by the negative gradients of chemical potential are given by X_1 and X_2 respectively. In the ternary isotopic solution of equal concentration, some of co-ion 2 is replaced by co-ion 4, which is chemically identical in all respects, such that the total co-ion concentration in the binary is the sum of the concentration of 2 and 4 in the ternary, Equation (9).

$$C_2 = c_2 + c_4 \tag{9}$$

The phenomenological equations for this ternary solution are given in Equation (10) in which forces, frictional coefficients and flows are represented by lower case symbols, x_i, r_{ik}, and j_i^3 respectively.

$$\begin{vmatrix} x_1 \\ x_2 \\ x_4 \end{vmatrix} = \begin{vmatrix} r_{11} & r_{12} & r_{14} \\ r_{21} & r_{22} & r_{24} \\ r_{41} & r_{42} & r_{44} \end{vmatrix} \begin{vmatrix} j_1^3 \\ j_2^3 \\ j_4^3 \end{vmatrix} \tag{10}$$

From Equations (8) and (10) the Onsager reciprocal relations, (O.R.R.) will be assumed, such that $R_{ik} = R_{ki}$ and $r_{ik} = r_{ik}$ for all i and k. (Extensions to relationships proved between R- and r-coefficients, due to these identities, will be represented by (+O.R.R.)).

In a comparison of a binary solution, Equation (8), and the same solution in which an isotopic form of 2 is present, Equation (10), the following identities exist, Equations (11), (12) and (13);

$$J_1^3 = j_1^3; \tag{11}$$

$$J_2^3 = j_2^3 + j_4^3; \tag{12}$$

$$X_1 = x_1 \tag{13}$$

and from the Gibbs-Duhem relationship for forces in both systems, Equation (14),

$$C_2 X_2 = c_2 x_2 + c_4 x_4 \tag{14}$$

Under conditions for isotopic diffusion of co-ion, Equation (10), $x_1 = 0 = j_1^3$ and from Equations (11), (12) and (13),

$$r_{12} = r_{14} \quad (+\text{O.R.R.})$$

Comparison of X_1 and x_1 in experiments without isotopic forces shows,

$$R_{11} = r_{11} \quad \text{and} \quad R_{12} = r_{12} = r_{14} \quad (+\text{O.R.R.})$$

The coefficients r_{22}, r_{24}, (r_{42}), and r_{44} of Equation (10) cannot be determined solely by comparison of coefficients in Equations (8) and (10) but are related to the isotopic diffusion coefficient for co-ion, D_{22} (and D_{44}).

If a purely electrical force is applied to both solutions,

$$X_2 = x_2 = x_4 = Z_c F \, (-\text{grad } \psi) \tag{15}$$

where $Z_c = Z_2 = Z_4$, is the co-ion valency (including sign), and grad ψ, the local gradient of electrical potential. Comparison of expansions for X_2 and x_2 (and x_4) give the equalities of Equations (16) and (17).

$$R_{22}J_2^3 = r_{22}j_2^3 + r_{24}j_4^3 \tag{16}$$

and

$$R_{22}J_2^3 = r_{42}j_2^3 + r_{44}j_4^3 \tag{17}$$

Since co-ions 2 and 4 are chemically identical, they will have the same electrochemical mobility, V, (cm s^{-1}) under unit electrical potential gradient. Since $J = CV$,

$$C_2 R_{22} = c_2 r_{22} + c_4 r_{24} = c_2 r_{42} + c_4 r_{44} \tag{18}$$

Under conditions for isotopic diffusion of co-ion, Equation (10), the total force and total flow of co-ion are separately zero and from Equations (14) and (12)

$$c_2 x_2 = - c_4 x_4 \quad \text{and} \quad j_2^3 = - j_4^3 \tag{19}$$

It is easily shown that the isotopic flows obey Fick's Law and that the isotopic diffusion coefficients of 2 and 4 are equal, D_{22} and D_{44} respectively, Equations (20).

$$D_{22} = \frac{RT}{c_2(r_{22} - r_{24})} \equiv \frac{RT}{c_4(r_{44} - r_{42})} \tag{20}$$

From Equation (18) these diffusion coefficients may be expressed in terms of the direct frictional coefficient of co-ion in the binary, R_{22} and the frictional coefficient between isotopes 2 and 4 in the ternary $r_{24}(r_{42})$ so that

$$D_{22} = D_{44} = \frac{RT}{C_2(R_{22} - r_{24})}$$

The relative concentrations of co-ion isotopes, c_2 and c_4, are not required in this equation and the isotope-isotope frictional coefficient $r_{24}(r_{42})$, may be obtained directly from the self- or isotopic diffusion coefficient of the co-ion in solution and the corresponding values of R_{22} and C_2 in the binary.

From Equations (18) and (21) explicit expressions for the frictional coefficients $r_{24}(r_{42})$, r_{22} and r_{44} of Equation (10) are obtained; Equations (22), (23), (24)

$$r_{24} = r_{42} = R_{22} - \frac{RT}{C_2 D_{22}} \tag{22}$$

$$r_{22} = R_{22} + \frac{c_4}{c_2}\left(\frac{RT}{C_2 D_{22}}\right) \tag{23}$$

$$r_{44} = R_{22} + \frac{c_2}{c_4}\left(\frac{RT}{C_2 D_{22}}\right) \tag{24}$$

These last two coefficients, r_{22} and r_{44}, depend upon the relative proportions of isotopes 2 and 4 in the solution.

From the identities

$$\sum_{i=1}^{3} C_i R_{ik} = 0 \qquad k = 1, 2, 3$$

and

$$\sum_{i=1}^{4} c_i r_{ik} = 0 \qquad k = 1, 2, 3, 4$$

applied to the phenomenological Equations (8) and (10) respectively, it is easily shown that the ion-to-water frictional coefficients in the isotopic ternary are equal to those in the parent binary, Equation (23).

$$R_{13} = r_{13} \quad \text{and} \quad R_{23} = r_{23} = r_{43}; \, (+ \text{O.R.R.}) \tag{25}$$

consequently $R_{33} = r_{33}$

3. Phenomenological Equations Relative to Ion 4

Equation (10) may be taken as a model for the exchanger membrane in which ion 4 represents the fixed charge on the polymer and 2, the co-ion imbibed by the membrane from the external solution. Membrane flows are measured relative to the stationary matrix and therefore relative to the ion, species 4. On a '4-fixed' frame of reference the phenomenological equations of Equation (10) become Equation (26)

$$\begin{vmatrix} x_1 \\ x_2 \\ x_3 \end{vmatrix} = \begin{vmatrix} r_{11} & r_{12} & r_{13} \\ r_{21} & r_{22} & r_{23} \\ r_{31} & r_{32} & r_{33} \end{vmatrix} \begin{vmatrix} j_1^4 \\ j_2^4 \\ j_3^4 \end{vmatrix} \tag{26}$$

The flows j_i^4 are the flows of ions 1, 2, and water, 3, relative to the fixed anion 4, where

$$j_i^4 = \left(j_i^3 - \frac{c_i j_4^3}{c_4} \right)$$

All frictional coefficients, r_{ik}, are defined by the binary, with the exception of r_{22}, which, from Equation (23), is dependent on the isotopic diffusion coefficient D_{22} and the concentration ratio, c_4/c_2. Thus the direct coefficient r_{22} is strongly dependent on the concentration of salt in the membrane and increases as co-ion uptake, c_2, diminishes; a feature observed in experimental studies of ion exchange membranes [2].

4. Scaling Factors for Transport Parameters, Relating to Those of the Solution Model to Practical Membrane Values

Equation (26) is a model of a non-tortuous membrane in which no account has been taken of the presence of polymer. The model makes the explicit assumption that the membrane function is determined by the ionogenic fixed groups on the polymer, that these are similar to simple aqueous anions and that the polymer matrix has no influence on the movement of ions or water other than, by its presence, to restrict movement by constraining mobile species to tortuous diffusional pathways.

In common practice concentrations in the membrane are expressed in moles cm^{-3} of total membrane volume, $\bar{C}$, where $\bar{C} = vc$ and v is the fractional aqueous volume (or pore-volume) of the membrane. Equally flow across the membrane is, in practice, referred to flow/unit area of exposed membrane, $\bar{J}$ so that $\bar{J} = jv'$ where v' is the ratio of 'pore' to geometric area at the membrane surface. The membrane may be defined as homogeneous in macroscopic terms if $v' = v$.

The presence of polymer in the membrane may be considered to increase the effective length of diffusional pathways across the membrane, such that a membrane of geometric thickness d may be considered to have a diffusional path length of $d\theta$ which corresponds to a solution of path length d, where $\theta > 1$.

Fick's equation for isotopic diffusion of co-ion may be chosen to illustrate these scaling effects, Equation (26).

$$j_2 = - D_{22} \Delta c_2/d \tag{27}$$

becomes

$$\bar{J}_2/v' = -\left(\frac{D_{22}}{\theta}\right)\left(\frac{\Delta \bar{C}_2}{d}\right)\frac{1}{v} \tag{28}$$

and if $v' = v$, as in a homogeneous membrane

$$\bar{J}_2 = -\frac{D_{22}}{\theta}\left(\frac{\Delta \bar{C}_2}{\bar{d}}\right) \tag{29}$$

so that

$$D_{22}/\theta = \bar{D}_{22} \tag{30}$$

The diffusion coefficient of co-ion in the tortuous membrane $\bar{D}_{22} = D_{22}/\theta$ using this salt model calculation and so is smaller than in free solution. Using barred symbols to represent the membrane,

$$\bar{D}_{22} = \frac{RT}{\bar{C}_2(\bar{R}_{22} - \bar{R}_{22*})} \tag{31}$$

and consequently

$$\bar{R}_{22} = \frac{r_{22}\theta}{v} \quad \text{and} \quad \bar{R}_{22*} = \frac{r_{22*}\theta}{v},$$

where 2* is the isotopic form of 2 used in membrane co-ion diffusion (it is easily shown that $r_{22*} = r_{24}$). Since this analysis may be applied to the forces and flows in the phenomenological Equations (26), it is generally true that,

$$\bar{R}_{ik} = r_{ik}\frac{\theta}{v} \tag{32}$$

or in inverse form as mobility coefficients l_{ik};

$$\bar{l}_{ik} = l_{ik}\frac{v}{\theta} \tag{33}$$

The value of the tortuosity coefficient calculated by theoretical estimations is dependent upon the statistical model of the exchanger phase chosen and values from independent theoretical models may not be consistent. In earlier papers Prager's estimate of the tortuosity factor θ_p and Meares' value of the *path* tortuosity θ_m have been used [2, 3, 4].

The salt model calculation, SMC, however defines θ as $D_{22}/\bar{D}_{22}$ and as θ_s, will be used in the comparisons of experimental membrane and SMC parameters given below.

An estimate of the frictional coefficients for an experimental membrane may therefore be obtained from a knowledge of its physical dimensions, the concentrations of ions and water in the membrane, and the co-ion diffusion coefficient. Before making comparisons of this sort it is useful to summarise the predicted correspondence between membrane, $\bar{R}_{ik}$, and solution frictional parameters, Equation (34), which is represented conveniently in matrix form,

$$
\begin{vmatrix}
\bar{R}_{11} & \bar{R}_{12} & \bar{R}_{13} & \bar{R}_{14} \\
\bar{R}_{21} & \bar{R}_{22} & \bar{R}_{23} & \bar{R}_{24} \\
\bar{R}_{31} & \bar{R}_{32} & \bar{R}_{33} & \bar{R}_{34} \\
\bar{R}_{41} & \bar{R}_{42} & \bar{R}_{43} & \bar{R}_{44}
\end{vmatrix}
=
\begin{vmatrix}
R_{11} & R_{12} & R_{13} & R_{12} \\
R_{21} & r_{22} & R_{23} & r_{24} \\
R_{31} & R_{32} & R_{33} & R_{32} \\
R_{21} & r_{42} & R_{23} & r_{44}
\end{vmatrix}
\times \theta/v
\tag{34}
$$

         'Membrane' (SMC)   Solution

where r_{22}, r_{44} and $r_{24}(=r_{42})$ are defined by Equations (23), (24) and (22) respectively.

5. Application of the SMC

Before making a detailed comparison between this simple model calculation and the observed properties of membranes, it is of interest to note that the model predicts that a value of $\bar{R}_{22}$ may be obtained directly from the isotopic diffusion coefficient of co-ion in the membrane $\bar{D}_{22}$ provided the ratio of fixed charge to co-ion concentrations is large, Equations (20), (22) and (23). For most binary electrolytes the function $RT/C_2 D_{22}$ is of the same order of magnitude as R_{22}. For example in aqueous sodium chloride at 3 M the function is some 20% smaller than R_{22}. If the co-ion to sulphonate friction in the experimental membrane is, even approximately, equal to r_{24} of the salt model calculation,

$$
\bar{R}_{22} \sim RT/(\bar{C}_2/\bar{D}_{22})
\tag{35}
$$

Again within the limits of applicability of the model the error in using Equation (35) would be approximately $-0.2x\%$ when the ratio of $\bar{C}_2/\bar{C}_4 \times 100 = x$ in the membrane (at these molalities).

There are certain consequences of the isotope model presented. The first is that, since co-ion and matrix fixed charge 4 are taken as chemically identical the transport number of co-ion, t_2 will, be defined as zero. The change of frame of reference to membrane or 4-fixed automatically requires co-ion to be stationary relative to 4 in an electrical experiment. Equally the value of the electro-osmotic transference number t_3 will be identical to that obtained by considering the flow of water relative to all co-ions, fixed, in the binary, so that

$$t_3 = -\frac{C_3}{Z_2 C_2} \; t_2^3 = \frac{C_3 t_2^3}{Z_1 C_1}$$

where t_2^3 is the transport number of co-ion in the binary solution. The specific conductivity κ is independent of the frame of reference chosen for the measurement and so the SMC predicts the specific conductivity of the membrane $\bar{\kappa}$ by Equation (36)

$$\bar{\kappa} = \kappa \frac{v}{\theta} \tag{36}$$

For these parameters it is therefore sufficient to know the co-ion transport number relative to water, t_2^3, and the specific conductivity of the model binary, κ, to obtain predicted membrane parameters.

6. Observed and Calculated Membrane Parameters

The frictional coefficients for the binary model electrolyte, sodium chloride, were obtained from Miller's tabulated data [16] and isotopic diffusion coefficients, for chloride co-ion, obtained by Mills [12]. Since molarity, C, of species relative to unit volume of aqueous pore solution is not defined unequivocally, this concentration was estimated by assuming the ratio of molarity to molality in the membrane to be the same as in equimolal aqueous sodium chloride at 25 °C.

Frictional coefficients obtained from experimental data and from the Salt Model Calculation, (SMC are given in Table IV. Tortuosity corrections, θ/v, have been estimated, using the ratio $D_{22}/\bar{D}_{22}$, as suggested by the model, and those of Meares and Prager in the manner discussed above. These three methods, designated (s), (m) and (p) respectively are shown in all tabulated data.

The agreement between calculated and experimental coefficients, $\bar{R}_{ik}$, is in general very good, with calculated values following in detail the trends and magnitudes found from data derived from experimental measurements [2, 4]. There is particularly good correspondence between calculated and experimental values of $\bar{R}_{22}$ and $\bar{R}_{44}$, which largely justifies the basic assumption of the SMC; that aqueous chloride and sulphonate-matrix anions have similar kinetic characteristics. For both normal and expanded membranes the salt uptake is small and $\bar{R}_{44}$ is given to a good approximation by the solution coefficient $R_{22}/(\theta/v)$, Table IV, Equation (24). Equally, the very large values for $\bar{R}_{22}$ are explained by the dominant contribution of the concentration ratio, c_4/c_2 in Equation (23). The lower value of $\bar{R}_{22}$ in the expanded membrane, C60E (0.1) is caused primarily by the greater uptake of co-ion, $\bar{C}_2$, since the value of the direct coefficient R_{22} in the model solution, is largely unaffected by the change in concentration from 2.87 m in C60N (0.1) to 2.11 m in C60E (0.1).

The direct frictional coefficients for sodium and water, $\bar{R}_{11}$ and $\bar{R}_{33}$, are somewhat over-estimated by the SMC particularly in the more concentrated normal membrane,

TABLE IV

Comparison of salt model calculations (SMC) with experimental values of $\overline{R}$-coefficients for C60N and C60E Membranes in sodium chloride solutions. (0.1M) at 25 °C

(a) Method	$\overline{R}_{22}$	$\overline{R}_{11}$	$\overline{R}_{44}$	$R_{22}{}^{a}$	$\overline{R}_{33}$ J cm s mole^{-2} × 10^{-12}	$-\overline{R}_{12}$	$-\overline{R}_{14}$	$-\overline{R}_{13}$	$-\overline{R}_{23}$	$-\overline{R}_{43}$
C60N										
Exptal.	349	1.01	0.88	—	0.0028	-1.57	0.442	0.030	0.144	0.023
SMC (s)	294	1.30	0.82	0.82	0.0045	0.204	0.204	0.057	0.032	0.032
(m)	329	1.46	0.92	0.92	0.0050	0.229	0.229	0.064	0.036	0.036
(p)	260	1.15	0.72	0.72	0.0040	0.181	0.181	0.050	0.028	0.028
SMC (uncorrected)	24.76	0.1101	0.0691	0.06895	0.000386	0.01725	0.01725	0.0048	0.00267	0.00267

$m_1\,m_2\,m_4$, the molalities of sodium, chloride and sulphonate were 2.87, 7.08 × 10^{-3} and 2.863 respectively, and the concentration ratio $m_4/m_2 = 404.38$

	$\overline{R}_{22}$	$\overline{R}_{11}$	$\overline{R}_{44}$	$R_{22}{}^{a}$	$\overline{R}_{33}$	$-\overline{R}_{12}$	$-\overline{R}_{14}$	$-\overline{R}_{13}$	$-\overline{R}_{23}$	$-\overline{R}_{43}$
C60E										
Exptal.	102	0.83	0.602	—	0.0013	-1.65	0.283	0.0218	0.102	0.0128
SMC (s)	103	0.88	0.619	0.619	0.0019	0.149	0.149	0.0320	0.0181	0.0181
(m)	103	0.98	0.621	0.621	0.0019	0.149	0.149	0.0320	0.0181	0.0181
(p)	82.9	0.79	0.501	0.501	0.0015	0.120	0.120	0.0258	0.0145	0.0145
SMC (uncorrected)	14.024	0.134	0.0848	0.0848	0.000261	0.0203	0.0203	0.00436	0.00247	0.00247

m_1, m_2 and m_4 are 2.12, 11.49 × 10^{-3}, and 2.113 respectively and $m_4/m_2 = 183.49$

In salt model calculations, (SMC), (s), (m), (p) refer to scaling factors θ/ν calculated using θ_s (experimental), θ_p, and θ_m respectively.
[a] R_{22} the aqueous binary frictional coefficient, for chloride, is retained for comparison with $\overline{R}_{44}$.

C60N (0.1). Ion-to-water coefficients $\bar{R}_{13}$ and $\bar{R}_{34}$ indicate that sodium and sulphonate-matrix have similar water interactions to sodium and chloride in the model. From a linear relationship between t_3 and t_1 observed in earlier studies on these membranes [2], it may be deduced that the ratio $-\bar{R}_{13}/\bar{R}_{33} \approx 10.8$ in C60N and 12.6 in C60E. Values of the ratio calculated by the SMC, Table IV, are 12.66 and 16.84 respectively. Apart from two obvious misfits, for co-ion to water friction $-\bar{R}_{23}$ and for counter-ion to co-ion friction $\bar{R}_{-12}$, the model calculation shows remarkably good agreement with experimental data; especially so when its conceptual simplicity is considered.

Mobility coefficients, l_{ik}, for this system are shown in Table V. The model, by assuming identical co-ion and fixed charge restricts the co-ion transport number (calculated) to zero so that for a 1:1 salt, $l_{12} = l_{22}$ in the SMC calculation, Equation (A3). Once more the agreement between experimental and calculated coefficients is good and tortuosity corrections (s) and (m) superior to the Prager estimate, (p).

The experimental value of l_{12} is small and cannot be calculated with confidence for these membranes which have such low co-ion uptake. Good agreement between calculated and observed coefficients, l_{22}, indicate that the coupling coefficient l_{12}, in the membrane, must be small. The most serious disagreement is observed for values of l_{33}, the direct mobility of water. The SMC underestimates this major parameter. In consequence, water flow, for example, osmotic flow in Equation (A6) which depends largely on the magnitude of l_{33} is underestimated.

In earlier papers [2, 4] approximations were made which involved the neglect of certain coefficients. In particular neglect of co-ion-to-water coupling l_{23} (such that

TABLE V

Mobility coefficients, l_{ik}, for C60N and C60E membranes in sodium chloride solutions. (0.1M) at 25°C.

Method	l_{11}	l_{12}	l_{22}	l_{13}	l_{23}	l_{33}
			mole^2 J^{-1} s^{-1} cm^{-1} $\times 10^{12}$			
C60N						
	1.47	–	0.0029	16.0	0.159	547
SMC (s)	1.65	0.0034	0.0034	20.7	0.065	475
(m)	1.49	0.0031	0.0031	18.6	0.059	427
(p)	1.84	0.0039	0.0039	23.6	0.075	540
SMC (Uncorrected)	19.9	0.0405	0.0405	247.6	0.784	5673
C60E						
	2.05	–	0.0103	33.6	1.07	1930
SMC (s)	2.25	0.0099	0.0099	33.3	0.257	1152
(m)	2.24	0.0099	0.0099	33.2	0.256	1149
(p)	2.78	0.0122	0.0122	46.4	0.328	1425
SMC (Uncorrected)	16.14	0.0722	0.0722	274.3	1.88	8423

$l_{23} \ll l_{13}$) was found to be valid. The SMC justifies this assumption for these membranes, since, by calculation, l_{23} is less than one per cent of the value of l_{13} and appears only in the expression for electro-osmotic transference number, t_3, Equation (A4) (which can normally be measured only to an accuracy $\pm 1\%$).

For the expanded membrane C60E (1.0) in which salt uptake is some 14%, neglect of this coefficient is no longer justified, Table II. The SMC is included in Table II, where calculated coefficients may be compared with experimental and ternary-model coefficients. Once more the agreement is good, but l_{33} (and l_{23}) are again underestimated by both calculation models.

7. Prediction of Experimental Measurements

Measured and predicted transport properties are given in Tables III and VI. In both sets the specific conductivity is estimated and particularly good agreement obtained with (s) and (m) tortuosity corrections. Electro-osmotic transference numbers calculated by the salt model are some 10–15% too large, but, since these are calculated from the transport number of co-ion, t_2^3, in the binary model, they do not take account of co-ion movement in the membrane which will tend to reduce electro-osmotic flow.

TABLE VI

SMC predictions for membranes in 0.1 M sodium chloride

Membrane		Specific conductivity $\kappa \times 10^2$ $\Omega^{-1}\,cm^{-1}$	t_1	t_3	J_s [a] mole cm^{-2} s^{-1} $\times 10^{10}$	J_w [a] $\times 10^7$
C60N						
	obs.	1.37	0.998	10.75	3.90	0.65
SMC	(s)	1.55			4.90	0.43
	(m)	1.39	1.00	12.47	4.40	0.39
	(p)	1.76			5.60	0.49
C60E						
	obs.	1.92	0.995	15.77	12.2	1.65
SMC	(s)	2.06			15.2	1.40
	(m)	2.06	1.00	16.68	15.2	1.40
	(p)	2.57			19.2	1.73

[a] J_s and J_w are flows of salt and osmotic flows of water observed when a concentration gradient 0.15/0.05 M salt is maintained across the membrane [2].

Salt and osmotic flows across the membrane were measured when a concentration gradient 0.15/0.05 m was maintained across the membrane [2]. For the SMC these flows, given in Equations (A5) and (A6), become, for a univalent form,

$$\mathbf{J}_s = (l_{22})\,\mathbf{X}_{12} + (l_{23})\,\mathbf{X}_3 \tag{37}$$

and

$$\mathbf{J}_w = (l_{32})\,\mathbf{X}_{12} + (l_{33} - t_3^2\alpha)\,\mathbf{X}_3 \tag{38}$$

Salt and water flows calculated by the SMC are respectively higher and lower than observed. (In each case the error is $\approx 20\%$). The lower value of J_w, Equation (38), is due primarily to underestimated l_{33}, and over-estimated t_3, in the dominant second term of that equation. Salt flow, J_s, is overestimated because the coupling coefficient, l_{12}, is overestimated, but in both cases the agreement may be said to be remarkable when it is considered that only concentrations of the species in the membrane phase are required for the model calculation.

8. Isotope-Isotope Friction

At the basis of the model is the requirement that co-ion diffusion in membrane and model electrolyte differ solely due to tortuosity effects. Consequently $\bar{R}_{22}$ and $\bar{R}_{22*}$ in the membrane are very similar to solution values. The agreement between SMC and experimental frictional coefficients largely justifies this assumption. It is of interest however to examine isotope-isotope friction for counter-ion and for tritiated water, $\bar{R}_{11*}$ and $\bar{R}_{33*}$ respectively. Since Equation (21) may be written for any isotopic species, $\bar{R}_{11*}$ may be calculated from the self-diffusion coefficient $\bar{D}_{11}$ and estimated values of $\bar{R}_{11}$, (which are in good agreement with the SMC, Table IV). From this membrane data it is easily shown that isotopic friction between counterions, $\bar{R}_{11*}$, is negative for all membranes studied [2, 4, 1b]. For all ionic solutions for which data are available, ion-to-ion isotopic friction R_{ii*} is positive [19]. The source of this particular effect is unknown but it is not to be expected, since even for non isotopic species i and k, R_{ik} is positive where i and k are ions of like-charge [15].

TABLE VII

Water friction in C60 membranes and in aqueous solutions of alkali chloride salts

Membrane (NaCl ext)	$\bar{f}_{33*}/\theta_s$ $\times 10^{-7}$ J s cm^{-2} mole^{-1}	Internal Molality m
C60E (0.1)	9.3	2.13
C60N (0.1)	9.7	2.87
C60E (1.0)	8.8	3.21
C60N (1.0)	8.2	4.00
Solutions	f_{-33*} $\times 10^{-7}$ J s cm^{-2} mole^{-1}	
NaCl	12.34	3.20
KCl	9.87	3.31
CsCl	9.27	3.45
Water	11.09	0

Solution data calculated from ref. [16] for NaCl and KCl and from ref. [20] for CsCl. Isotopic diffusion coefficients for tritiated water were obtained from Brun's data [13] and Anderson and Paterson [19].

Few data are available for isotopic diffusion coefficients of water in concentrated aqueous electrolytes. The data of Brun [13] using trititated water in sodium and potassium chlorides and of Anderson and Paterson [19] for caesium chloride were combined with frictional coefficients from Miller [16] and from Dunsmore et al. [20] to provide Spiegler frictional coefficients f_{33*}, $(-C_3R_{33*})$. In pure water C_3R_{33} is zero and so $D_{33}^0 = RT/(-C_3R_{33*})$ with $f_{33*} = 11.09 \times 10^7$ J s cm^{-2} mole^{-1} (using Mills value of $D_{33}^0 = 2.236 \times 10^{-5}$ cm^2 s^{-1} [21].

The effect of salts on f_{33*} is relatively small even at 3 M. Sodium chloride results show increasing values as concentration increases, while the more order-destroying salts, potassium and caesium chlorides reduce f_{33*} as their concentration is increased. Data for the membranes, Table VII, indicate a tendency for f_{33*}/θ_s to decrease with increasing membrane molality and have magnitudes which are in keeping with potassium rather than sodium chloride. It appears therefore that the water-to-water friction in the membranes is in keeping with a more order destroying salt than sodium chloride [21]. The effect is not sufficiently large to seriously undermine the model.

9. Electro-Osmotic Coefficients, t_3, for Various Ionic Forms

The C60E membrane was converted into a number of ionic forms in 0.1 M solutions. Water content, capicity, and electro-osmotic transference number t_3 were then determined. To test the predictions of the SMC calculated values of $t_3 = (\bar{C}_3/\bar{C}_1) t_2^3$ for the corresponding aqueous chloride salts were plotted against $\bar{C}_3/\bar{C}_1$, Figure 6. The slopes of these lines correspond to the transport numbers of co-ion chloride in each salt solution. Experimental values of t_3 are shown on the same figure.

For lithium, sodium, potassium and hydrogen forms the SMC gives agreement which is excellent for the hydrogen form and same 10–15% high for the remaining ions. For the caesium and rubidium forms the SMC gives results which are much lower than observed. Indeed, the observed values are close to the limiting maximum for t_3, which would occur when $t_3 = \bar{C}_3/\bar{C}_1$ and both ion and water would have identical velocities relative to the membrane. In this condition the electrical field would cause the whole pore solution to be transported in electro-osmosis. There is no obvious explanation for this effect, which has no parallel in solution behaviour and must therefore be considered a polymer effect.

To illustrate the calculation for a quite different membrane data for Zeo-Karb 315 (Na-form, 0.05 M NaCl) [24] is included with those for C60 membranes of varying degrees of expansion and salt uptake [2, 23], Figure 7. The SMC predicts a t_3 of 54.4 compared with observed, 45.7 and the salt model is shown to follow observed electro-osmotic transference over an order of magnitude range from 50 to 5.

10. Hyperfiltration

A series of hyperfiltration experiments have been carried out by Burke and Paterson [23]. Calculated and observed desalination characteristics are shown in Table VIII.

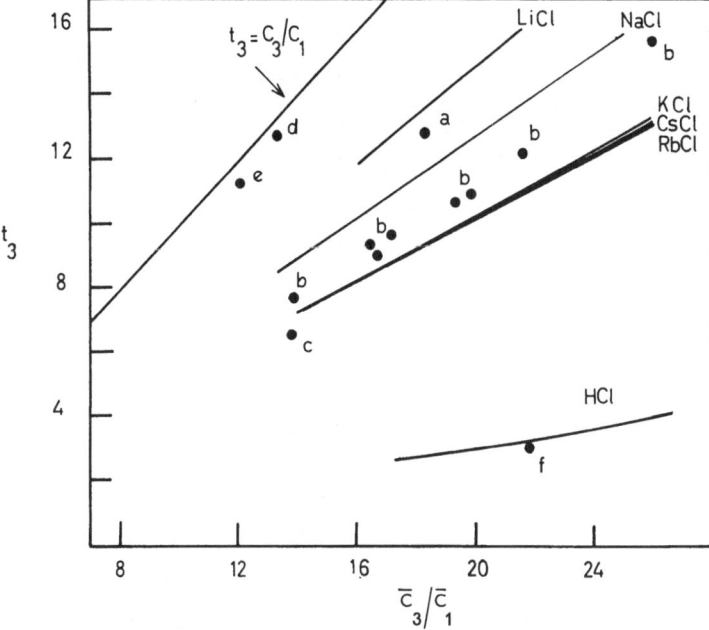

Fig. 6. Electro-osmotic transference numbers for ionic forms of C60E. Solid lines represent pre-dicted behaviour, from the SMC, circles experimental points for ionic forms. Lithium, sodium, potas-sium, rubidium, caesium and hydrogen forms are denoted by *a, b, c, d, e* and *f* respectively.

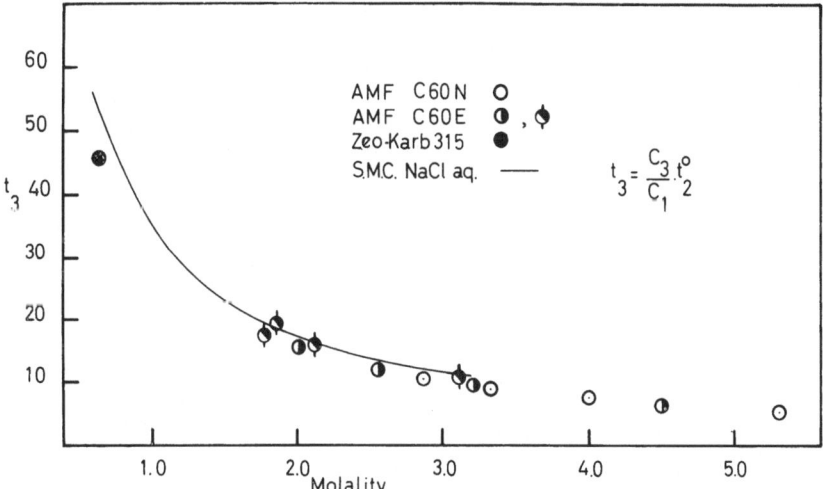

Fig. 7. Comparison of transference numbers for water in the sodium form of C60 membranes with varying degrees of expansion ○ [23] and for Zeo-Karb 315. [24], ●.

Salt and water flows were calculated, using Equations (A5) and (A6) using guessed values for product concentration, C_P. Since $(J_s/J_w) \cdot C_3{}^*$ also defines the product con-centration C_P, C_P (guessed) was varied until self consistent results were obtained. The

---

* C_3 is the molarity of water in the product solution and is required when J_s and J_w are defined in their usual unit mole $cm^{-2}s^{-1}$.

TABLE VIII

Hyperfiltration results and model calculations: Feed 0.1 M NaCl and pressure difference across the membrane 27.57×10^5 Nm^{-2}. (400 p.s.i.)

Membrane	Method	Salt flow $J_s \times 10^{10}$ mole cm^{-2} s^{-1}	Water flow $J_w \times 10^7$	Product C_p mole l^{-1}	Rejection $\dfrac{C_f - C_p}{C_f} \times 100\%$
C60N	obs.	(4.01)	4.80	0.042	58%
	l_{ik}-[2]	4.12	5.17	0.044	56%
	SMC (s)	3.70	3.40	0.060	40%
C60E	obs.	(16.00)	21.80	0.041	59%
	l_{ik}-[2]	14.50	19.90	0.041	59%
	SMC (s)	10.96	10.35	0.059	41%

Salt flow J_s was not measured directly but obtained from J_w and C_p.

results are given in Table IX where observed performance may be compared with that calculated using l_{ik} coefficients obtained by independent experimental studies [2] and the model calculation. The former accurately reproduce membrane performance and justify the approximations used in these earlier studies [2]. The salt model estimates are in poorer agreement. Predicted flows are somewhat smaller than observed and salt rejection under estimated by some 10%. It is however significant that the general effect of expansion on membrane behaviour is reproduced and the model predicts that expansion and concomitant increase in salt uptake will have little effect on rejection but will increase flows of salt and water by a factor of three: not a conclusion to be reached intuitively.

The examples above serve to indicate the degree to which predictions by this simple model are valid and so in turn the validity of the basic assumption that ion exchange polymers are similar to electrolyte solutions. The calculated results give good reason to believe that there are indeed close similarities and that specific polymer effects are rather fewer than might have been expected. Tortuosity correction remains a serious theoretical problem although the model defines θ_s, which may be superior to theoretical calculations of this parameter.

It is to be hoped that this method will be tested and expanded in future studies, both as a quick and simple method of estimating probable conductance or t_3 values in new membranes and for detailed comparison of transport coefficients, using familiar, if not fully understood references, ionic solutons.

Appendix

From the phenomenological equations, Equations (3), measured transport parameters may be expressed in terms of mobility coefficients, l_{ik} [2]
Specific Conductivity, κ:

$$\begin{aligned} \kappa &= [Z_1^2 l_{11} + Z_2^2 l_{22} + Z_1 Z_2 (l_{12} + l_{21})] F^2 \\ &= \alpha F^2 \end{aligned} \tag{A1}$$

Transport and transference number for water:

$$t_1 = (Z_1^2 l_{11} + Z_1 Z_2 l_{12})/\alpha \tag{A2}$$

$$t_2 = (Z_2^2 l_{22} + Z_2 Z_1 l_{21})/\alpha \tag{A3}$$

$$t_3 = (Z_1 l_{31} + Z_2 l_{32})/\alpha \tag{A4}$$

Salt flow, $\bar{\mathbf{J}}_s$, and osmotic flow of water, $\bar{\mathbf{J}}_3$ under local gradients of chemical potential $\bar{X}_{12}$ and $\bar{X}_3$;

$$\bar{\mathbf{J}}_s = l_{ss}\bar{X}_{12} + l_{sw}\bar{X}_3 \tag{A5}$$

$$\bar{\mathbf{J}}_3 = l_{ws}\bar{X}_{12} + l_{ww}\bar{X}_3 \tag{A6}$$

where

$$l_{ss} = \frac{-Z_1 Z_2}{r_1 r_2}\left(\frac{l_{11}l_{22} - l_{12}l_{21}}{\alpha}\right)$$

$$l_{sw} = l_{ws} = \frac{1}{r_1}\left(l_{13} - \frac{t_1 t_3 \alpha}{Z_1}\right)$$

and

$$l_{ww} = \left(l_{33} - \frac{t_3^2 \alpha}{Z_1}\right)$$

Z_1, Z_2 are the signed valencies of the salt ions and r_1, r_2 their stoichiometric coefficients in the salt molecule.

References

1. (a) Katchalsky, A. and Curran, P.: *Non Equilibrium Thermodynamics in Biophysics*, Harvard University Press, Cambridge, Mass., 1965.
   (b) Meares, P., Thain, J. F., and Dawson, D. G.: in G. Eisenman (ed.), *Membranes – A Series of Advances*, Dekker, New York, 1972, Chapter 2.
   (c) Caplan, S. R. and Mickulecky, D. C.: in J. A. Marinsky (ed.), *Ion Exchange*, Dekker, New York, 1966, Chapter 1.
2. Paterson, R. and Gardner, C. R.: *J. Chem. Soc. A* 2254 (1971).
3. Ferguson, H., Gardner, C. R., and Paterson, R.: *J.C.S. Faraday I* **68**, 2021 (1972).
4. Gardner, C. R. and Paterson, R.: *J.C.S. Faraday I* **68**, 2030 (1972).
5. Onsager, L.: *Ann. N.Y. Acad. Sci.* **46**, 241 (1945).
6. Spiegler, K. S.: *Trans. Faraday Soc.* **54**, 1409 (1958).
7. Meares, P., Dawson, D. G., and Sutton, A. H.: *Ber. Bunsenges f. Physik. Chem.* **71**, 765 (1967).
8. Pikal, M. J.: *J. Phys. Chem.* **75**, 3124 (1971).
9. Glueckauf, E.: *Proc. Roy. Soc. (London) A* **214**, 207 (1952).
10. Arnold, R. and Koch, D. F. A.: *Austral. J. Chem.* **19**, 1299 (1966).
11. Mackie, J. S. and Meares, P.: *Proc. Roy. Soc. (London), A* **232**, 510 (1955).
12. Mills, R.: *Rev. Pure Appl. Chem.* **11**, 78 (1961).
13. Brun, B.: *Ph. D. Thesis, Montpellier University*, 1967.
14. Prager, S.: *J. Chem. Phys.* **33**, 122 (1960).
15. Miller, D. G.: *J. Phys. Chem.* **71**, 616 (1967).
16. Miller, D. G.: *J. Phys. Chem.* **70**, 2639 (1966).
17. Laity, R. W.: *J. Phys. Chem.* **63**, 80 (1959).

18. Kedem, O. and Essig, A.: *J. Gen. Physiol.* **48**, 1047 (1965).
19. Anderson, J. and Paterson, R.: *J. C. S. Faraday I* **71**, 1335 (1975).
20. Dunsmore, H. S. Jalota, S. K., and Paterson, R.: *J. Chem. Soc. A.* 1061 (1969).
21. Mills, R.: *Ber. Bunsenges f. Physik. Chem.* **75**, 195 (1971); *J. Phys. Chem.* **77**, 685 (1973).
22. Gurney, R. W.: *Ionic Processes in Solution*, McGraw-Hill, New York, 1953.
23. Burke, I. S. and Paterson, R.: in preparation.
24. Mackay, D. and Meares, P.: *Trans. Faraday Soc.* **55**, 1221 (1959).

MEASUREMENT OF FLUXES AND FORCES AT THE SURFACE OF CATION EXCHANGE MEMBRANES UNDER CONDITIONS OF CONTROLLED POLARIZATION

I: *Methodology*

ERIC SÉLÉGNY and CHRISTIAN BOURDILLON*

Laboratory of Macromolecular Chemistry, ERA 471, Faculty of Sciences and Technology of the University of Rouen, 76130 Mont Saint Aignan, France

Abstract. Methods for the measurement of fluxes and forces at the interfaces of a cation-exchange membrane in dilute (0.1 M) sodium chloride have been developed. Instead of neglecting or eliminating inasfar as possible both polarization layers, one of them was stabilized and expanded by the use of a contiguous agarose gel film. This method makes for a high level of uniformity over the interface and affords easy control of concentration profiles. Consequently, a well-defined polarographic plateau is observed in current-voltage data. From determinations of the relaxation time of the overpotential, one can calculate both the interfacial concentrations and the forces which prevail, namely, the potential ΔE between the two membrane interfaces. Volume flows were measured in the conventional manner but new methods were employed to measure the ionic fluxes. All of this methodology allows one to obtain directly the nine coupling coefficients of the membrane without resort to pressure measurements as shown in the second part of this contribution.

List of Symbols

$a\pm^n$	mean activity of NaCl (see C^n for notation)
C	molar concentration, side (1)
C^n	interfacial concentration at polarization film interface with $n = 1, 2, 3, 4$ denoting the profiles of Figure 1.
C_0^n	molar concentration, side (2)
C_r	concentration of regulating solution in measurement of Js.
$C^n(x, t)$	concentration at distance x from the interface at time t during diffusion-relaxation.
$C^n(t)$	interfacial concentration at time t of the diffusion-relaxation process
D	diffusion coefficient
θ	membrane thickness
E_{J1}, E_{J2}	junction potentials
E_d	diffusion potential in polarization layer
$E_s, \bar{E}, E_s^n, E_{s0}^n$	potential due to IR drop in solution (1), membrane, polarization layer, solution (2).
$E_M(t)$	potential after current interruption at time t
E_0	Nernst potential
F	Faraday
i	current
I	current density
J_v	general case, volumetric flux through membrane
J_v^0	volumetric flux at $I = 0$, osmosis
J_s	NaCl flux, general case
J_s^0	NaCl flux at $I = 0$

* Present address: Université Technologique de Compiègne, Lab. Technologie Enzymatique, Compiègne, France.

Eric Sélégny (ed.), Charged Gels and Membranes I, 183–206. All rights reserved.
Copyright © 1976 by D. Reidel Publishing Company, Dordrecht-Holland.

P	hydrostatic pressure
p	slope of relaxation graph
R	gas constant
$R_s, \bar{R}, R_s{}^n, R_{s0}{}^n$	ohmic resistances of solution (1), membrane, polarization layer, solution (2)
R_1, R_2, R_3, R_4	resistances defined in (35)
S	membrane surface area
t	time
t_i	real transport number of species i in membrane, with $i = 1, 2, 3$ for cation, anion, and water.
$t_{i(\text{ap})}$	apparent transport number of species i in membrane.
t_v	volume transport number
$t_1{}^0$	transport number of Na^+ in solution
$\bar{V}_3, \bar{V}_{Ag}, \bar{V}_{AgCl}$	partial molar volumes (water, Ag, AgCl)
X	distance to the membrane
Δ	difference between sides (1) and (2) of membrane
$\overline{\Delta E}$	membrane potential
ΔE	potential between the two membrane interfaces
$\Delta\Pi_s$	difference in osmotic pressure between sides (1) and (2) of membrane
$\Delta\mu_s$	difference in chemical potential of salt between sides (1) and (2) of membrane
ϕ	osmotic coefficient (molar scale)
$\gamma\pm$	activity coefficient (molar scale)
$\alpha_s, \alpha_{s0}{}^n$	dimensional constants of the cell
δ_m	thickness of boundary layer
χ_c	conductance of NaCl of concentration C.
Λ_C	equivalent conductivity of NaCl of concentration C (or $C_0{}^n$).

1. Introduction

Due to the discontinuity of the mobility of ionic species, all fluxes through an ion-exchange membrane give rise to diffusion-convection or diffusion-convection-electro-migration phenomena at interfaces, consequently influencing the interfacial concentrations there and, in turn, the functional behavior of ion-exchange membranes. In order to minimize or even eliminate concentration polarization, forced convection which has been more efficient [1–3] or less so [4, 5] has been employed, the latter coupled with the use of low current densities to minimize further the effects of concentration polarization. It was often suggested that these measurements were valid only when transport numbers were independent of current density and (sometimes) also of stirring. Nevertheless, some authors have reported on variations of transport numbers, particularly that of water [6–8] which were justified by hydrodynamic considerations (vertical natural convection).

Brun [9] has shown that the independence of the water transport number on current density is not, in fact, a criterion of the absence of polarization. We recently confirmed this result and proposed the EQUI method, one which is truly independent of the uncontrolled influence of polarization [6, 10]. Further, we show in this communication that it is possible to use the phenomenon of polarization itself to obtain definitive information about the membrane. This procedure requires a good knowledge of polarization, in particular as it is related to hydrodynamic conditions.

In this contribution we evaluate the necessary measurements and the previously available methodology, describe new procedures and the theory which obtains and then present some typical experimental results, as an example, with a cation-permeable membrane.

2. Theoretical-Rationale of Methodology

The first requirement is a precise method for measuring interfacial concentrations.

The concentration overpotential method introduced by Cooke [11] that we have recently modified and improved [10] allows the necessary precision. The second is control of the thickness and uniformity of the polarization layers. Under usual conditions of forced or of vertical, natural convection the thickness of the layer is different at different points. After a study of the methods adapted to natural convection [13] and to forced convection, we concluded that by mechanical stabilization of the layer one could control local concentrations with precision, and used for this purpose a layer of agarose gel of constant thickness (about 1 mm) cast on one face of the membrane. The other interface is stirred vigorously so its polarization layer is negligible in thickness to that materialized by the gel. We thus define a bilayer with which it is possible to obtain stable and uniform concentration profiles in an effective manner.

Moreover, this polarization layer has further theoretical interest because in the stationary state the equivalence of fluxes in each layer, each characterized by a different set of forces, allows for a more fundamental understanding of the phenomena which obtain, particularly for those occurring at the interface.

In the stationary state at a given current density across the ion-exchange membrane – agarose gel bilayer, the polarization layer in the gel remains stable if certain external controls are maintained. A number of concentration profiles can be investigated. Figure 1 shows *four profiles* that are reported herein with the corresponding notation.

Profiles 1 and 2 are classical examples of polarization produced by a current in either direction, with one polarization layer artificially enlarged to make the other negligible by comparison. These are complex cases because none of the forces acting on the membrane are zero (except that there is no hydrostatic pressure difference) so all of the fluxes of electro-osmosis, osmosis, diffusion or retrodiffusion and electro-migration are present. Profiles 3 and 4 show equal concentrations at both interfaces of the membrane; they correspond to the Equi method [10] where a certain current density creates this interfacial situation, one that can exist for two series of an infinite number of concentrations, one with $C_0^n > C$, the other with $C_0^n < C$ depending on the direction and intensity of current. All of the symbols used herein are defined at the beginning of this communication.

The notation we adopt is to designate the solution concentration at the (left hand or lh) non-polarized side as C, as that on the (right hand or rh) polarized side as C_0^n where the n used herein is 1, 2, 3, or 4, referring to the profile. At the gel-membrane interface the concentration is C^n. The corresponding mean activities are $a_\pm$, $a_\pm^n$ etc. The IR drop in the non-polarized solution is E_s, it is $\bar{E}$ across the membrane, E_s^n in the (gel) polarization layer and E_{s0}^n in the other solution place. The dimensional constraints defined later are α_s in the non-polarized solution and α_{s0}^n in the gel layer and the adjacent solution. The corresponding non-ohmic potential drops are E_{jl} due to the lh electrode, $\overline{\Delta E}$ the membrane potential, E_d^n the diffusion potential in the gel and E_{j2} at the rh electrode. The membrane thickness is e and that of the gel layer is δ_m. The positive scalar direction x is lh to rh as are those of all fluxes J for water and both ionic species and for the positive current I. This type of profile makes it possible to make $\Delta \mu_s$ and $\Delta \Pi_s$ the gradients of the chemical potential and osmotic

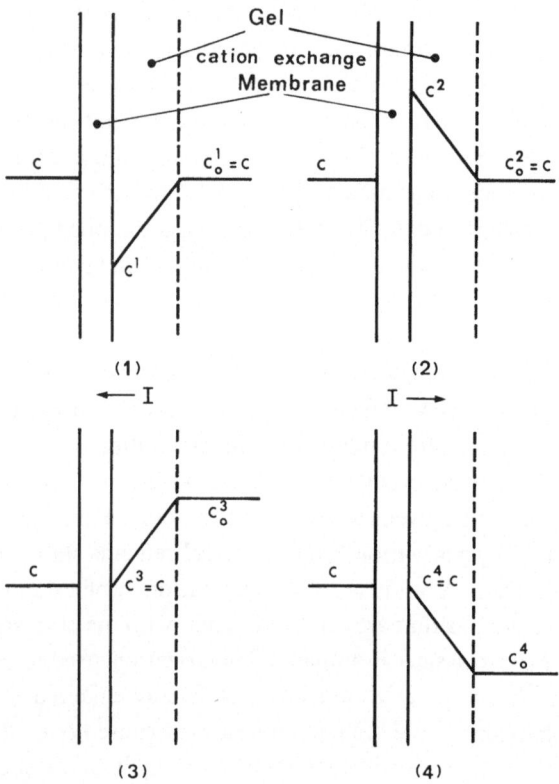

Fig. 1.   Concentration profiles under four polarizing conditions showing direction of positive current. *Profiles (3) and (4) are under the Equi condition.* Concentration on non-polarized side is C, in the solution on the gel side C_0^n where $n = 1, 2, 3, 4$ are referring to the profile, and where C^n is the concentration at the membrane-gel interface.

pressure across the membrane (as opposed to the gel layer) both zero, but with a finite electric potential acting on the system.

3. Relaxation Phenomena

A technique introduced by Cooke [11] measures the concentration overpotential as soon as possible after the interruption of current. We have shown [10] that this potential is more easily determined by measuring its relaxation as a function of time.

The overpotential is composed [11] of that of the membrane responding reversibly to its interfacial concentrations and the diffusion potentials in the boundary layers, both functions of the concentration. The potentials of measuring electrodes and any junction potentials must also be added where they are present.

The membrane potential is expressed conveniently using the transport number of the cation in the membrane $t_{1\,(ap)}$, a value which can be measured at any given concentration gradient independently without current, as will be shown later.

Neglecting the contribution of polarization on the side without gel and with the operator Δ expressing differences between the two membrane sides for processes going from 1 to 2:

$$E_M(t) = \overline{\Delta E} + E_d^n \tag{1}$$

$$\overline{\Delta E} = \frac{RT}{F} [(1 - 2t_1(ap)] \ln \frac{a_\pm}{a_+^n(t)} \tag{2}$$

$$E_d^n = \frac{RT}{F} (1 - 2t_1^0) \ln \frac{a_\pm^n(t)}{a_{\pm 0}^n} \tag{3}$$

Since the activity $a_\pm = C\gamma_\pm$ and at $C < 0.3$ M for NaCl,

$$- \log\gamma_\pm = \frac{0.5115\sqrt{C}}{1 + 1.316\sqrt{C}} - 0.055C \tag{4}$$

Since the relationship connecting activity and concentration is complex, there is not a simple analytic solution giving $C^n(t)$ (the interfacial concentration) as a function of the measured potential $E_M(t)$. Numerical methods are used here.

At zero time of current interruption, the polarization layers still present the profile corresponding to the state of quasi-stationary equilibria and the potential $E_m(0)$ is for the system preceding current interruption. Its relaxation depends on the properties of the membrane system, so the slope of the relaxation curve depends upon the characteristics of the diffusion layer and thus on the assumptions we make concerning this layer.

We assume *first* that: the membrane is an impermeable wall throughout the relaxation of diffusion; the concentration profile in the diffusion layer is linear and not influenced by convection through the membrane during the passage of the current; the diffusion coefficient of the salt is constant.

Under these conditions the interfacial concentration as a function of time is given by [10, 16]:

$$C^n(t) = C^n + (C_0^n - C^n)(2/\delta_m)(Dt/\Pi)^{1/2} \tag{5}$$

Since $C^n(t)$ varies linearly with the interfacial concentration this allows one to obtain C^n by a linear extrapolation to zero time. This has already been verified for all types of convections [13, 14].

The slope (p) is given by,

$$p = \frac{dC''(t)}{d\sqrt{t}} = \frac{2(C_0^n - C^n)\sqrt{D}}{\delta_m \sqrt{\Pi}} \tag{6}$$

through which δ_m is obtained if the assumptions taken are valid. This is a good approximation for nearly ideal membranes, but inadequate for *more permeable ones*:

The electro-osmotic flux J_v caused by the passage of current through the membrane modifies the concentration profile in the diffusion film and in turn influences the relaxation process. The gel which provides a well-defined layer can make this phenomenon evident, as follows. With current, the flux of the cation (J_1) at any point in the diffusion layer is now,

$$J_1 = -D\frac{dC''}{dx} + \frac{t_1^0 I}{F} + J_v C''(x) \tag{7}$$

Since

$$dC/dt = -dJ/dx \tag{8}$$

and assuming D and t^0 invariable with x in the stationary state,

$$\frac{dC}{dt} = D\frac{d^2C}{dx^2} - J_v\frac{dC}{dx} = 0 \tag{9}$$

The boundary conditions are: $x=0$, $C=C''$; $x=\delta_m$, $C=C_0^n$. Then,

$$C''(x) = \frac{C_0^n - C^n}{\exp(J_v \delta_m/D) - 1}\left[\exp\left(\frac{J_v x}{D}\right) - 1\right] + C^n \tag{10}$$

and at both zero time and x (subscript 0, 0),

$$\frac{dC''(0,0)}{dx} = \frac{J_v(C_0^n - C_n^n)}{D[\exp(J_v \delta_m/D) - 1]} \tag{11}$$

From this profile we can now calculate the diffusion relaxation law [16]. For small values of t,

$$C''(t) = C^n + \frac{2J_v(C_0^n - C^n)t^{1/2}}{[\exp(J_v \delta_m/D) - 1](\Pi D)^{1/2}} \tag{12}$$

As in (5), the interfacial concentration C^n changes with the square root of time, and its slope p is,

$$p = \frac{dC''(t)}{d\sqrt{t}} = \frac{2J_v(C_0^n - C^n)}{[\exp(J_v \delta_m/D) - 1](\Pi D)^{1/2}} \tag{13}$$

The assumptions of this second calculation correspond to the relaxation process for profiles 3 and 4, but are too restrictive for the more general case (1 and 2) because osmosis and retrodiffusion are not accounted for.

In the *general case*, the less restrictive assumptions we make are: D and t_1^0 are constant in the diffusion layer; the osmotic volume flow J_v^0 in the system without current is independent of t throughout the relaxation process; its numerical value is equal to that measured for the profile at $t=0$; the salt flux J_s^0 without current follows the same assumptions. These are approximations since we know that J_v^0 and J_s^0 are dependent on the interfacial concentration which varies during the relaxation process. Nevertheless, the perturbation of diffusion due to these fluxes is weak (about a few percent so it can be neglected at small times. A later calculation will consider these perturbations at larger times or for different kinds of membranes [16].

The calculation of the diffusion relaxation is an involved one and is presented elsewhere [16]. Here too the interfacial concentration is a function of $t^{1/2}$ and for p also,

$$C''(t) = C'' + \left[\frac{2J_v(C_0'' - C'')}{\exp(J_v\delta_m/D) - 1} + 2J_s^0 - 2C''J_v^0 \right] \frac{t^{1/2}}{(\Pi D)^{1/2}} \qquad (14)$$

$$p = \left[\frac{2J_v(C_0'' - C'')}{\exp(J_v\delta_m/D) - 1} + 2J_s^0 - 2C''J_v^0 \right] \frac{1}{(\Pi D)^{1/2}} \qquad (15)$$

Accordingly, *the potential measured throughout the relaxation of the diffusion layer can be related to the interfacial concentration*. The change in this concentration is always a function of $t^{1/2}$ and extrapolation to $t = 0$ gives the concentration profile with current. Thus, we can describe the perturbations in the relaxation-diffusion process by the electro-osmotic volume flux J_v with current and by the salt and volume fluxes during the relaxation via convective terms containing J_s^0 and J_v^0. This treatment is used later to compute interfacial concentrations, the salt flux with current and the effective thickness of the gel layer.

4. Fluxes and Forces

We use the definitions of fluxes and forces given by Kedem and Katchalsky [17] for a cation-exchange membrane separating two solutions of a 1-1 electrolyte, where $(\Delta P - \Delta \Pi)$, $\Delta \mu$, ΔE are the forces and J_v, J_s, I the fluxes. While one can readily measure volume and salt fluxes through an ion-exchange membrane, this is not so for the forces acting at the boundaries of the membrane. No attempt has heretofore been made in this direction. The means proposed to obtain the phenomenological coefficients [2, 18–20] use certain conditions for which combined coefficients can be calculated from standard membrane measurements, *i.e.*, conductivity, transport number, etc. Here it is not necessary to know the forces at the membrane boundaries, so the possibility of verifying this treatment for direct experiment does not exist.

Recently Demarty and Sélégny [21] examined *the direct measurement of these phenomenological coefficients* from a knowledge of the forces at the level of the control or measurement electrodes, and obtained coefficients characteristic of the membrane – solution system between the measuring electrodes, not of the membrane only. Now we propose a new methodology which allows one to determine and to control $\Delta \Pi_s$,

$\Delta\mu_s$, and ΔE in an effective manner at the boundaries of the membrane itself. This does not apply to the hydrostatic pressure which will be treated later.

The measurement of *current density* is straightforward. No corrections were made for 'edge effects'; by a standard method [22] it can be shown that they are negligible under our experimental conditions.

The *volume flux density* J_v with reference to the membrane is usually measured by volume change [7, 23–25].

We use our technique [25] wherein the volume change of one compartment is recorded automatically by a photo-electric device which follows the displacements of a meniscus. Then, after making corrections for volume changes due to the Ag/AgCl working electrodes, the volume change ΔV for the membrane surface S is,

$$J_v = \frac{\Delta V}{St} + \frac{I}{F}(\bar{V}_{AgCl} - \bar{V}_{Ag})$$ (16)

The *osmotic flux* J_v^0 without current is measured in the same way, and the volume transport number t_v is defined as $t_v = J_v F / V_0 I$.

The *cation flux* through the membrane J_s (or J_1) is usually measured using isotopic traces [28, 29], a measurement not subject to the usual restrictions [26, 27]. A method of the Hittorf type can also be used, but since it requires changes in solution concentration at the membrane to obtain sufficient precision and can result in modification of membrane properties, it is not the method of choice [30]. We use here the method [21] which by regulating the concentration of ambient solutions avoids the use of corrections inherent in the use of isotopic tracers and maintains the system in a stationary state. As will be shown later, it allows one to analyze polarization phenomenon and to calculate J_s from the relaxation of the overpotential.

The measurement of J_s is on the one hand accomplished with the *automatic addition of a solution* of concentration C_r, one more or less concentrated than C_n^0. This keeps the conductivity of the solution constant in each compartment, if necessary. The rate of volume addition of C_r is constant under a stationary state, one that can be maintained for a long time with considerable advantage in measuring ΔE, for example. J_s is now calculated from concentration and V, the volume of solution C_r added during time t,

$$J_s = \frac{V(C_0^n - C_r)}{St} + J_v C_0^n$$ (17)

with the sign of J_v and J_s defined by the conventions stated earlier. Note that $t_1 = J_s F / I$. The experimental details of these methods are given later.

On the other hand for the measurement of J_s by the *analysis of the diffusive processes during relaxation*, in the general case of polarization with current (profiles 1 or 2) one can express the contributions of diffusion, electromigration, and convection to the flux at the membrane/gel interface. At $x = 0$,

$$J_s = -D\left(\frac{dc}{dx}\right)_{x=0} + \frac{t_1^0 I}{F} + J_v C^n$$ (18)

and from (11),

$$D\left(\frac{dc}{dx}\right)_{x=0} = \frac{J_v(C_0^n - C^n)}{\exp(J_v\delta_m/D) - 1}$$

The general case of relaxation leads to

$$\frac{J_v(C_0^n - C^n)}{\exp(J_v\delta_m/D) - 1} = \frac{p\sqrt{\Pi D}}{2} - J_s^0 + J_v^0 C^n \qquad (19)$$

and

$$J_s = -\frac{p\sqrt{\Pi D}}{2} + J_s^0 + C^n(J_v - J_v^0) + \frac{t_1^0 I}{F} \qquad (20)$$

Since J_v, J_v^0, and J_s^0 are measured, the only complementary determination required to obtain J_s is that of the slope p.

To *determine* $\Delta\Pi_s$, since our apparatus did not measure or create hydrostatic pressure differences, since only NaCl was present, $\Delta P - \Delta\Pi = \Delta\Pi_s$,

$$\Delta\Pi_s = \frac{2RTM_3}{\bar{V}_3}(\phi C - \phi^n C^n) \qquad (21)$$

where at the values of the osmotic coefficient ϕ used, there is but a maximum deviation of 0.2% from experimental values [31]. Thus, $\Delta\Pi_s$ is calculated from the interfacial concentrations C and C^n.

The value of $\Delta\mu_s$, *the chemical potential difference* is calculated from the interfacial concentrations using the activity coefficient,

$$\Delta\mu_s = 2RT \ln \frac{a_\pm}{a_\pm^n} \qquad (22)$$

In the formalism of Kedem and Katchalsky, *ΔE is the potential difference between two electrodes reversible to the anions* (for example Ag/AgCl) *supposedly placed at the membrane surfaces.* Since our electrodes are placed outside the polarization layers, we compute ΔE from the ohmic or not ohmic resistance in the different layers of solution: membrane; gel; solution. With calomel electrodes, the potential E_{st} in the stationary state is,

$$E_{st} = \Delta E_j + \overline{\Delta E} + E_d^n + E_s + \bar{E} + E_s^n + E_{s0}^n \qquad (23)$$

We wish to obtain $\Delta E = \bar{E} + \Delta\bar{E} + E_0$, with E_0 being the potential difference of two Ag/AgCl electrodes in NaCl of concentration C and C^n,

$$E_0 = \frac{RT}{F} \ln \frac{a_\pm^n}{a_\pm} \qquad (24)$$

Then from the geometry of the system and its concentration profiles,

$$\Delta E = E_{st} - \Delta E_j - E_d^n - E_s - E_{s0}^n - E_s^n + E_0 \qquad (25)$$

To calculate the IR drop, let E and E_{s0}^n be the IR drop in the stirred solution and E_s^n be that in the diffusion layer. This type of calculation was employed by Spiegler [32] for interpreting conventional current data. His assumptions were approximations too extensive for our purposes, particularly for the evaluation of E_s^n.

The IR drop E_s^n in the polarization layer is calculated from its thickness and the concentrations at its boundaries,

$$E_s^n = \int_0^{\delta_m} \frac{i\, dx}{S\chi_c} \tag{26}$$

with the conductivity χ_c defined by

$$\chi_c = \Lambda\, C \tag{27}$$

and if one assumes that Λ is linear with $C^{1/2}$ [31],

$$\Lambda = \Lambda_0 + k\sqrt{C} \tag{28}$$

where we use a negative value for k. Since a linear concentration profile in the boundary layer is a reasonable approximation,

$$C_x = C^n + \frac{1}{\delta_m}(C_0^n - C^n)\, x \tag{29}$$

$$\Lambda_c = \Lambda_0 + k\sqrt{C^n + \frac{1}{\delta_m}(C_0^n - C^n)\, x} \tag{30}$$

and from [26],

$$E_s^n = \int_0^{\delta_m} \frac{i\, dx}{S\left[\Lambda_0 + k\sqrt{C^n + \dfrac{1}{\delta_m}(C_0^n - C^n)\, x}\right]\left[C^n + \dfrac{1}{\delta_m}(C_0^n - C^n)\, x\right]} \tag{31}$$

Complete integration leads to:

$$E_s^n = \frac{2i\delta_m}{S\Lambda_0(C_0^n - C^n)} \ln \frac{\sqrt{C_0^n}\,(\Lambda_0 + k\sqrt{C^n})}{\sqrt{C^n}\,(\Lambda_0 + k\sqrt{C_0^n})} \tag{32}$$

For the calculation of E_{s0}^n and E_s,

$$E_{s0}^n = \frac{(\alpha_{s0}^n - \delta_m/S)i}{\Lambda_{C_0^n}C_0^n} \tag{33}$$

$$E_s = \alpha_s i / \Lambda_c C \tag{34}$$

where α_{s0}^n and α_s are parameters characteristic of the volume and solution. C and C_0^n being known, Λ_c and ΛC_0^n are calculated from [28].

The measurements of α_s and α_{s0}^n are the most difficult part of the procedure. There must be a rigorous positioning of all elements relative to one another, in particular,

of the capillaries of the probe electrodes. The complex cell geometry does not necessarily insure a good parallelism of field lines in the measurement zone, even if they
are parallel at the membrane. The resistance of the solutions are, therefore, not
calculable with sufficient precision by the conventional formula, $R = l/S\chi$.

The so-called substitution cell used for AC membrane conductivities [33] could
not be employed here because the two solutions adjacent to the membrane (ex.
profiles 3 and 4) were not identical. A new method was developed. If we make all
elements of the cell occupy a strictly reproducible position and the temperature
constant so that α_{so}^n and α_s are not dependent on operating conditions, one can make
the following measurement without gel and with vigorous stirring at the two membrane interfaces. The potential drop between two Luggin capillaries placed a few mm
from each membrane surface is measured at a low constant current density (2 ma cm^{-2}
in 0.1 M NaCl), for maximum reduction of polarization in the chain $C_1/\text{Memb}/C_2$,
with C_1 and C_2 at all combinations and permutation of 0.05 and 0.1 M.

At different concentrations on each side of the membrane, it is necessary to consider separately membrane and junction potentials to calculate the resistance R,

$$
\begin{aligned}
R_1 &= \alpha_s/\chi_{0.1} + \bar{R} + \alpha_{so}^n/\chi_{0.1} \\
R_2 &= \alpha_s/\chi_{0.1} + \bar{R} + \alpha_{so}^n/\chi_{0.05} \\
R_3 &= \alpha_s/\chi_{0.05} + \bar{R} + \alpha_{so}^n/\chi_{0.05} \\
R_4 &= \alpha_s/\chi_{0.05} + \bar{R} + \alpha_{so}^n/\chi_{0.1}
\end{aligned}
\tag{35}
$$

Since the membrane resistance $\bar{R}$ varies little in these relatively dilute solutions,

$$
R_2 - R_1 = R_3 - R_4 = \alpha_{so}^n \left(1/\chi_{0.05} - 1/\chi_{0.1} \right)
\tag{36}
$$

$$
R_4 - R_1 = R_3 - R_2 = \alpha_s \left(1/\chi_{0.05} - 1/\chi_{0.1} \right)
\tag{37}
$$

A subsequent comparison of $R_2 - R_1$ and $R_3 - R_4$ or $R_4 - R_1$ and $R_3 - R_2$ verified
that R was nearly constant within the limit of experimental error. Moreover, $\bar{R}$ could
be recalculated from the α thus determined from (35). It is evident that these experimental procedures are difficult and limited to mean concentrations > 0.03 M.
The principal sources of error will be discussed later.

The membrane potential ΔE is calculated from the interfacial concentration (2)
where the apparent transport number t_1 (ap) depends on the concentration ratio
C/C^n. The potential E_d^n in the diffusion layer is calculated using (3). The cation solution
transport number t_1^0 varies little with concentration. It is presumed constant in the
layer and defined at the concentration C^n.

The electrode potential difference $\Delta E_j = E_{j1} - E_{j2}$ was measured using SCE with
gelified capillaries leading to the solutions. ΔE_j was calibrated using a pair of Ag/AgCl
electrodes and of capillary probe electrodes, and a comparison of potentials measured
with each type for different concentrations gave $t_{1\,(ap)}$ and ΔE_j as a function of C^n.

Figure 2 shows a diagram of the system of measurements leading to the calculation
of forces and fluxes. In order to acquire data and make calculations in reasonable
times an automated and computer-connected measuring system was devised.

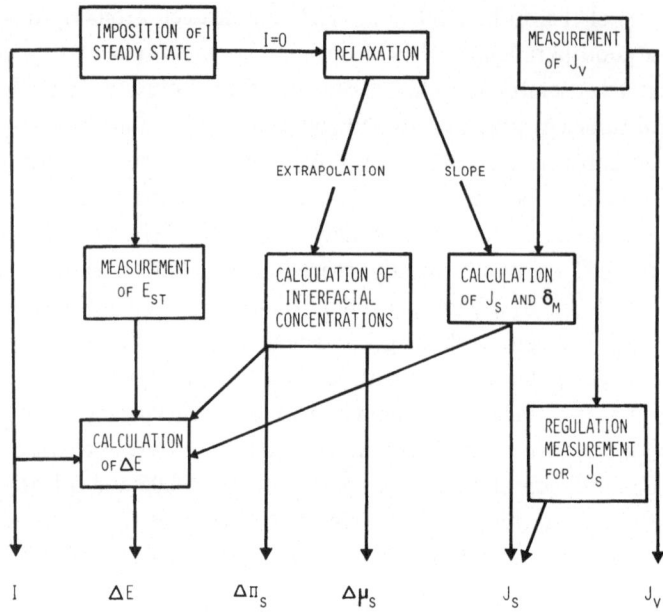

Fig. 2.   Diagram of the measurements and their combinations for the determination of fluxes and forces.

5. Experimental

The bi-layer material was prepared by coating a 3% agrose gel onto the ion-exchange membrane. This gel had the same electrochemical characteristics as the saline solution in which it was prepared and did not introduce any discontinuity at the gel/solution interface. A 3% agarose gel fits these qualifications well, with an absence of tortuosity as shown by conductivity and diffusion coefficient measurements [35]. To assure a reasonable life to the bi-layer, the ion-exchange membrane was well stretched on a support to minimize its deformations under use, render its surface more planar and increase adhesion to the gel. A disc of membrane equilibrated in dilute LiCl and washed with water to achieve a swollen state was mounted in a holder and its edges immobilized (Figure 3). Equilibration in 0.1 M NaCl then shrinks and stretches the membrane, a technique practical only with non-reinforced membranes whose dimensional variations are not negligible. The membrane used had a degree of swelling, (100 g water/g dry) of 90% in the Li form and 65% in the Na form. Since one might modify membrane properties by this tension, all measurements were made on one unique sample, immobilized permanently in the frame. The non-commercial cation-exchange membrane K used [37], is made by vapor phase chlorosulfonation of polyethylene followed by hydrolysis in water. Its thickness was 0.026 cm, capacity was 1.37 meq per dry g and swelling was 65% in the Na state. It was virtually transparent, indicative of good homogeneity at the microscopic level.

The hot solution of 3% agar in 0.1 M NaCl was cooled to 50° and poured into the

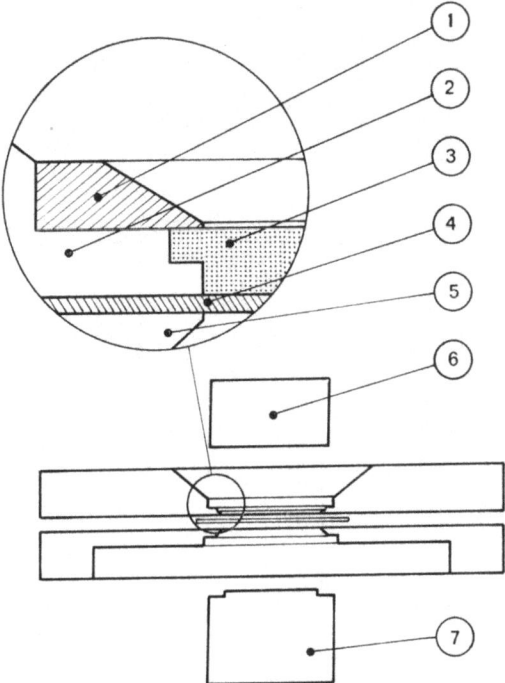

Fig. 3.   Membrane holder from altuglass with: 1, clip partly covering the gel; 2, membrane holder, side (2); 3, gel; 4, membrane; 5, membrane holder, side (1); 6 and 7, wedges B and A. The clamping system including the two half-membrane holders is not represented here. It includes screws made of altuglass mounted on '○' rings which insure sealing.

horizontal membrane supported by a precisely fabricated wedge A (see Figure 3). Another wedge (B) insures elimination of excess hot solution. After cooling and formation, the two wedges and excess agar were removed and the altuglass clamp having an internal diameter exactly equal to that of the membrane was placed above the gel to avoid displacement due to stirring during measurements. These agar discs show maximum thickness irregularities of 0.02 mm with a life span and adherence to the membrane longer than one month.

As measurements of fluxes and forces (except ΔP) were made virtually simultaneously, the experimental constraints were rather significant. Two half-dialyzers were used (Figure 4), each containing a different solution. For the half-electrodialyzer on the side of the membrane with no gel, strong convection made the thickness of the boundary layer but a few % that of the gel. This apparatus [14] at a rate flow of 25 cm^3 s^{-1} gave a boundary 0.02 mm thick, satisfying our requirements. The flow was forced tangentially to the surface by a peristaltic pump whose pulsations were damped.

Current was introduced by a Pt electrode separated by auxiliary compartments wherein NaCl five times more concentrated than that in the compartment and 5 l in

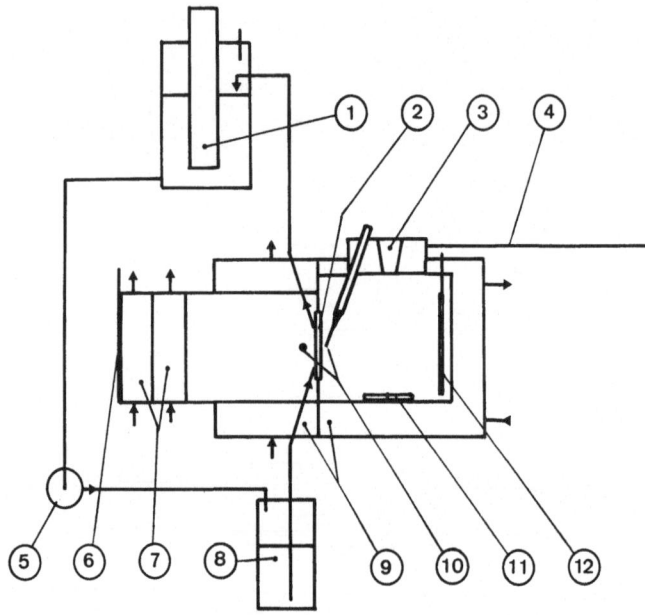

Fig. 4. Schematic diagram of electrodialyser made of altuglass: 1, storage flask for solution recycled on (1); 2, membrane holder and membrane; 3, standard taper stop cock for calibrating capillary (to measure volume fluxes) or conductivity electrode (for concentration regulation); 4, calibrated capillary (1 mm id) to measure volume variations with the meniscus automatically followed by differential photocell; 5, peristaltic pump – 1.5 l/min; 6, Pt injection electrode, side (1); 7, auxiliary compartments with ion-exchange membranes protecting central compartment; 8, storage flask; 9, thermostating circuits; 10, extremities of potential-measuring probes on both sides of membrane; 11, magnetic stirrer; 12, Ag/AgCl injection electrodes.

volume was circulated [36]. Thus, variations in concentration at all current densities used were $< 0.3\%$/day.

The other half-electrodialyzer allowed the masurement of fluxes; the hydro-dynamic requirements here were less restrictive, the only aim being a uniform composition easily achieved by magnetic stirring at 600 rpm. The volume here was limited to 100 cm^3. The large working electrode (Ag/AgCl) supported currents of 30 mÅ for several hours.

The measurement of E_{st} with current and the overpotential $E_{m(t)}$ with current interruption used the same pair of probe SCE electrodes fitted with silicone tubes and a capillary to allow a precise measurement. Junctions were made with saturated KCl in 4% agar. The tips of the capillaries of 0.8 mm id were ground so the exterior surface of the gel was perpendicular to the field lines. Shafts allowed the positioning of the two probes in a highly reproducible manner. The distance between the capillary tips was about 12 mm and the electrical asymmetry between the two measuring electrodes placed in the same solution was < 0.05 mV.

A constant temperature (to 0.01°) in the electrodialyzer necessary, was obtained partly by having the apparatus in a room at $25 \pm 0.2°$; this also helped to stabilize the electronic equipment. The electrodialyzer was kept at $25 \pm 0.01°$ by circulation

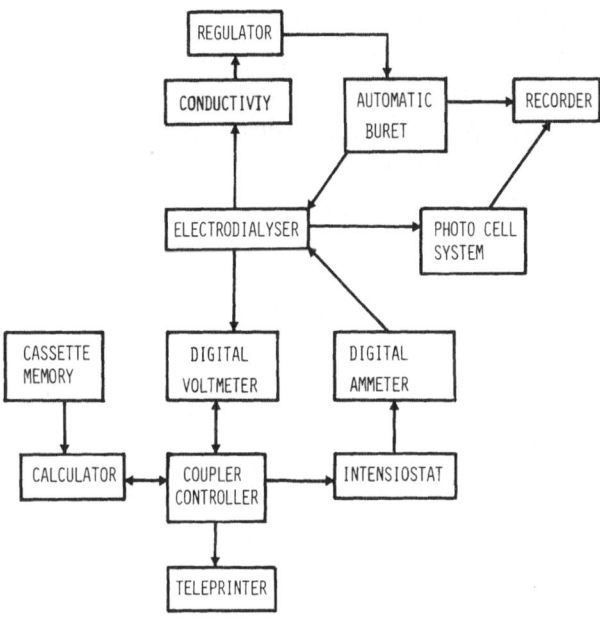

Fig. 5.   Diagram of connections between the measuring and data acquisition systems, not including
the thermostatic and temperature measuring systems.

through a double jacket of constant temperature water. A Pt probe electronic thermometer precise to $0.01°$ (at 23–27°) was used.

A general diagram of the electronic measuring equipment is given in Figure 5. Three circuits provided: current and measurement of potentials; regulation of the concentration in (2); measurement of volume variations in (2). These three were not always simultaneously for it was not possible to regulate concentrations during the volume flux measurement because the latter required the addition of solution. On the other hand, the measurement of potentials was done with regulation of concentration, and allowed an indefinite maintenance of the system in a stationary state and good reproducibility of the measurement.

The current injection and potential measurement circuit already described [14], allowed determination of interfacial concentrations and the boundary layer thickness, extended to measurements and calculations of $\Delta\mu_s$, $\Delta\Pi_s$ and ΔE. The order of operation during a measurement was:

the intensiostat imposed a fixed current density on the working electrodes of the electrodialyzer; once a stationary state was attained, the computer through the coupler-controller had the voltmeter measure the potential with current (E_{st}) and store this information; the piloted intensiostat then reduced the current to zero and the internal clock of the coupler-controller assured the measurement of overpotential by command of the voltmeter every Δt seconds (the relaxation process with a gel thickness of 1 mm was rather slow, so $\Delta t = 3$ s was selected); the stored data was used for automated calculations. This program of calculation and monitoring represented about 2000 steps and 50 constants (Hewlett Packard calculator, model 9810). For each measurement, the following sets were printed: p, C^n, $\Delta\Pi_s$, $\Delta\mu_s$, E_{st}, $\bar{E}$, $\bar{R}$, E_s, E_{s0}^n, E_s^n, E_d^n, $\Delta\bar{E}$ and ΔE. J_s was calculated separately because this required J_v which could not be measured at the same time.

A statistic linear regression calculation allowed the evaluation of the degree of correlation between the experimental points of the overpotential relaxation and the appropriate relationship as a function of $t^{1/2}$. For those experiments used, this correlation was always 0.999. The principal advantage of this program is that it can be adapted to all types of profiles and any interfacial concentration. Part of the program was used for assembling tables of the parameters and variables, together with the concentration, t^0, D, $t_{1\,(ap)}$ and Λ. A conditional test during the program insured, with no intervention of the operator, that the values used are truly those corresponding to the experimental situation.

The regulation circuit of (2) used a Urectron 4 Solea which commanded, step by step, the motor of an automated buret which delivered the regulating solution to (2). The delivered volume of solution was taken from the analog repeater of the buret and recorded with time.

An electronic problem in the conductivity measurement in the presence of another electric field and of the potential measurement appeared. It was solved by using an intensiostat and a differential voltmeter that were entirely floating. Grounding was then accomplished by one of the conductivity cell electrodes. Under these conditions the regulation of concentration, current imposition and measurement of E_{st} were simultaneously accomplished.

6. Results

The results presented here are largely on the consequences of hydrodynamic control of the boundary layer on membrane behavior. The current-voltage curves, variation with current density of water and ion transport, the evolution of interfacial concentrations and the potentials interacting in this system are also examined.

It was shown earlier that certain parameters (α_s, α_{s0}^n) should be standardized or known as a function of concentration $C^n(t_{1\,(ap)}, \Delta E_j)$. The calculation of α_s, α_{s0}^n was from data obtained without gel and with vigorous stirring at the two membrane interfaces of boundary layers $<20\,\mu$ thick at 3 ma cm^{-2} (Table I). The results allowed

TABLE I

Experimental values of E_{st} used to calculate dimensional constants α_s and α_{s0}^n

Soln. (1) (M)	Soln. (2) (M)	E_{st} in mV		Mean E_{si} (mV)	R (Ω)
		$I = -3\,\text{mA cm}^{-2}$	$I = +3\,\text{mA cm}^{-2}$		
0.10	0.10	-83.61	83.7_2	83.6^6	$R_1 = 27.89$
0.10	0.05	-138.7	106.3	122.5	$R_2 = 40.83$
0.05	0.05	-158.9	158.7	158.8	$R_3 = 52.93$
0.05	0.10	-103.9	136.1	120.0	$R_4 = 40.00$

the experimental verification of (35). The uncertainties here were $\pm 0.2\%$, a precision justified by the following. First, the temperature was measured to be $25°\pm 0.02°$ which led to an uncertainty of $<0.1\%$ in the conductivity. The solutions were prepared by weight in precision flasks. The distance between the two probes was about 1.2 cm, so the two solution layers contributed only 0.5% of the total resistance. The current density used caused quite small interfacial concentration variations of 2–3%, so corresponding changes in the resistance were negligible. A surprising degree of reproducibility of about 0.1% was obtained after complete disassembly and assembly of the electrodialyzer and renewal of the equilibrating solutions. In view of this experimental precision, there is a precision of the order of $\pm 0.2\%$ in α_s and α_{so}^n. Since one can calculate the membrane resistance under these experimental conditions from (35) where $S=2.584$ cm^2, $\bar{R}=0.77\pm 0.03$ Ω. Calculations based on IR drops from $(R_3-\bar{R})/(R_1-\bar{R})=\chi_{0.1}/\chi_{0.05}$ leads to an agreement better than 0.1%.

It is evident that the relative precision in $\bar{R}$ and equally in ΔE are not as good, since of the total IR drop, that due to membrane resistance represents but 2–3%. For this method, this is the principal limitation, as it is for substitution techniques [33], making precise measurements in dilute solutions not possible. Bringing the probes

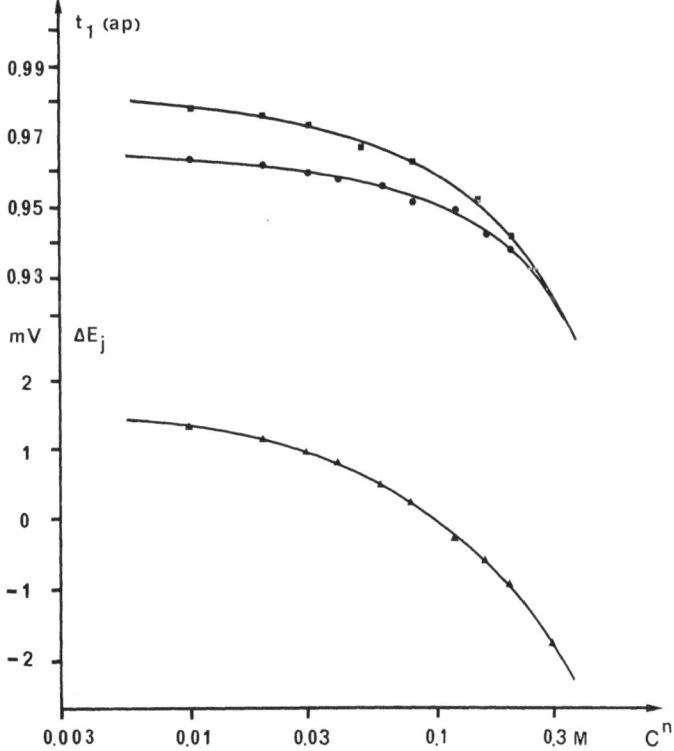

Fig. 6. Variation of apparent transport number of cation, as measured via the membrane potential for $[C=0.1$ M in (1)$/C^n$ in (2)] for stretched membrane (■), unstretched membrane (●) and standardization of junction potential of measuring probes (▼) with membrane K at 25°.

closer to the membrane can lead to a screening effect (38); moreover, the relative uncertainty due to the positioning of the probes and/or electrophoretic or mechanical deformation of the membrane tends to increase the error. It is believed that the precision can be increased 5-fold by improved cell construction, a problem of current study.

The variation of $t_{1\,(ap)}$ and ΔE_j with C^n is given in Figure 6. The uncertainties in $t_{1\,(ap)}$ were about 0.2% and of ΔE_j not more than 0.1 mV.

The variation of $t_{1\,(ap)}$ measured under the same conditions for a non-stretched membrane is also shown. There is a significant difference between the two curves in that the stretched membrane has a greater permselectivity. This was a reversible phenomenon for the relaxed membrane showed again its former transport number. This effect of mechanical tension on ion-exchange membranes has not been reported previously and justifies further study. By taking the precaution to always work with the same sample of stretched membrane, membrane potentials were constant over several months.

Figure 7 shows interfacial concentration variations and membrane resistance, as well as the different, interrelated potentials as a function of current density in the case of profiles (1) and (2), obtained in the presence of the gel layer. The variation in interfacial concentration C^n with current density is virtually linear up to high degrees of polarization.

This latter result is never obtained under non-stirred vertical convection [10, 11, 39]. (Figure 8). It is observed under forced convection [14, 40] but only at weak polarization. At strong polarization the disparity in diffusion layer thickness (and current distribution) at different points on the membrane explains these deviations from linearity [40]. The gel, therefore, does regulate the polarization layer thickness and, as a consequence, the local interfacial concentration distribution is narrow and the current is uniform.

The current-voltage curves obtained show an almost ideal plateau, one usually difficult to observe except in natural, horizontal convection [41] or when the measurement is taken with industrial electrodialyzers at a large distance from the point of fluid entry [42, 43]. The width of the polarographic plateau is usually explained by a non-uniform current density where the limiting current is reached at some points before others, so an average is observed. Variations in pH in the critical zone were not studied because of the chemical instability of agar. The use of a stable gel would allow a study of concentration polarization as related to water decomposition, phenomena which may not always be interrelated [45].

A high degree of stability of interfacial concentrations and potentials was observed at prolonged times, near the critical current density. This was due partly to efficient regulation of solution concentration and partly to the absence of convective oscillations that are sometimes observed in the limiting current region [46], especially under natural convection. This stability favored reproducibility of measurement. For example, the spread of interfacial concentrations did not exceed ±0.2%. The membrane resistance remained practically constant outside of the critical zone,

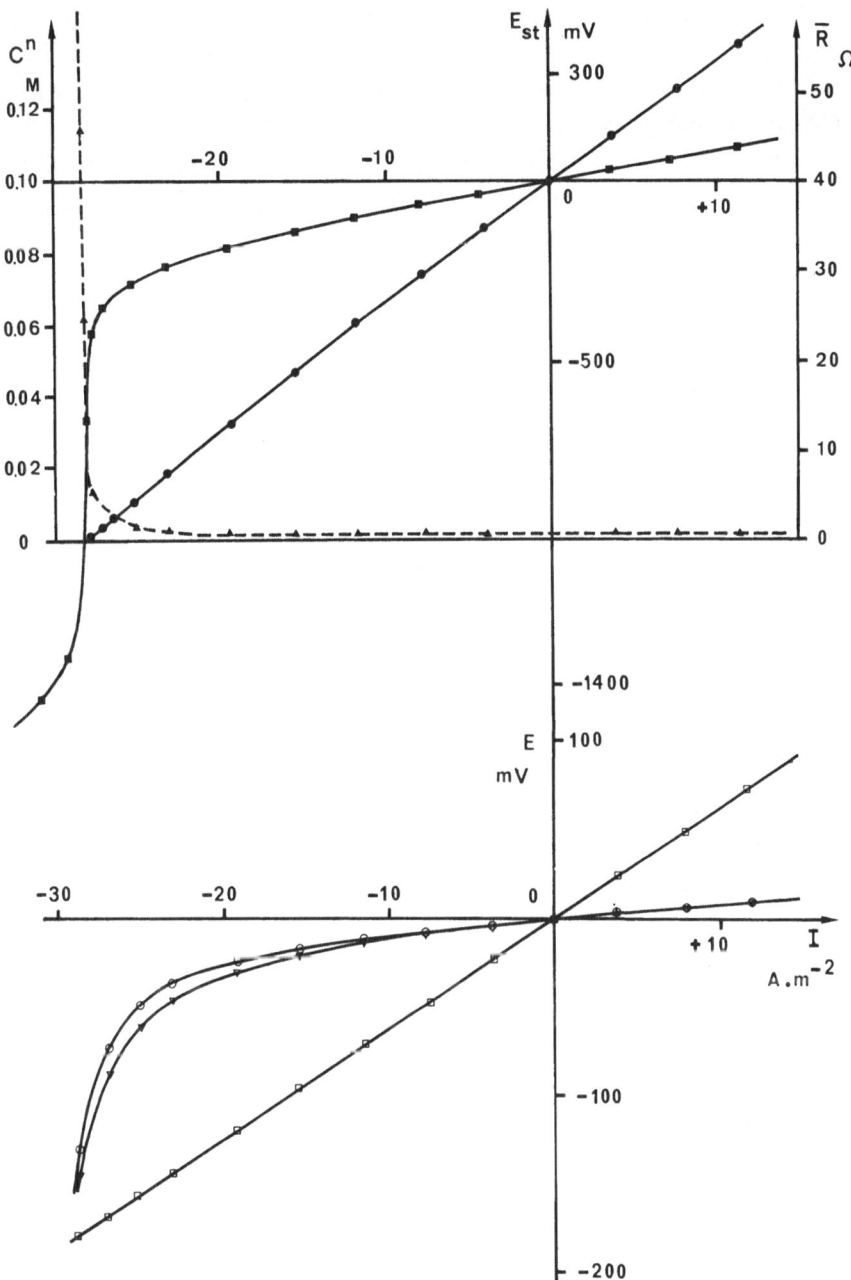

Fig. 7.   Results with profiles (1) and (2) with membrane K in 0.1 M NaCl, gel thickness 1.09 mm at 25°. The symbols represent: (■), current-voltage data for the two measuring probes (E_{st}); (●) variation of interfacial concentration C^n with I; (▼) membrane resistance $\bar{R}$; (□), IR drop in solution ($E_{so}{}^n + E_s$); (○), IR drop in polarization layer ($E_s{}^n$); (△), non-ohmic voltages ($\Delta E + E^n{}_d$).

independent of the direction of the current ($\bar{R}=0.78\pm0.03\ \Omega$). This value agrees with that obtained without gel. This measurement serves as a test of reliability; (23) and (25) show that the calculation of $\bar{E}$ and therefore of $\bar{R}$ is quite sensitive to different experimental errors.

The region corresponding to the critical zone (Figure 7) is interesting because it shows the appearance of overvoltage due to water decomposition which adds to the

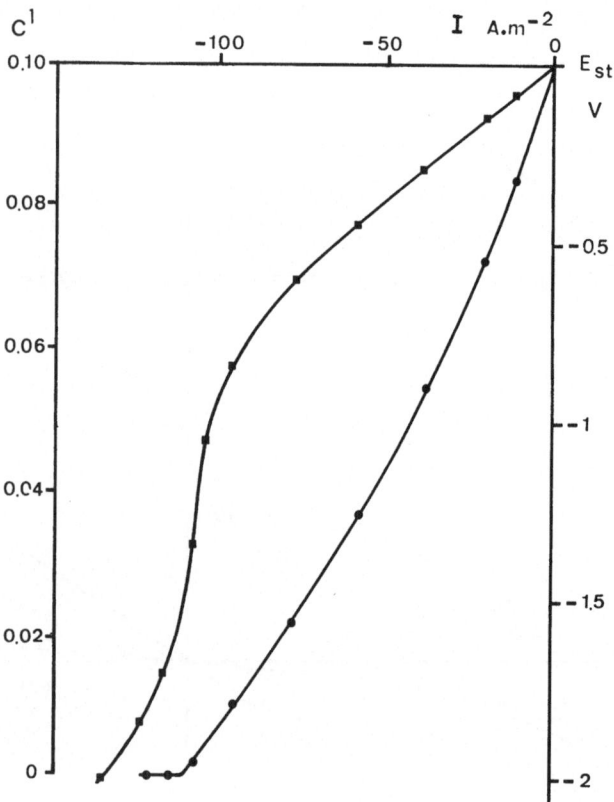

Fig. 8. Examples of current-voltage curves (■) and of the variation in the interfacial concentration, (●) in natural, vertical convection. Only interface (2) is polarized; interface (1) is strongly stirred by flowing solution at 25 m s^{-1}; profile (1); membrane K without gel in 0.1 M NaCl at 25°.

membrane resistance. This overvoltage appears suddenly and helps prove that the local interfacial concentrations tend towards zero almost simultaneously at all the points on the membrane.

The non-ohmic potentials E_d^n, $\Delta\bar{E}$ and E_s^n constitute a small part of the total potential drop and show a nearly ohmic character at low degrees of polarization. We can thus conclude somewhat differently from Meares $et\ al.$ [19] that the observation of ohmic behavior of the system cannot justify by itself the conclusion that negligible concentration polarization exists.

Finally, it was not difficult to obtain from these measurements values of ΔE, $\Delta\Pi_s$

and $\Delta\mu_s$ as a function of current density, and with gratifying precision. Further results are given in the following communication (48).

The influence of current density on fluxes, more particularly on water and ion transport numbers has been reviewed [30], but often the concentration polarization phenomena are neglected or receive only qualitative attention. In all cases the independence of t_v and t_1 with I were evident (Figure 9), yet the numerical values obtained are quite different according to the profile studied. One usually seeks to determine transport numbers at both $\Delta\Pi_s=0$ and at $\Delta P=0$, usually obtained using the Equi method. It must be restated that the independence of transport numbers of I cannot be considered proof of the absence of polarization during flux measurements. Our study carried out under well defined hydrodynamic conditions confirms previous results on water transport numbers [6, 9, 47]. For a more complete interpretation of

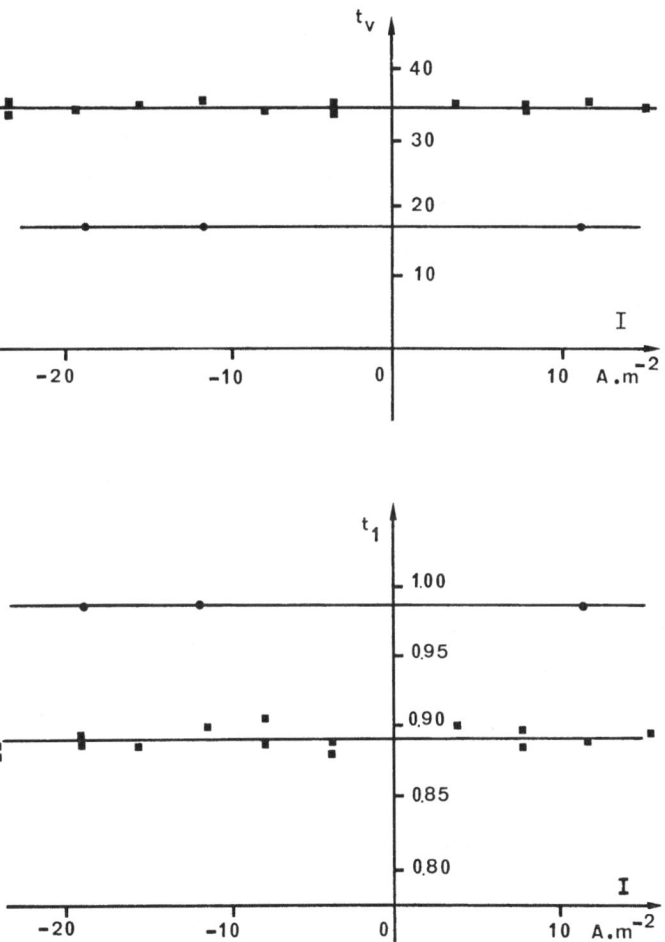

Fig. 9.   Volumic and cation transport numbers with: profiles (1) and (2) (■); profiles (3) and (4) of the Equi method (●) with $C_0{}^n = 0.03, 0.05$ and 0.15 M. The relative error in t_1 is 1%, in t_v 2%; membrane K, gel thickness 1.09 mm at $25°$.

these results, one needs the phenomenological coefficients; these are reported on in our next contribution [48].

As was observed earlier, the slope of the interfacial concentration – time curve is dependent upon fluxes through the membrane before as well as after current interruption (20). Table II compares results obtained by concentration regulation and by relaxation for two different types of profile. The uncertainties in p measured by relaxation are $<0.1\%$ and in C^n about 0.2%. Values of J_s^0, J_v, J_v^0 are given with a precision of $\pm 5\%$, leading to J_s values of a precision about $\pm 1\%$.

TABLE II

Comparison of J_s obtained independently from relaxation and regulating measurements

I	t_1^0	p	$10^9 D$	$10^8 J_v$	C^n	$10^8 J_v^0$	$10^5 J_s^0$	$10^5 J_s^a$	$10^5 J_s^b$
				Profile (1), $C = C_0^n = 0.1\ M$					
-3.87	0.386	0.72	1.49	-2.4	86.6	-1	0.4	-3.54	-3.57
-7.74	0.386	1.34	1.49	-4.9	73.6	-1.8	0.8	-7.11	-7.14
-11.6	0.387	2.03	1.50	-7.3	60.1	-2.6	1.1	-10.8	-10.7
-15.5	0.388	2.74	1.51	-9.7	46.8	-4.0	1.5	-14.4	-14.3
-19.3	0.389	3.43	1.52	-12.2	32.9	-5.0	2	-17.9	-17.8
-23.2	0.390	4.01	1.53	-14.6	18.9	-6.3	2.3	-21.1	-21.4
-25.2	0.392	4.40	1.54	-15.8	11.7	-6.7	2.5	-23.1	-23.2
				Profile (4), $C_0^n = 0.05\ M$, $C = 0.1\ M$					
-1.93	0.388	0.335	1.50	-3.5	52.3	-3.4	1.3	-0.64	-0.70
$+3.87$	0.387	-0.693	1.49	-0.55	73.4	-1.8	0.7	$+4.71$	$+4.80$
$+7.74$	0.386	-1.35	1.49	1.5	87.1	-1.0	0.3	8.24	8.25
$+11.61$	0.385	-2.00	1.48	3.7	100.7	0	0	11.8	11.8
$+15.48$	0.385	-2.66	1.48	5.3	114.5	0.8	-0.5	15.7	15.5

[a] by relaxation
[b] by regulation, with mean square values reported

The excellent agreement between the two methods proves that the assumptions made in the calculation of the diffusion-relaxation process are reasonable. In particular, the changes in J_s^0 and J_v^0 with t after current interruption did not perceptibly influence the slopes.

This new method has several advantages over the usual measurements of J_s in that it is rapid enough to follow the relaxation of the overpotential after current interruption once J_s^0, J_v, J_v^0 are known. *It can be used even at very low concentrations and therefore at very low fluxes since the measured potential depends only on the concentration profile and not on the magnitude of the flux.* However, if a precision of 1% is desired, this method is valid at present only with the use of a gel.

7. Conclusions

The methodology for the control and measurement of an ion-exchange membrane – polarization film system, one which allows the measurement of forces (except ΔP) at the membrane boundaries, are described and tested by experiment. The use of an agar gel allows for the precise definition of the polarization layer and of interfacial concentrations and eliminates the usual hydrodynamic limitations which obtain. All of the experimental results obtained show that a high degree of uniformity in the conditions prevailing at the membrane surface was attained as attested by: current-voltage curves with a distinct polarographic plateau; a linear variation of interfacial concentration with current density; the sudden appearance of an overvoltage showing water-splitting. We conclude, therefore, that each point on and in the membrane undergoes the same fluxes and forces.

A theory of the diffusion-relaxation process is presented, one which takes account of both the influence of an electro-osmotic flux on the concentration profile in the polarization layer and of the osmotic retrodiffusion during relaxation. This treatment leads not only to values of the interfacial concentration but also to the thickness of the polarization layer and the salt flux through the membrane. The profile analysis method reported previously [10] was developed herein and our results confirm that the Equi method does allow the measurement of fluxes and, therefore, of volume and ionic transport numbers when there is no difference in osmotic pressure across the membrane.

This methodology leads to values of the forces and, in particular, of the potential difference between the membrane faces. It can be used in any condition of the concentration profile. Further, detailed results and their employment in the general framework of irreversible thermodynamics are given in Paper II [48].

References

1. Mackay, D. and Meares, P.: *Trans Farad. Soc.* **55**, 1221 (1959).
2. Paterson, R. and Gardner, C. R.: *J. Chem. Soc.* A 2254 (1971).
3. Wills, G. B. and Lightfoot, E. N.: *Ind. Eng. Chem.* **5**, 114 (1966).
4. Lakshminarayanaiah, N. and Brennen, K. R.: *Electrochim. Acta* **11**, 949 (1966).
5. George, J. H. B. and Courant, R. A.: *J. Phys. Chem.* **71**, 246 (1967).
6. Bourdillon, C., Demarty, M., and Sélégny, E.: *J. Chim. Phys.* **71**, 6 (1974).
7. Bary, P. H. and Hope, A. B.: *Biophysical J.* **9**, 700–729 (1969).
8. Kobatake, Y. and Kamo, N.: *4th International Symposium on Fresh Water from the Sea* (ed. by E. and A. Delyannis), Vol. 3, p. 91 (1973).
9. Brun, T. S. and Vaula, D.: *Ber. Bunsenges Physik Chem.* **71**, 824 (1967).
10. Metayer, M., Bourdillon, C. and Sélégny, E.: *Desalination* **13**, 2 (1973).
11. Cooke, B. A.: *Electrochim. Acta* **3**, 307 (1961), **4**, 179 (1961).
12. Levich, V. G.: *Physicochemicals Hydrodynamics*, Prentice Hall, New York 1962.
13. Bourdillon, C., Metayer, M. and Sélégny, E., *J. Chim. Phys.* **5**, 722 (1973).
14. Bourdillon, C., Metayer, M. and Sélégny, E., *J. Chim. Phys.* **5**, 788 (1974).
15. Crank, J.: *The Mathematics of Diffusion*, Oxford at the Clarendon Press, 1967.
16. Metayer, M. and Bourdillon, C.: *J. Chim. Phys.*, to be published.
17. Kedem, O. and Katchalsky, A.: (1963).

18. Meares, P., Thain, J. F., and Dawson, D. G.: in G. Eisenman (ed.), *Membranes*, Marcel Dekker, New York, 1972.
19. Foley, T., Klinowski, J., and Meares, P.: *Proc. Roy. Soc.* **A 336**, 327 (1974).
20. Spiegler, K. S.: *Trans Farad. Soc.* **54**, 1408 (1958).
21. Demarty, M. and Sélégny, E.: *C. R. Acad. Sci.* **C 276**, 1549 (1973).
22. Barrer, R. M. Barie, J. A., and Rogers, M. G.: *Trans Farad. Soc.* **58**, 2473 (1962).
23. Sosipatrov, N. and Krapivin, P.: *Izv. Sibiv., Otdel. Akad. Nauk* **3**, 150 (1967).
24. Mc-Hardy, W. J., Meares, P., Sutton, A. H., and Thain, J. F.: *J. Colloid Interfare Sci.* **29**, 116 (1969).
25. Demarty, M., Maurel, A., and Sélégny, E.: *J. Chim. Phys.* **6**, 811 (1974).
26. Koefoed-Johnsen, V. and Ussing, H. H.: *Acta Phys. Scand.* **28**, 60 (1953).
27. Nims, L. F.: *Science* **137**, 130 (1962).
28. Kedem, O. and Essig, A.: *J. en Physiol.* **48**, 1047 (1965).
29. Meares, P. and Sutton, A. H.: *J. Colloid Interface Sci.* **28**, 118 (1968).
30. Lakshminarayanaiah, N.: *Transport Phenomena in Membranes*, Academic Press, New York 1969.
31. Robinson, R. A. and Stockes, R. H.: *Electrolyte Solutions*, Butterworth, London 1959.
32. Spiegler, K. S.: *Desalination* **9**, 367 (1971).
33. For example Lorimer, J. W.: *Disc. Faraday Soc.* **21**, 198 (1956).
34. Kwak, J. C. T.: *Desalination* **11**, 61 (1972).
35. Olsztajn, M.: *Thesis*, Paris 1969.
36. Kressman, T. R. E. and Tye, F. L., *Disc. Faraday Soc.* **42**, 279 (1959).
37. Kindly furnished by Dr. E. Korngold, Negev Institute.
38. Guillou, M.: Thesis, Paris 1968.
39. Boari, G., Lacava, G., Merli, C., Passino, R., and Tiravanti, G.: *Proceedings of the 4th International Symposium on Fresh Water from the Sea* (ed. by A. and E. Delyannis), Athens 1973.
40. Cooke, B. A. and Van der Valt, S. J.: *Electrochim. Acta* **5**, 216 (1961).
41. Khedr, G. and Varoqui, R.: *C. R. Acad. Sci.* **C 275**, 1185 (1972).
42. Mas, L. J., Pierrard, P. M., Prax, P. A., and Sohm, J. C., *Desalination* **7**, 285 (1970).
43. Sonin, A. A., and Probstein, R. F.: *Desalination* **5**, 293 (1968).
44. Block, M. and Kitchener, J. A.: *J. Electrochim. Soc.* **113**, 947 (1966).
45. Oda, Y. and Yawataya, T.: *Desalination* **5**, 129 (1968).
46. Takemoto, N.: *Nippon Kagaka Kaishi* **1**, 44 (1973).
47. Ripoll, C., Demarty, M., and Sélégny, E.: *J. Chim. Phys.* **6**, 828 (1974).
48. Bourdillon, C. and Sélégny, E.: this volume, p. 207.

MEASUREMENT OF FLUXES AND FORCES AT THE SURFACE OF CATION-EXCHANGE MEMBRANES UNDER CONDITIONS OF CONTROLLED POLARIZATION

II: *Application of Measurement – Use of Phenomenological Coefficients*

CHRISTIAN BOURDILLON* and ERIC SÉLÉGNY

Laboratory of Macromolecular Chemistry, ERA 471, Faculty of Sciences and Technology of the University of Rouen, 76130 Mont Saint Aignan, France

Abstract. The nine differential phenomenological coefficients of a non-commercial cation-exchange membrane were measured in $0.1\ M$ sodium chloride and the Onsager reciprocal relations verified. Since these results were obtained from measurements of fluxes and forces at the membrane interface without the intervention of a hydrostatic pressure, the technical difficulties connected with the compression of the membrane were thus overcome. Hydrodynamic influences at these interfaces upon volumetric and ionic transport numbers could readily be simulated thereby, and it was shown that the existence of a transport number independent of current density does not constitute full proof of complete depolarization.

List of Symbols

$a_{\pm}, a_{\pm}^{n}$	mean activity of NaCl solutions of concentration C and C^{n}.
C	concentration on side (1)
C^{n}	interfacial concentration on polarization side, where $n = 1, 2, 3, 4$ according to profiles
C_0^{n}	concentration, side (2)
C_s	mean thermodynamic concentration defined by (10)
D	diffusion coefficient in solution
e	membrane thickness
ΔE	potential difference across two membrane interfaces
I	current density
J_i	flux density of species i (1, 2, or 3 for counter-ion, co-ion or water)
J_s	flux density of NaCl through membrane
J_v	volumetric flux through membrane
K, K'	constants defined in (24) and (27)
l_{ij}	'direct' phenomenological coefficients
L_{ij}	'practical' phenomenological coefficients
M, N, Q	constants defined in (20) and (23)
P	hydrostatic pressure
r_{ij}	generalized friction coefficients
R_{ij}	practical friction coefficients
t_i	transport number of species i
t_v	volumic transport number
V_3, V_s	partial molar volume of water and salt
x	distance to the membrane
X_i	negative gradient of electrochemical potential of species i.
Δ	operator, difference between sides 1 and 2
$\Delta \Pi_s$	difference in osmotic pressure
$\Delta \mu_s$	difference in chemical potential of salt
ϕ	osmotic coefficient (molar scale)

* Present address: Université Technologique de Compiègne, Lab. Technologie Enzymatique, Compiègne, France.

δ polarization layer thickness
η_i electrochemical potential of i
θ constant from (24)

1. Introduction

Meares *et al.* [1, 2] noted recently the lack of experiments on ion-exchange membranes which employed the formulations of irreversible thermodynamics. Although the relevant theoretical developments are now quite complete (see for example [3–6]), this situation is explained by the experimental difficulties which are found; the methodology for all these coefficients has appeared but recently. Even so, one must recognize those techniques which do not necessitate any specific assumptions in the calculations of these coefficients. In this contribution, we are interested only in those methods by which one makes specific measurements of coefficients.

Two approaches are possible: to measure the coefficients of the membrane itself; to consider the membrane and its solutions together as a thermodynamic entity whose parameters are to be determined. In the first case, it is absolutely necessary to obtain the interfacial parameters and, in particular, the forces. By the use of combinations of coefficients Foley *et al.* [2] avoided the need to measure certain interfacial forces, but were hampered by hydrostatic pressure variations which necessarily introduced experimental complications, in particular, a compression of the membrane. In the second case, one must be content to measure the parameters of the entire system between measuring electrodes [7]. Unfortunately, in this case comparisons between two membranes or even of two solution concentrations across the same membrane are not straightforward. However, by varying the osmotic pressure by the addition of a non-ionic solute the use of a hydrostatic pressure can be avoided [7, 11] provided ionic activities are appropriately corrected.

We have proposed [12] methods which allow one to measure fluxes and forces at the membrane interfaces. These quantities were, in this study, measured under hydrodynamic conditions such that the boundary layer was uniform across the entire membrane surface. It is demonstrated herein that the nine L_{ij} coefficients can be obtained directly from these experimental results.

2. Mathematical Formalism

The system studied is composed of a cation-exchange membrane interposed between two solutions of a $1-1$ electrolyte (here NaCl of concentration C'' and C) at the membrane-solution boundary and in the solution phase. The assumptions for calculations are: the fluxes across the membrane are constant; the system is isotropic and isothermal; a mechanical equilibrium exists which allows the application of the Prigogine principle [13], the chosen reference frame being the membrane; fluxes are perpendicular to the membrane, making the system one-dimensional; the system is sufficiently near to equilibrium so that fluxes are linear functions of forces.

Under these conditions the phenomenological relations between fluxes of each species and the conjugated forces take the form:

$$
\begin{aligned}
J_1 &= l_{11}X_1 + l_{12}X_2 + l_{13}X_3 \\
J_2 &= l_{21}X_1 + l_{22}X_2 + l_{23}X_3 \\
J_3 &= l_{31}X_1 + l_{32}X_2 + l_{33}X_3
\end{aligned}
\tag{1}
$$

Subscripts 1, 2, 3 correspond to the counter-ion, co-ion and water, respectively, and

$$
x_i = -\frac{d\eta_i}{dx}
\tag{2}
$$

The transformation to generalized friction coefficients may be obtained by matrix inversion [13, 14]. Let

$$
x_i = \sum_{k=1}^{3} r_{ik}J_k
\tag{3}
$$

If the reciprocal relations of Onsager are verified, the number of coefficients is reduced from 9 to 6.

$$
l_{ik} = l_{ki} \quad \text{and} \quad r_{ik} = r_{ki} \quad (i \neq k)
$$

The conditions of entropy production require that the l_{ii} and r_{ii} coefficients as well as the principal minors must be positive.

Since it is impossible to control each electrochemical gradient, the experimental determination of coefficients l_{ik} is never based directly upon these equations.

A certain number of indirect methods have been tested as outlined in a recent review [9]. By appropriate combinations of coefficients, measurable quantities appear but the validity of the reciprocal relations and also certain complementary assumptions must be assumed from the start so as to be able to complete the system [5, 10].

Kedem and Katchalsky [4] proposed the use of phenomenological equations in which fluxes and forces have physical meanings obtainable from direct experiments. Let us introduce, also, the thickness of the membrane to make comparisons with other measured coefficients easier.

$$
\begin{aligned}
J_v &= L_{11}\frac{(\Delta P - \Delta\Pi)}{e} + L_{12}\frac{\Delta\Pi}{eC_s} + L_{13}\frac{\Delta E}{e} \\
J_s &= L_{21}\frac{(\Delta P - \Delta\Pi)}{e} + L_{22}\frac{\Delta\Pi}{eC_s} + L_{23}\frac{\Delta E}{e} \\
I &= L_{31}\frac{(\Delta P - \Delta\Pi)}{e} + L_{32}\frac{\Delta\Pi}{eC_s} + L_{33}\frac{\Delta E}{e}
\end{aligned}
\tag{4}
$$

Then the practical friction coefficients R_{ij} are obtained by inversion, thus expressing forces as functions of fluxes

$$
\frac{\Delta P - \Delta\Pi}{e} = R_{11}J_v + R_{12}J_s + R_{13}I
$$

$$\frac{\Delta\Pi}{eC_s} = R_{21}J_v + R_{22}J_s + R_{23}I \tag{5}$$

$$\frac{\Delta E}{e} = R_{31}J_v + R_{32}J_s + R_{33}I$$

We then have for the fluxes,

$$J_v = \bar{V}_s J_s + J_3 \bar{V}_3 \tag{6}$$

It should be noted that $\bar{V}_s$ and $\bar{V}_3$ are not strictly independent of concentration; however, in practice these corrections can be neglected [6].

$$I = F(J_1 - J_2) \tag{7}$$

$$J_s = J_1 \tag{8}$$

For the forces in the particular case where $\Delta P = 0$, we obtain:

$$\Delta P - \Delta\Pi = -\Delta\Pi_s = -\frac{2RTM_3}{\bar{V}_3}(\phi C - \phi''C'') \tag{9}$$

$$\frac{\Delta\Pi_s}{C_s} = \Delta\mu_s = 2RT \ln \frac{a_{\pm}}{a'^n_{\pm}} \tag{10}$$

$$\Delta E = -\frac{\Delta\eta_2}{F} \tag{11}$$

with ΔE defined as the potential difference between two electrodes reversible to the anions placed at the membrane interface.

Knowing a set of coefficients l_{ij}, r_{ij}, L_{ij} and R_{ij} permits one, in theory, to obtain all of the others by matrix inversion. We will see in a later section that this procedure is strongly limited by error duplication.

The coefficients previously defined are dependent upon solution concentration, with a series of coefficients corresponding to each. A problem arises in obtaining certain coefficients (for example L_{2j}) which require that a concentration gradient be imposed across the membrane. Thus, one may distinguish between two types of coefficient variations: a general variation with C where the membrane is not subjected to a concentration gradient; the dependence of the coefficients upon the mean concentration C_s.

In order to measure coefficients corresponding to a given concentration C, one can either assume that these are constant in the domain of C_s [9, 10] (mean coefficients), or introduce differential coefficients [2, 6, 7] where C_s tends toward the interfacial concentration C.

3. Experimental

All calculations make use of fluxes and forces measured according to the techniques described in detail in the previous paper [12] and in the absence of hydrostatic pressure. For membrane K [15], the results of Figures 1–4 correspond to the four types of concentration profiles previously defined in Figure 1 of [12] and recalled in the circles of Figures 1–4. The concentration was 0.1 M NaCl. The precision obtained for the measured fluxes and forces are (with all % in ±): J_v, 2%; J_s, 1%; I, 0.5%; C^n, 0.2%; ΔE, 2%.

4. Discussion

Values of L_{ij}: The solution of the three systems of equations is possible, in theory, only if the coefficients are independent of C_s and where knowledge of all fluxes and forces at the interfaces is available. We found this way impractical because of error propagation. We prefer to create situations where the number of unknowns is reduced to only one or two.

Approach to L_{i3} Values: These coefficients are considered the ones most easily attainable since it suffices if $\Delta\mu_s$, ΔP and $\Delta\Pi=0$. Given the case where the influence of concentration polarization upon flux can be considered negligible (see subsequent discussion), it is sufficient to measure the volume flux, salt flux and the conductivity

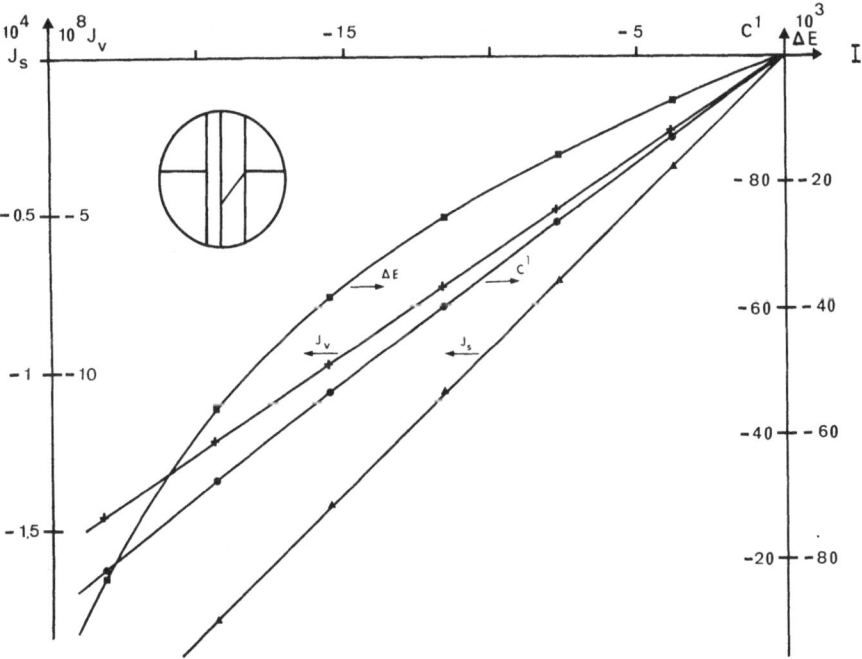

Figs. 1–4. Fluces and forces as a function of current I(amp m^{-2}) with membrane K having a gel thickness 1.09×10^{-3}m and at 25° *for four concentration profiles*: 1 and 2, 0.1 M sodium chloride; 3, 0.10 M on (1), 0.15 M on (2); 4, 0.10 M on (1), 0.05 M on (2). The fluxes are: J_v, volume flux in m s^{-1}; J_s flux of cation in moles m^{-2} s^{-1}; ΔE, potential difference at membrane boundary in volts. The interfacial concentration C^n (n designating the profile type) is on the polarized side in moles m^{-3}. The scales of the ordinates and abcissae are designated at the origins; for example, for the left side of the rt. hand ordinate, the scale is in C^1, on its rt. side it is $\Delta E \cdot 10^3$; arrows indicate the ordinates corresponding to the curve.

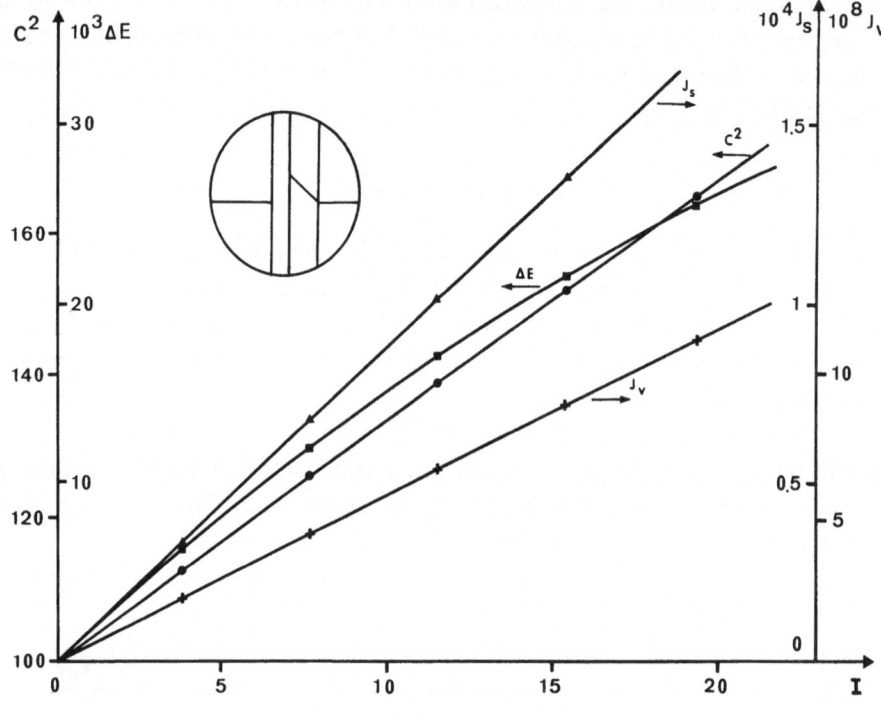

Fig. 2.   See caption to Figure 1.

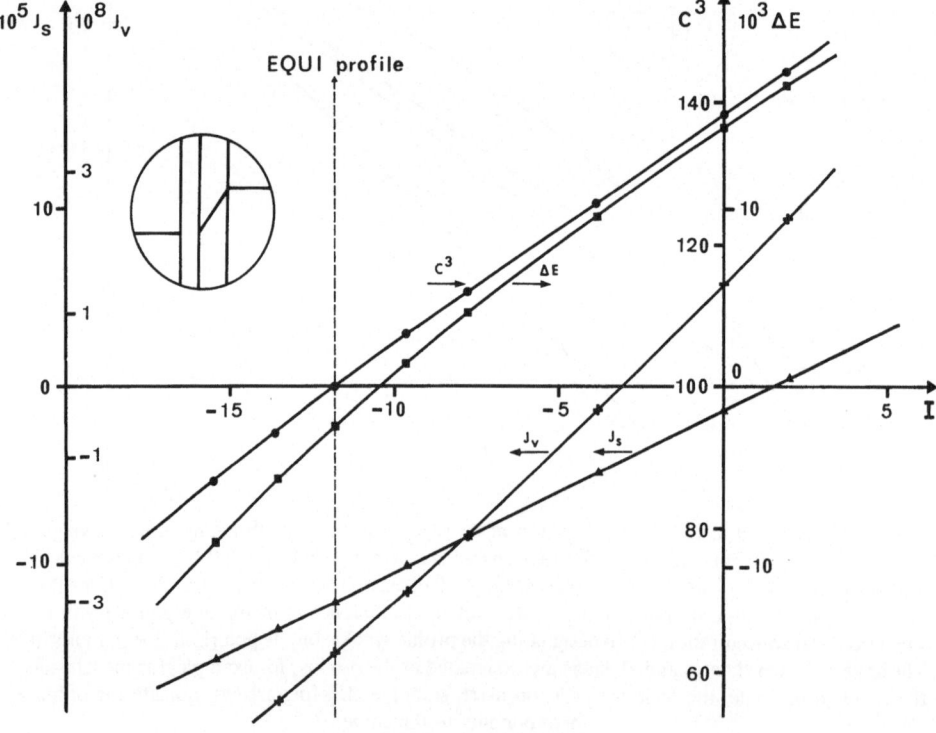

Fig. 3.   See caption to Figure 1.

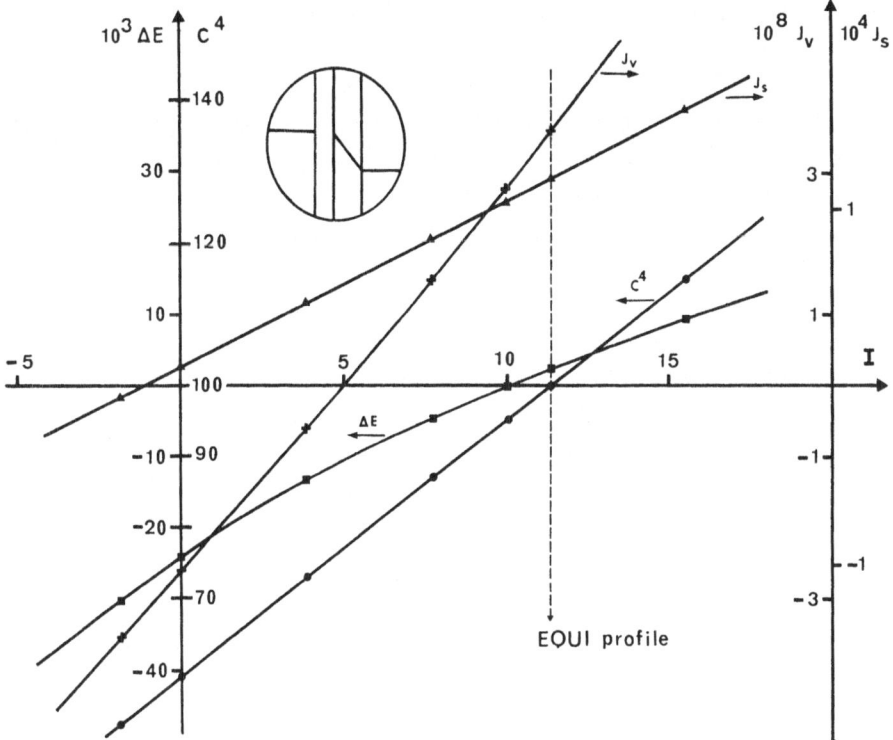

Fig. 4. See caption to Figure 1.

of the membrane in order to obtain the values of L_{i3}. *Here, the use of the Equi method* [12] *creates situations where $\Delta\Pi_s$ and $\Delta\mu_s$ are equal to zero at the membrane boundaries.* Figures 3 and 4 give two examples of such situations in which the following, simplified relations hold, all under the Equi condition:

$$J_v = L_{13}\,\Delta E/e; \quad J_s = L_{23}\,\Delta E/e; \quad I = L_{33}\,\Delta E/e \qquad (12)$$

From these results:

$$L_{13} = 4.03 \pm 0.06 \times 10^{-9}\ \mathrm{V^{-1}\ m^2\ s^{-1}};$$
$$L_{23} = 1.30 \pm 0.02 \times 10^{-5}\ \mathrm{mole\ V^{-1}\ m^{-1}\ s^{-1}};$$
$$L_{33} = 1.27 \pm 0.03\ \Omega^{-1}\ \mathrm{m^{-1}}.$$

The close agreement between these results on the two types of Equi profiles and the results of the first study [12] show a good control of concentration profile by this method.

Obtaining L_{i1} and L_{i2}: We use here the results of studies wherein $\Delta\Pi_s$ and $\Delta\mu_s$ are varied (profiles 1 and 2). From (4) one can easily demonstrate that if $\Delta P=0$,

$$\frac{J_v - L_{13}\Delta E/e}{\Delta\mu_s} = -\frac{L_{11}C_s}{e} + \frac{L_{12}}{e} \qquad (13)$$

$$\frac{J_s - L_{23}\Delta E/e}{\Delta\mu_s} = -\frac{L_{21}C_s}{e} + \frac{L_{22}}{e} \tag{14}$$

$$\frac{I - L_{33}\Delta E/e}{\Delta\mu_s} = -\frac{L_{31}C_s}{e} + \frac{L_{32}}{e} \tag{15}$$

Here L_{i3} are known and by plotting these three curves as functions of C_s, L_{i1}/e can be obtained from the slope of the tangent at $C_s = C$, and L_{i2}/e from its value on the ordinate at the origin of these tangents (see Figure 5). The variation of these coeffi-

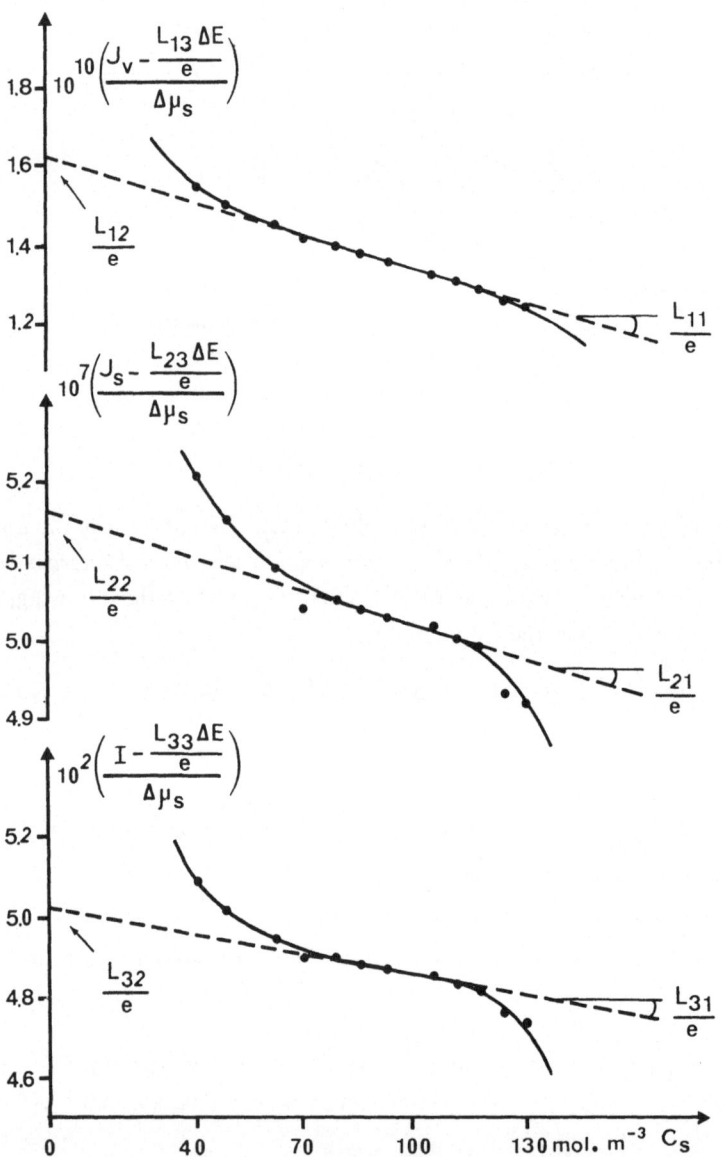

Fig. 5.   Graphical determination of the L_{i1} and L_{i2} coefficients. Tangents at $C_s = 100$ are practically confounded with curves between $70 < C_s < 120$ moles m⁻³.

cients with the imposed concentration gradient is clearly established in the divergence between tangent and curve at extreme C_s values. The coefficients measured under these conditions are differential coefficients, characteristic of $C = 0.1 \ M$. Nevertheless, one is able to establish that these coefficients remain a good approximation for $0.07 \ M < C_s < 0.12 \ M$.

The slopes and the ordinates at the origin lead to the following results:

$$L_{11} = 6.8 \ \pm 0.6 \quad 10^{-17} \ \text{m}^4 \ \text{N}^{-1} \ \text{s}^{-1}$$
$$L_{12} = 4.15 \pm 0.1 \quad 10^{-14} \ \text{mole m N}^{-1} \ \text{s}^{-1}$$
$$L_{21} = 3.8 \ \pm 0.4 \quad 10^{-14} \ \text{mole m N}^{-1} \ \text{s}^{-1}$$
$$L_{22} = 1.34 \pm 0.02 \ 10^{-10} \ \text{mole}^2 \ \text{m}^{-2} \ \text{N}^{-1} \ \text{s}^{-1}$$
$$L_{31} = 4.2 \ \pm 0.2 \quad 10^{-9} \ \text{m}^2 \ \text{v}^{-1} \ \text{s}^{-1}$$
$$L_{32} = 1.30 \pm 0.02 \ 10^{-5} \ \text{mole v}^{-1} \ \text{m}^{-1} \ \text{s}^{-1}$$

Comparing these values with L_{i3} shows that *Onsager's reciprocal relations apply to these results in the limits of experimental error.* In agreement with the general treatment of Snell and Stein [16], the membrane can be thus considered to be *phenomenologically macroscopically homogeneous* if one examines the fluxes perpendicular to its surface.

In Table I the six values of L_{ij} show that the conditions of positive entropy production are respected. Of course, if one *assumes* the Onsager relations, the determination of the six coefficients can be simplified and it is not necessary to use the Equi method.

Determination of Other Coefficients: Here there are two possibilities; to utilize new kinds of experimental data or to establish the necessary matrix transformations for the calculation of R_{ij}, l_{ij} and r_{ij} from the previously measured L_{ij} values. For the experimental determination of R_{ij}, Figure 3 and 4 show data previously unexploited where one of the fluxes J_v, J_s or I cancel. In these cases (5) reduces to:

$$
\begin{aligned}
J_v &= 0 & \Delta E &= R_{32}J_s + R_{33}I \\
J_s &= 0 & \Delta E &= R_{31}J_v + R_{33}I \\
I &= 0 & \Delta E &= R_{31}J_v + R_{32}J_s
\end{aligned}
\tag{16}
$$

By knowing the parameters for the corresponding situations and recognizing their independence of C_s in the stipulated zones, it is possible to calculate the mean terms for R_{3i}. Unfortunately, it has been conceded that fluctuations of the order of 2% in the numerical values for fluxes and forces induce, in turn, considerable fluctuations in the calculated values of the coefficients, ones that can produce errors of more than 100%. Therefore, this method is applicable only when the precision of measured fluxes and forces is high.

Transformations of L_{ij}: The transformation equations are well known [6]. In Table I the results of R_{ij}, l_{ij} and r_{ij} are presented as calculated by the direct method from L_{ij}. The fluctuations caused by the use of the minimum or maximum value of each co-

TABLE I

Numerical values (in S.I. Units) and Uncertainties in differential coefficients, with R_{ij}, l_{ij} and r_{ij} values calculated from L_{ij} by matrix transforms, for membrane K in 0.1 M sodium chloride at 25°

ij	11	12	13	22	23	33
L_{ij}	$6.8 \pm 0.6 \times 10^{-17}$	$4.15 \pm 0.1 \times 10^{-14}$	$4.03 \pm 0.06 \times 10^{-9}$	$1.34 \pm 0.02 \times 10^{-10}$	$1.30 \pm 0.02 \times 10^{-5}$	1.27 ± 0.03
R_{ij}	$1.8 \pm 0.4 \times 10^{16}$	$-1.5 \pm 0.8 \times 10^{13}$	$9 \pm 8 \times 10^{7}$	$1.09 \pm 0.05 \times 10^{12}$	$-1.17 \pm 0.05 \times 10^{7}$	$1.14 \pm 0.05 \times 10^{2}$
l_{ij}	$1.34 \pm 0.02 \times 10^{-10}$	$-7 \pm 8 \times 10^{-13}$	$2.20 \pm 0.05 \times 10^{-9}$	$9 \pm 7 \times 10^{-12}$	$2 \pm 3 \times 10^{-11}$	$2.0 \pm 0.2 \times 10^{-7}$
r_{ij}	$9.1 \pm 0.2 \times 10^{9}$	$8 \pm 3 \times 10^{9}$	$-9.8 \pm 0.8 \times 10^{8}$	$1.06 \pm 0.01 \times 10^{12}$	$-1.6 \pm 0.9 \times 10^{8}$	$5.9 \pm 0.6 \times 10^{6}$

efficient upon others are specified. A systematic study of error duplications in the matrix calculations is left for a later report. The variation in certain coefficients is quite great; thus, little interest is to be found in the transformations. However, even though one continues to improve the precision of the coefficient L_{ij}, it appears doubtful that the precision could be increased by a factor of 20 or 30 as seems necessary. The errors produced depend upon the numerical values themselves and therefore on the type of membrane.

This calculation applied to the L_{ij} coefficients of Foley et al. [2] led to still greater discrepancies* in the values of certain coefficients due to experimental inaccuracies. It should be pointed out that the passage from l_{ij} to L_{ij} is achieved with virtually no additional error. Thus, in the future it will be necessary to define the succession of optimal operations and how to express the experimental results.

Applications of These Data: A series of differential coefficients L_{ij} capable of application in experimental situations where $C=0.1$ M was established. Several examples of these, like the calculation of membrane potentials or fluxes from forces can be made; a simulation of the influence of concentration polarization upon the fluxes is also possible. The membrane potential when $I=0$ as measured with reversible chloride electrodes is readily expressed from [4] by

$$(\Delta E)^{I=0}_{\Delta P=0} = \frac{\Delta \Pi_s}{L_{33}}\left(L_{31} - \frac{L_{32}}{C_s}\right) \tag{17}$$

Membrane potentials measured with $C=0.1$ M on side (1) and variable concentrations C'' of 0.01–0.3 M on side (2) are listed in Figure 6. Good agreement with values calculated from (17) confirms that the experimental situation corresponded to a linear relationship between fluxes and forces. This relationship was tested elsewhere also with success, but using transport numbers of water and ions [6, 17].

It is important to point out that variations in the coefficients L_{3i} with C_s do not appear in this case. All this would be justified if the three coefficients, without being constant, were to vary proportionately to C_s; however, a lack of sufficient experimental evidence does not as yet allow us to confirm this assumption.

With the exception of the Equi condition, the results of Figure 3 and 4 have not been used for the calculation of coefficients. It is possible to do so by comparing measured to calculated fluxes using Equation (4). A plot of three fluxes as a function of C_s was made in order to study the region in which the differential coefficients were valid (Figure 7). The gap between the calculated and experimental values is narrowed as C_s approaches 0.1 M. One cannot be certain whether these differences are due to variations in the coefficients with C_s or to a duplication of errors in the calculations.

---

* Before our calculation, an evident typographical error was corrected for the order of magnitude of the values of L_p and L_π of this paper.

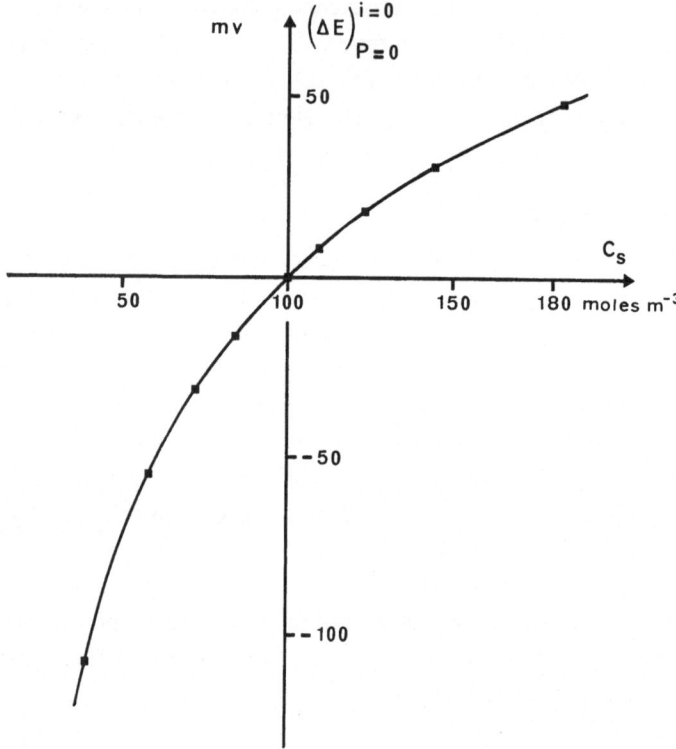

Fig. 6. Comparison of measured potential values with ones calculated (solid curve) from (17), with the bulk sodium chloride concentration on side (1) at 0.1 M, that on side (2) from 0.01–0.3 M with membrane K and without a gel layer, at 25° and with vigorous stirring ($\delta < 210^{-5}$ m).

These numerical errors, as previously observed in the matrix inversion of L_{ij} seem to constitute a limitation of the use of relations (4).

Influence of Polarization on Transport Numbers: Two phenomena influence water and transport numbers. First there is concentration polarization acting via the concentration gradient to produce retrodiffusion and osmosis. Second, there is a reduction in the migration of salt ions caused by the decomposition of water [18–20]. Our knowledge does not allow a treatment of the latter phenomena at this time, but we can consider the effects of concentration polarization. We have already presented a treatment [21] relating volume transport numbers to concentration polarization. That treatment was qualitative because we could not dispose of combinations of coefficients, but now we can derive expressions for t_v and t_1. From [4], J_v is expressed as a function of I when $\Delta P = 0$,

$$J_v = \left(\frac{L_{13}L_{31}}{L_{33}} - L_{11} \right) \frac{\Delta\Pi_s}{e} + \left(L_{12} - \frac{L_{13}L_{32}}{L_{33}} \right) \frac{\Delta\Pi_s}{eC_s} + \frac{L_{13}}{L_{33}} I \tag{18}$$

where

$$t_v = J_v F / V_3 I \tag{19}$$

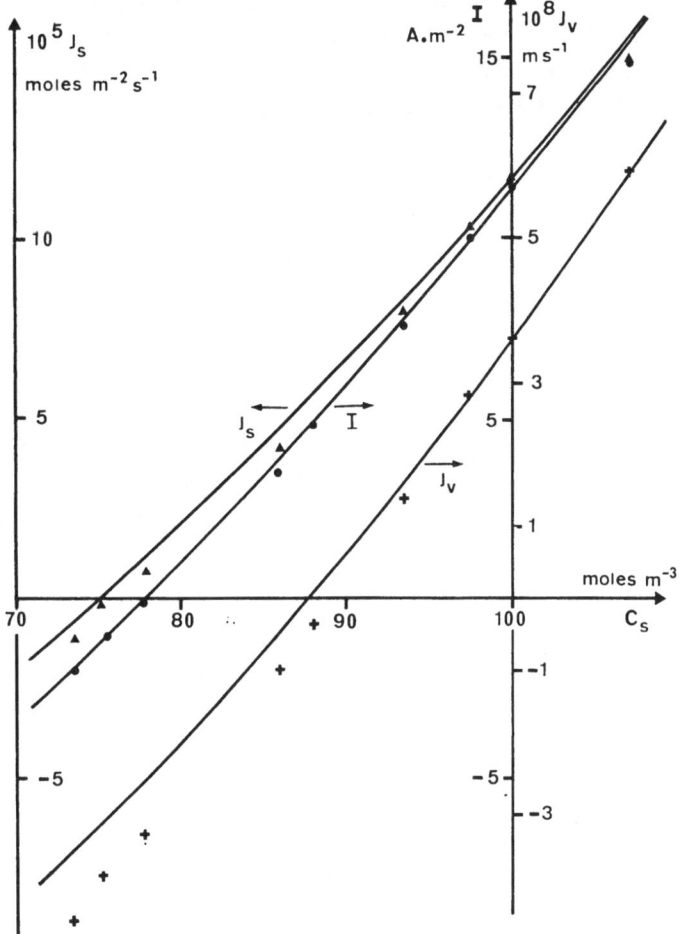

Fig. 7. Comparison of measured fluxes (full lines) and those calculated (points) from the L_{ij} coefficients of Table I and measured forces of Figure 4 for profile 4 with membrane K, having a gel thickness of 1.09×10^{-3} m at 25°.

$$t_v = \frac{F}{e\bar{V}_3}\left[\left(\frac{L_{13}L_{31}}{L_{33}} - L_{11}\right) + \frac{L_{12} - L_{13}L_{32}/L_{33}}{C_s}\right]\frac{\Delta\Pi_s}{I} + \frac{L_{13}F}{L_{33}\bar{V}_3} \qquad (20)$$

or else

$$t_v = M\frac{\Delta\Pi_s}{I} + N \qquad (21)$$

Here M is a parameter which is a function of both concentration C and the concentration gradient. On the other hand, N changes only with C. The 'real' transport number at zero concentration gradient given by parameter N is quite independent of current density (experimentally, this corresponds to the Equi condition).

The $M \Delta\Pi_s/I$ term correcting polarization is related to the type of convection at the interface. Except when there was a very high level of stirring, this term appeared in every profile situation other than Equi. This agrees with our previous conclusions [21, 22] which pointed out the importance of the term $(C-C'')/I$; $\Delta\Pi_s$ is proportional to $C-C''$ except for the osmotic coefficient corrections.

Counter-ion transport numbers can also be expressed, with

$$t_1 = J_s F/L \tag{22}$$

$$t_1 = \frac{F}{e}\left[\left(\frac{L_{23}L_{31}}{L_{33}} - L_{21}\right) + \frac{L_{22} - L_{23}L_{32}/L_{33}}{C_s}\right]\frac{\Delta\Pi_s}{I} + \frac{L_{23}F}{L_{33}} \tag{23}$$

regrouping the terms.

$$t_1 = \theta\frac{\Delta\Pi_s}{I} + Q. \tag{24}$$

As already seen, θ is dependent on both C and C_s while Q depends only on C. The magnitude of the term $\Delta\Pi_s/I$ of (21) and (24) is determined by the convection at the interfaces. Several types of convection were studied wherein $\Delta\Pi_s/I$ took different values.

The use of an agarose layer on the membrane [12] acts to maintain a constant boundary layer and thus amplifies polarization phenomena, increasing $\Delta\Pi_s/F$. Since the thickness of the boundary layer on the stirred, non-gel side is negligible (profile 1, [12]), C'' is proportional to the current density so M and θ remain constant with t_v and t_1 independent of current density. This has indeed been observed, with $t_v = 33.7$ and $t_1 = 0.89$ (cf. Figure 11 of [12]).

Table II shows excellent agreement in t_v but one not as good for t_1. Variations in

TABLE II

Calculated transport numbers and related parameters for membrane K, with gel thickness 1.09×10^{-3} m in 0.1 M sodium chloride and profile 1

I	$10^3 M$	$10^6\theta$	t_v	t_1
-3.87	-1.089	3.6	34.0	0.93
-7.74	-1.078	3.9	33.7	0.93
-11.61	-1.072	4.3	33.8	0.92
-15.68	-1.064	4.8	33.5	0.91
-19.35	-1.053	5.6	33.7	0.90
-23.21	-1.032	7.0	33.5	0.88

t_1 are due to ones in θ which is more sensitive to concentration gradients at the two interfaces of the membrane.

Vertical Convection: Here no external stirring is imposed and only convection induced by concentration gradients takes place [23], so the concentration gradient can be expressed as

$$C_0^n - C'' = KI^{4/5} \tag{25}$$

TABLE III

Caculated values for gel-free membrane under natural vertical convection

I	$10^3 M$	$10^6 \theta$	t_v	t_1
−6.05	−1.086	3.37	33.3	0.937
−14.0	−1.086	3.41	31.1	0.944
−23.0	−1.086	3.47	29.9	0.946
−33.5	−1.083	3.57	28.8	0.949
−44.5	−1.082	3.70	28.0	0.950
−56.0	−1.082	3.90	27.5	0.950
−67.4	−1.074	4.20	27.2	0.948
−79.5	−1.067	4.66	26.8	0.945

In Table III a previously measured value for K was used [21]. Calculated t_v and t_1 values are found to be a function of current density, reflecting a variation in $\Delta\Pi_s/I$ due to convection and because M and θ are influenced by concentration gradients.

Forced Convection: Under these conditions [24, 25] the mean thickness of the polarization layer is independent of current density and variations in interfacial concentration are proportional to I, at small values of I. Here $\Delta\Pi_s/I$ is virtually constant. Table IV shows t_v and t_1 calculated for an arbitrarily chosen thickness of

TABLE IV

Calculated values for gel-free membrane under forced convection ($\delta = 3 \times 10^{-5}$m)

I	C	C^n	$10^3 M$	$10^6 \theta$	t_v	t_1
−10	101.1	98.9	−1.087	3.36	18.1	0.985
−30	103.5	96.5	−1.087	3.36	18.1	0,985
−50	105.8	94.2	−1.087	3.36	18.1	0.985
−70	108.2	91.8	−1.087	3.36	18.1	0.985
−90	110.5	89.5	−1.086	3.37	18.1	0.984

3×10^{-5} m, corresponding to a convection velocity of 12 cm s^{-1}. The two interfaces are assumed equally polarized. At these current densities C_s hardly varies, M and θ are constant and constant transport numbers are found, as is evident also from the literature [21, 26, 27]. However, the fact that the transport number is independent of I is not sufficient evidence to assert that polarization has no influence on the fluxes.

Errors Due to Boundary Layers: Figure 8 shows that certain interpretations in the literature must be revised. If the Equi method is not used, it heretofore was necessary to ascertain the systematic error in calculated transport parameters introduced by polarization. A calculation of the systematic error is now proposed, relating it to the boundary layer thickness by the use of L_{ij} values. Here θ and M are postulated to be

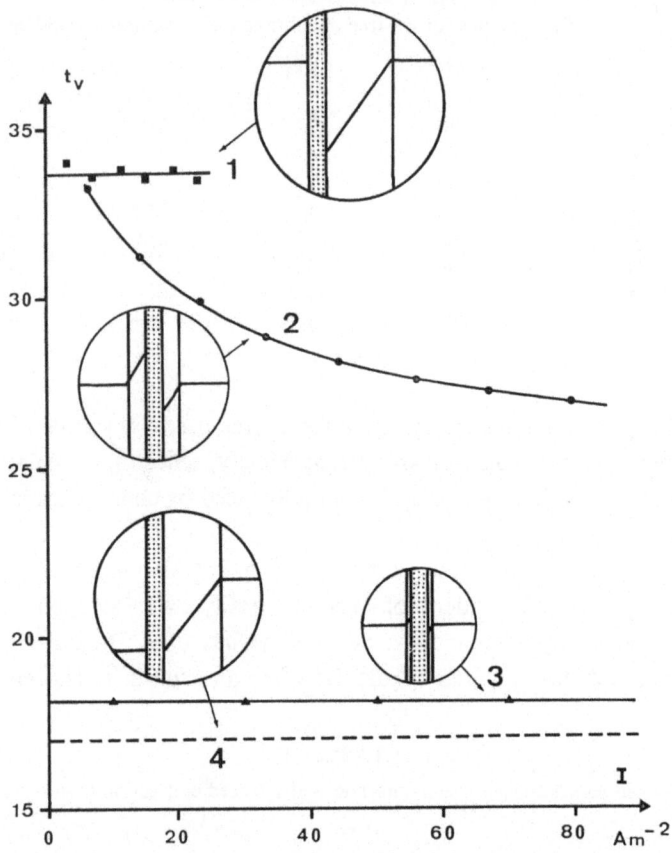

Fig. 8. Variation of volumic transport number with current density for different interfacial hydro-dynamic conditions with the boundary layer due to: 1, a gel layer 1.09×10^{-3} m thick ($\Delta\Pi_s/I$ constant); 2, both controlled by natural convection ($\Delta\Pi_s/I$ changes with I); 3, both controlled by forced convection ($\Delta\Pi_s/I$ constant); 4, Equi profile ($\Delta\Pi_s = 0$). Membrane K in 0.1 M sodium chloride used at 25°.

constant and the osmotic coefficient is also assumed independent of concentration. Let

$$\Delta\Pi_s = \frac{2\,RTM_3}{\bar{V}_3}\,\phi\,(C - C^n) \tag{26}$$

Then after the simplified summation of fluxes,

$$C^n - C = \frac{2I\delta\,(t_1 - t_1^0)}{DF} \tag{27}$$

Then let

$$\frac{\Delta\Pi_s}{I} = -\frac{4RTM_3\phi\,(t_1 - t_1^0)}{\bar{V}_3DF}\,\delta = K'\delta \tag{28}$$

The relative error on each transport number is then given by,

$$\frac{\Delta t_1}{t_1} = \frac{\theta K' \delta}{Q} \tag{29}$$

$$\frac{\Delta t_2}{t_2} = \frac{\theta K' \delta}{1 - Q} \tag{30}$$

$$\frac{\Delta t_v}{t_v} = \frac{M K' \delta}{N} \tag{31}$$

Since these errors are independent of current density, it is no longer necessary to employ very low current densities. These calculations were applied to our results and also to those of Foley et al. ([2], Table V). The uncertainties obtained depend strongly

TABLE V

Values of membrane parameters and systematic errors

	M	N	θ	Q	K'	$\Delta t_1/t_1$	$-\Delta t_v/t_v$	$-\Delta t_2/t_2$
Foley et al. [2]	-3.40×10^{-3}	41.6	2.25×10^{-5}	0.937	-3.76×10^7	$8.7 \times 10^4 \, \delta$	$3.1 \times 10^5 \, \delta$	$3.1 \times 10^6 \, \delta$
This study	-1.09×10^{-3}	17.0	3.36×10^{-6}	0.988	-3.89×10^7	$1.3 \times 10^4 \, \delta$	$2.5 \times 1 \times 10^5 \, \delta$	$1.1 \times 10^6 \, \delta$

upon the type of membrane. They are larger for the Zeo-Carb 315 membrane which is much more permeable, where the minimum boundary layer thickness (for which the uncertainties in t_v and t_1 are less than 1%) are: for t_v, $\delta < 0.3 \times 10^{-5}$ m; for t_1, $\delta < 1.2 \times 10^{-5}$ m.

5. Conclusions

The employment of a new experimental methodology for membrane studies, specially adapted to the application of the thermodynamics of non-equilibrium processes has led to a measurement of the 9 differential phenomenological coefficients L_{ij} with a margin of error of ± 1–2% (except for L_{11} where it was ± 10%). This method has the advantage of not requiring assumptions as to the numerical values of certain coefficients as in the past [5, 9, 10]. It permits experimental verification of the applicability of the Onsager relations while avoiding errors due to interfacial concentration polarization, and does not require the intervention of a hydrostatic pressure.

Although a knowledge of the L_{ij} coefficients permits one, in theory, to use matrix calculations to obtain values of R_{ij}, l_{ij} and r_{ij}, it is shown that when a direct numerical calculation was employed, a large duplication of errors makes these coefficients of little use.

The influence of polarization on transport numbers is interpreted simply from the L_{ij} coefficients. The volume and ion transport numbers are functions of a term independent of the concentration gradient and of corrections due to polarization. A means for estimating the systematic error due to the presence of the boundary layer thickness is proposed.

The Equi [12, 21] method which permits the measurement of fluxes under controlled polarization conditions proved to be particularly suitable for avoiding systematic errors with extremely permeable or fragile membranes.

Acknowledgments

The authors are most indebted to Professor H. P. Gregor for reading, correcting and simplifying the English manuscript.

They also which to thank the 'Délégation Générale à la Recherche Scientifique' for financial support during the Dr. Sc-thesis of C. B.; these articles are an integral part of this thesis.

References

1. Meares, P.: this volume, p. 123.
2. Foley, T., Klinowski, J., and Meares, P.: *Proc. Roy. Soc.* A **336**, 327 (1974).
3. Staverman, A. J.: *Trans Faraday Soc.* **48**, 176 (1952).
4. Kedem, O. and Katchalsky, A.: (1963).
5. Spiegler, K. S.: *Trans Faraday Soc.* **54**, 1408 (1958).
6. Ramer, H. K. and Meares, P.: *Biophys. J.* **9**, 1006 (1969).
7. Demarty, M. and Sélégny, E.: *C.R. Acad. Sci.* C **276**, 1549 (1973).
8. Dorst, W. and Staverman, A. J.: *Rec. Trav. Chim.* **86**, 61 (1967).
9. Meares, P., Thain, J. F., and Danson, D. G.: in G. Eisenman (ed.) *Membranes*, Marcel Dekker, New York, 1972.
10. Paterson, R. and Gardner, C. R.: *J. Chem. Soc.* A, 2254 (1971).
11. Thau, G., Block, R., and Kedem, O.: *Desalination* **1**, 129 (1966).
12. Sélégny, E. and Bourdillon, C.: this volume, p. 183.
13. Katchalsky, A. and Curran, P.: *Non-Equilibrium Thermodynamics in Biophysics*, Harvard Univ. Press, Boston, 1965.
14. Kirkwood, J. G.: *Ion Transport Across Membranes*, Academic Press, New York, 1954.
15. Kindly furnished by Dr. E. Korngold (Negev Institute).
16. Snell, F. M. and Stein, B.: *J. Theoret. Biol.* **10**, 177 (1966).
17. Oda, Y. and Yawataya, T.: *Bull. Chem. Soc. Japan* **29**, 673 (1956).
18. Block, M. and Kitchener, J. A.: *J. Electrochem. Soc.* **113**, 947 (1966).
19. Gregor, H. P. and Peterson, M. A.: *J. Phys. Chem.* **68**, 2201 (1964).
20. Oda, Y. and Yawataya, T.: *Desalination* **5**, 129 (1968).
21. Bourdillon, C., Demarty, M., and Sélégny, E.: *J. de Chim. Phys.* **71**, 6, 819.
22. Ripoll, C., Demarty, M. and Sélégny, E.: *J. de Chim. Phys.* **71**, 828 (1974).
23. Bourdillon, C., Metayer, M., and Sélégny, E.: *J. de Chim. Phys.* **3**, 91 (1973).
24. Cooke, B. A. and Van der Valt S. J.: *Electrochim. Acta* **5**, 216 (1961).
25. Bourdillon, C., Metayer, M., and Sélégny, E.: *J. de Chim. Phys.* **5**, 788 (1974).
26. Brun, T. S. and Vaula, D.: *Ber Bunsenges Physik, Chem* **71**, 824 (1967).
27. Kabatake, Y. and Kamo, N.: *4th International Symposium on Fresh Water from the Sea* (ed. by E. and A. Delyannis), Athens, Vol. 3, 91 (1973).
28. Mackay, D. and Meares, P.: *Trans Faraday Soc.* **55**, 1221 (1959).
29. Meares, P. and Sutton, A. H.: *J. Colloid Interface Science* **28**, 118 (1968).
30. McHardy, W. J., Meares, P., Sutton, A. H., and Thain, J. F.: *J. Colloid Interface Science* **29**, 116 (1969).
31. Lorimer, J. W., Boterenbrood, E. I., and Hermans, J. V.: *Disc. Faraday Soc.* **21**, 141 (1956).
32. Scatergood, E. M. and Lightfoot, E. N.: *Trans Faraday Soc.* **64**, 1135 (1968).
33. Levich, V. G.: *Physicochemical Hydrodynamics*, Prentice Hall, New York 1962.
34. George, J. H. B. and Courant, R. A.: *J. Phys. Chem.* **71**, 246 (1967).
35. Lakshminarayanaiah, N.: *Transport Phenomena in Membranes*, Academic Press, New York 1969.

TRANSCONFORMATION SURFACE REACTION
AND HYDRODYNAMIC STABILITY

A. SANFELD and A. SANFELD-STEINCHEN

Faculté des Sciences, Université Libre de Bruxelles

Abstract. In the present paper, the authors have tried to give a physico-chemical model of a membrane in which the ion flow is regulated by a transconformation reaction.

The membrane is modelized by a surface layer of carrying molecules spread at the interface between two immiscible liquids.

The influence of the chemical reaction on the mechanical stability of the interface is studied as well as the influence of a hydrodynamic instability on the stability of the chemical reaction. The results show that a deformation of the interface can be induced by a chemical instability and that a space structure of the chemical species at the interface can be induced by a hydrodynamic instability.

1. Introduction

It is well known that during phagocytosis the membrane of the lymphocytes is deformed by pseudopodes formation. The chemical mechanism that induces such a deformation is still unknown, but we can reasonably assume that a protein of the membrane is able to regulate the flow of a permeant molecule by a change of conformation [1]. Such a change of conformation might be taken as responsible for the change of elasticity of the membrane and to affect by consequence the boundary conditions of the hydrodynamic equations [2].

Cytoplasmic motion in the axon is also detected during the propagation of the action potential. A similar mechanism of transconformation of the ionophores in the membrane could be invoked to explain the inset of the cytoplasmic flow.

The aim of the present paper is not to deal with such complicated systems as the biological systems, but to show that the behavior of these biological systems can be modelized by physico-chemical systems.

As it is now generally assumed that the cell membrane is a fluid assembly of proteins moving in lipids [3], we thought that a good physical picture of the membrane could be given by a monolayer of proteins or polypeptides free to move and to change conformation at the interface between tow immiscible fluids.

It was recently observed [4] that a transconformation reaction at the interface changes the surface tension. Our purpose was then to analyze how the flow of ions through an interface, regulated by a transconformation reaction of a carrier, could affect the hydrodynamic stability of the interface and how a mechanical instability could destabilize the chemical reaction.

2. The Chemical Model

The system consists of two immiscible liquid phases separated by a spread monolayer of insoluble macromolecules (proteins or polypeptides). Two ionic double layers are

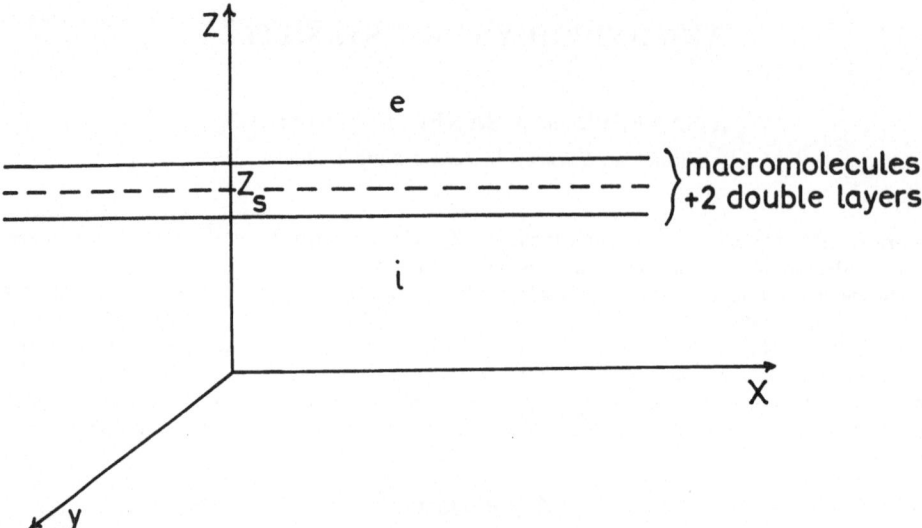

Fig. 1. The surface layer model.

adjacent to both sides of the protein layer. The layer of macromolecules together with the double layer will be considered as a pseudo membrane of width d.

We assume as Changeux and Lefever [5] and Van Roten [6] that the following sequence of reactions takes place in the layer:

$$\underset{C_3 \; k'_c \; C_1}{\bigcirc \overset{k_c}{\rightleftharpoons} \square} \qquad \text{(A) transconformation}$$

closed   open

$$\underset{C_1 \qquad k^i_d \; C_2}{\square + A \overset{k^i_a}{\rightleftharpoons} \boxdot} \qquad \text{(B)}$$

$$\underset{C_1 \qquad k^e_d \; C_2}{\square + B \overset{k^e_a}{\rightleftharpoons} \boxdot} \qquad \text{(C)}$$

fixation

$$\underset{k^i_{-p}}{A_i \overset{k^i_p}{\rightleftharpoons} A} \qquad \text{(D)}$$

$$\underset{k^e_{-p}}{B_e \overset{k^e_p}{\rightleftharpoons} B} \qquad \text{(E)}$$

diffusion migration from the bulk

where C is the concentration (in mass per unit volume) of the macromolecules in the layer; the subscripts 3, 2 and 1 refer respectively to the molecules in the closed conformation, and to the molecule in the open conformation, filled with an ion or free from ion. A and A_i are the concentrations of the ions respectively in the internal double layer and in the internal bulk phase, B and B_e the ionic concentration in the external layer and bulk phase.

All the kinetic constants depend on the electric field in the layer.

It is also assumed that the transformation reaction (A) is cooperative i.e. that the free energy (or the negative affinity $\mathscr{A}$) of the reaction depends on the concentration of the open macromolecules and on the electric field E

$$\mathscr{A} = \mathscr{A}^* + (\mu_1 - \mu_3)\, E + \eta\, (c_1 + c_3) \tag{1}$$

where μ_1 and μ_3 are respectively the mean dipole moment in the open state and in the closed state, η is the cooperativity parameter.

Taking into account the conservation equation of the total concentration of macromolecules, and assuming that the concentration C_2 of the molecules bounded to an ion is quasi stationary, we obtain the following kinetic equations

$$\frac{\partial C_R}{\partial t} = k_c [C_T - C_R - l^* e^{\Delta \mu E/RT} \Lambda^{C_R}(C_R - C_2^0)] - \operatorname{div} v^s C_R + \mathscr{D}_R \Delta^s C_R \tag{2}$$

$$\frac{\partial A}{\partial t} = k_p^{i*} [A_i e^{-px/4} - A e^{px/4}] + k_d^{i*} C_2^0 e^{(1-p)x/4} - k_a^{i*} A (C_R - C_2^0)\, e^{(p-1)x/4}$$
$$- \operatorname{div} v^s A + \mathscr{D}_A \Delta^s A \tag{3}$$

$$\frac{\partial B}{\partial t} = k_p^{e*} [B_e e^{px/4} - B e^{-px/4}] + k_d^{e*} C_2^0 e^{(p-1)x/4} - k_a^{e*} B (C_R - C_2^0)\, e^{(1-p)x/4}$$
$$- \operatorname{div} v^s B + \mathscr{D}_B \Delta^s B \tag{4}$$

where

$$C_R = C_2 + C_1 \tag{5}$$

$$k_a^i = k_a^{i*} e^{(p-1)\,x/4} \tag{6}$$

$$k_d^i = k_d^{i*} e^{(1-p)\,x/4} \tag{7}$$

$$k_a^e = k_a^{e*} e^{(1-p)\,x/4} \tag{8}$$

$$k_d^e = k_d^{e*} e^{(p-1)\,x/4} \tag{9}$$

$$k_p^i = k_p^{i*} e^{-px/4} \tag{10}$$

$$k_{-p}^i = k_{-p}^{i*} e^{px/4} \tag{11}$$

$$k_p^e = k_p^{e*} e^{px/4} \tag{12}$$

$$k_{-p}^e = k_{-p}^{e*} e^{-px/4} \tag{13}$$

with $\chi = z\mathscr{F}\Delta V/RT$ (reduced electric potential difference) and p the reduced width of the double layers (p is the parameter of location of the fixation site) v^s is the surface velocity, $\mathscr{D}_R$ the diffusion coefficient of the open macromolecules, $\mathscr{D}_A$ and $\mathscr{D}_B$ the diffusion coefficients of the ions, Δ^s is a two dimensional Laplace operator. The superscript ^{0} refers to the stationary state.

and where

$$\frac{k_c'}{k_c} = l\Lambda^{C_R} = l^* e^{\Delta\mu E/RT}\Lambda^{C_R} \tag{14}$$

with

$$l^* = e^{-\mathscr{A}^*/RT}$$

and

$$\Lambda = e^{-\eta/RT}.$$

In the absence of diffusion-migration and convection, the Equations (2), (3) and (4) read at the steady state (superscript ^{0})

$$C_2^0 = \frac{k_a^*[A^0 e^{(p-1)\chi/4} + B^0 e^{(1-p)\chi/4}] C_R^0}{k_a^*[A^0 e^{(p-1)\chi/4} + B^0 e^{(1-p)\chi/4}] + 2k_d^* \cosh(1-2)\chi/4} \tag{15}$$

$$k_c[C_T - C_R^0 - l^* e^{\Delta\mu^*\chi}\Lambda^{C^0 R}(C_R^0 - C_2^0)] = 0 \tag{16}$$

$$k_p^{i*}[A_i e^{-p\chi/4} - A^0 e^{p\chi/4}] + k_d^* C_2^0 e^{(1-p)\chi/4} - k_a^* A^0 (C_R^0 - C_2^0) e^{(p-1)\chi/4} = 0 \tag{17}$$

$$k_p^{e*}[B_e e^{p\chi/4} - B^0 e^{-p\chi/4}] + k_d^* C_2^0 e^{(p-1)\chi/4} - k_a^* B^0 (C_R^0 - C_2^0) e^{(1-p)\chi/4} = 0 \tag{18}$$

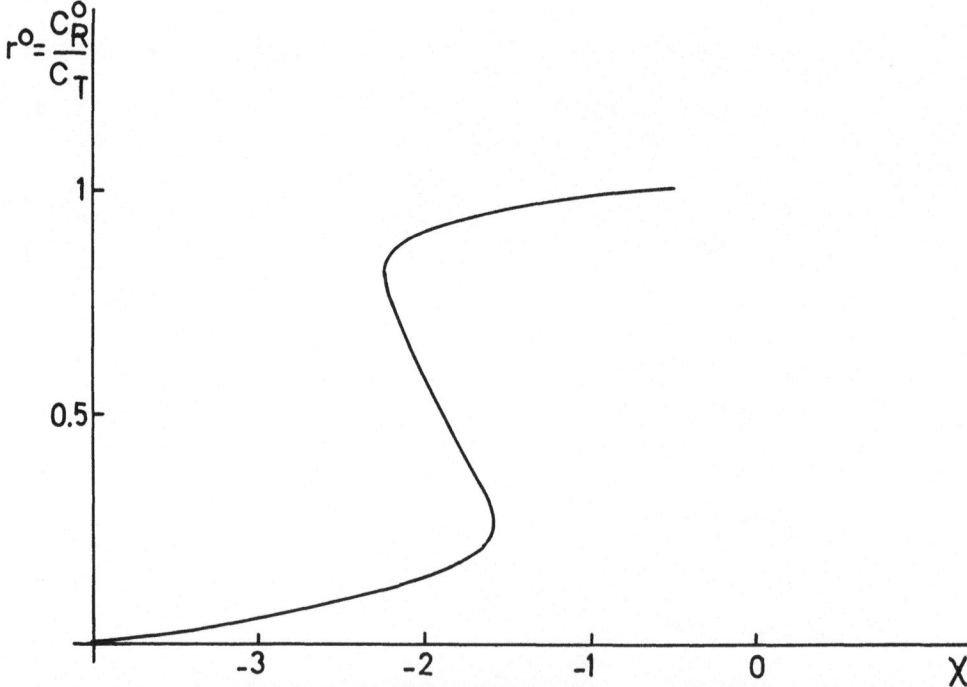

Fig. 2.   Multiple steady states: mole fraction $r^0 = C_R^0/C_T$ of open macromolecules versus potential χ for $C_T\eta=5$, $A_i(k_a^*/k_d^*)=1.26$, $B_e(k_a^*/k_d^*)=11.81$, $C_T k_a^*=1$, $(k_p^*/C_T k_a^*)=5\times10^{-3}$, $p=0.95$, $(\mu_3-\mu_1)/z\mathscr{F}d=2$, $l^*=4$, $k_c=1$.

with

$$\Delta\mu^* = \frac{\mu_1 - \mu_3}{zJd} \quad \text{(reduced dipole moment difference)} \tag{19}$$

$$k_a^{i*} = k_a^{e*} = k_a^* \tag{20}$$

and

$$k_d^{i*} = k_d^{e*} = k_d^* \tag{21}$$

M. Van Roten [6] shows that for peculiar values of the potential difference and for given parameters of the system, multiple steady states are obtained (see Figure 2)

3. Hydrodynamic Flow

The hydrodynamic motion in the two bulk phases, where the electric field is zero, is governed by the Navier-Stokes equations.

For an infinitesimal perturbation δ around the reference steady state (at mechanical equilibrium) these equations read (see for example [7] chap. X)

$$\varrho \frac{\partial v_x}{\partial t} = -\frac{\partial \delta p}{\partial x} + \mu \Delta v_x \tag{22}$$

$$\varrho \frac{\partial v_y}{\partial t} = -\frac{\partial \delta p}{\partial y} + \mu \Delta v_y \tag{23}$$

$$\varrho \frac{\partial v_z}{\partial t} = -g\delta\varrho - \frac{\partial \delta p}{\partial z} + \mu \Delta v_z \tag{24}$$

where μ is the viscosity coefficient, p the pressure and ϱ the density.

To test the stability of the reference steady state, we will use the normal mode analysis [7]. The boundary conditions associated to Equations (22) to (24) are the following (see for example [7] chap. X).

$$v_z^i = v_z^e = v_z^s \quad \text{at} \quad z = z_s \tag{25}$$

$$Dv_z^i = Dv_z^e = (Dv_z)^s \quad \text{where} \quad D = d/dz \tag{26}$$

(The superscripts i and e refer respectively to the internal phase $(z<z_s)$ and to the external phase $(z>z_s)$.)

$$\left\{ \varrho^e - \frac{\mu^e}{\omega}(D^2 - k^2)(Dv_z)^s \right\} - \left\{ \varrho^i - \frac{\mu^i}{\omega}(D^2 - k^2)(Dv_z)^s \right\} =$$
$$-\frac{k^2}{\omega}g(\varrho^e - \varrho^i)v_z^s + \frac{k^4}{\omega}\sigma^0 v_z^s - 2\frac{k^2}{\omega}(Dv_z)^s(\mu^e - \mu^i) \tag{27}$$

$$\mu^e\left(\frac{\partial v_x}{\partial z} + \frac{\partial v_z}{\partial x}\right)^e - \mu^i\left(\frac{\partial v_x}{\partial z} + \frac{\partial v_z}{\partial x}\right)^i = \frac{\partial \delta\sigma}{\partial x} \tag{28}$$

(the same for the y coordinate) (Landau-Levich conditions [8]).

To make the coupling between the hydrodynamic Equations (22) to (24) and the kinetic equations of the chemical species at the surface (2) to (4) (written for an infinitesimal perturbation around the steady state), we will use a state equation that gives the fluctuation of surface tension $\delta\sigma$ in terms of the fluctuations of concentrations in the surface i.e.

$$\delta\sigma = a_1\delta C_R + a_2\delta A + a_3\delta B \tag{29}$$

where a_1, a_2, a_3 are constants.

After injection of the solutions of v_z in the boundary conditions (25) to (28), together with equations (29), (2), (3) and (4), and assuming the incompressibility of layer (div $v^s = 0$) we obtain a system of algebraic equations.

The determinant of the system has to be zero. That condition gives us the dispersion equation which reads

$$D_H\left(\omega + k_c[1 - \eta^*(1 - r^0) + l^*\Lambda^{*r^0}e^{\Delta\mu^*\chi}] + \frac{\mathscr{D}_R}{\lambda^2}\right) \times$$

$$\times\left(\omega + k_p^*e^{p\chi/4} + C_Tk_a^*\frac{(1 - r^0)\,e^{(p-1)\chi/4}}{l^*\Lambda^{*r^0}e^{\Delta\mu^*\chi}}\right) \times$$

$$\times\left(\omega + k_p^*e^{-p\chi/4} + C_Tk_a^*\frac{(1 - r^0)\,e^{(1-p-)\chi/4}}{l^*\Lambda^{*r^0}e^{\Delta\mu^*\chi}}\right). \tag{30}$$

where ω is the frequency of the normal mode considered and $\lambda = 1/k$ the wavelength. D_H is the determinant of the hydrodynamic system (see for example [7] chap. X 'Rayleigh-Taylor Instability') and the three other terms under brackets give the dispersion equation of the chemical system.

4. Stability Analysis

If we restrict our analysis to an inviscid system, the pure hydrodynamic dispersion equation reads (see [7] p. 435)

$$\omega^2 = \frac{g}{\lambda}\left\{\frac{\varrho^e - \varrho^i}{\varrho^e + \varrho} - \frac{\sigma^0}{\lambda^2 g\,(\varrho^e + \varrho^i)}\right\} \tag{31}$$

if $\varrho^e < \varrho^i$ the arrangement is stable; if $\varrho^e > \varrho^i$ the arrangement is stable for $\lambda < \lambda_c$ where

$$\lambda_c = \left[\frac{(\varrho^e - \varrho^i)}{\sigma^0}g\right]^{-1/2} \tag{32}$$

while for $0 > \lambda > \lambda_c$ with $\varrho^e > \varrho^i$, the system is unstable.

Moreover there exists a mode of maximum instability corresponding to a wavelength

$$\lambda_* = \sqrt{3}\,\lambda_c \tag{33}$$

On the other hand, the only chemical contribution to the instability arises from the solution of the dispersion Equation (30) given by

$$\omega = -k_c\left[1 - \eta^*\left(1 - r^0\right) + l^*\Lambda^{*r^0}e^{\Delta\mu^*\chi}\right] - \frac{\mathscr{D}_R}{\lambda^2} \qquad (34)$$

if

$$k_c\eta^*\left(1 - r^0\right) > k_c\left(1 + l\Lambda^{*r^0}\right) + \frac{\mathscr{D}_R}{\lambda^2} \qquad (35)$$

the system is unstable.

From (31) and (35), one can see that even if $\varrho^e < \varrho^i$ the system may become unstable if (35) is realized. It means that the chemical instability induces an hydrodynamic instability.

Reversely, if condition (35) is not satisfied but if $\varrho^e > \varrho^i$ with $0 > \lambda > \lambda_c$ then the mechanical instability induces a chemical instability that will grow up to a wavelength λ_*. It means that the concentrations of the chemical species in the layer will become space dependent with a wavelength λ_* given by (33) for the mode of maximum instability.

5. Conclusions

The present analysis shows clearly that a chemical reaction occurring at the interface between to immiscible liquids is able to induce a mechanical instability i.e. a deformation of the interface. Such a process could perhaps explain how the cell membranes undergo deformations when they receive a chemical 'message'.

We can see also that a mechanical instability is able to induce a space distribution of the open and closed sites in the surface. Such a process could be invoked to explain the space distribution of the conducting channels in the axon membrane.

Acknowledgements

This work was sponsored by CNRS.

We wish to thank Prof. J. Guastalla (director of Laboratoire de Physico-chimie colloidale CNRS Montpellier) and Prof. H. Colson Guastalla (director of Laboratoire de Biologie physico-chimique, Faculté des Sciences, Montpellier) for the very interesting discussions we had together during our stay in their laboratories. We had also many fruitful discussions with Dr Lefever and M. Van Roten (Faculté des Sciences, Université Libre de Bruxelles).

References

1. Blumenthal, R., Changeux, J.-P., Lefever, R.: *J. Membr. Biol.* **2**, 351 (1970).
2. Sanfeld, A. and Sanfeld-Steinchen, A.: *Biophys. Chem.* **3**, 99 (1975).
3. Singer, S. J. and Nicolson, G. L.: *Science* **175**, 720 (1972).
4. Caspers, J.: Thèse de doctorat, Université Libre de Bruxelles, 1973.
5. Lefever, R. and Changeux, J. P.: submitted for publication to *J. Membr. Biol.*
6. Van Roten, M.: Mémoire de Licence, Université Libre de Bruxelles, 1972.
7. Chandrasekhar, S.: *Hydrodynamic and Hydromagnetic Stability*, Clarendon Press, Oxford, 1961.
8. Levich, V.: *Acta Physicochim. S.S.S.S.R.* **14**, 308 (1941).

IV

ULTRA- AND HYPER-FILTRATION MEMBRANES

FIXED-CHARGE ULTRAFILTRATION MEMBRANES

HARRY P. GREGOR

Dept. of Chemical Engineering and Applied Chemistry Columbia University,
New York, N.Y. 10027, U.S.A.

Abstract. The properties of a series of sulfonic acid fixed-charge membranes of graded porosity were described in terms of their fixed-charge molality, hydraulic permeability, concentration potential and rejection to a number of charged and uncharged solutes of differing molecular weight. Their use in ultrafiltration processes employing a number of test solutions and also solutions from the sewage-treatment, pulp and paper, and cheese industries was also described. The salient property of these membranes was their non-fouling character as exemplified by their ability to maintain their hydraulic flux and solute rejection properties with solutions known to foul conventional ultrafiltration membranes.

1. Introduction

This paper summarizes the results of ultrafiltration membrane studies that started about fifteen years ago. We had examined the electrodialysis of some rather complex solutions. Some were fermentation broths which contained substances which could be recovered in a high state of purity by electrodialysis, with the products also concentrated thereby. Others were blackstrap molasses and other sugar process streams from which inorganic salts could be removed electrodialytically to allow recovery of more sugar at considerable economic advantage. Both programs achieved their scientific objectives in terms of the separations achieved; they were failures from a practical point of view because the electrodialysis membranes deteriorated in a matter of hours or days.

In the course of these investigations, all of the commercially available electrodialytic membranes were subjected to these solutions for from two or three hours up to two or three days, if possible. The cation-permeable membranes of sulphonated polystyrene were always intact, smooth and apparently unchanged in their physical and chemical properties. The quaternary ammonium anion-permeable membranes which were originally light amber in color all became black and corroded and their ohmic resistances rose by orders of magnitude.

These results emphasized the critical importance of fouling in membrane phenomena, invariably by polyanions present in natural systems along with hydrophobic molecules and, particularly, hydrophobic anions. They taught that sulphonated polystyrene membranes had enormous advantages over quarternized membranes in their resistance to fouling. It was obvious also that anionic molecules of intermediate and high molecular weight were responsible for fouling.

2. Sulphonic Ultrafiltration Membranes

We proceeded to prepare ultrafiltration membranes which had sulphonic acid groups on their outer surface as well as inner pore surfaces. No other fixed-charge group would serve, for we examined all which were available, including the carboxylic, phospho-nous, phosphonic, etc. Only the fixed charge, sulphonic acid membranes were useful. A number of ultrafiltration membranes composed primarily of cross-linked polysty-renesulphonic acid and polyvinylsulphonic acid and copolymers thereof were then prepared.

The terms hydrophilic and hydrophobic are of necessity relative ones. The usual, classical physicalchemical means of characterizing surfaces in terms of wetting angle and the like are not applicable here. Rather a functional definition is required, one which entails the actual use of a surface in a given series of environments to determine the extent of adverse reactions.

In terms of membrane phenomena, the definitive test of hydrophilic or non-fouling character consists in employing the membrane in either an electrically driven (as in electrodialysis) or pressure driven (as in reverse osmosis or ultrafiltration) process. It is only under these conditions wherein there is a flux of both solvent and solute into the membranes that significant differences in the ability of membranes to withstand fouling are evident.

In the course of these investigations, a large number of membranes of ionic and non-ionic character were prepared and subjected to test. These included membranes cast from a number of different hydrophobic polymers and then coagulated in water in accordance with the classical procedures of Elford and of Sollner. The base poly-mers employed here included those from polyvinylidine fluoride, polyvinylchloride, copolymers of vinyl chloride and acrylonitrile, polyvinylacetate, cellulose acetate, cellulose nitrate and regenerated cellulose acetate and nitrate (cellulose). In addition, membranes were cast from a water-soluble polymers (and a suitable cross-linking agent) such as polyvinylalcohol, polyacrylamide and polyacrylic acid.

When each of these membranes was subjected to ultrafiltration, employing a natural solution containing high molecular weight anions and highly colored, non-polar anionic materials, in every case fouling ensued rapidly. For membranes having dia-meters in the range from 10 to 30 AU, the hydraulic permeability fell rapidly, the membranes became discolored and the generalized phenomena of fouling was ex-hibited. With coagulated membranes of high porosity, with pore diameters of about 100 AU, hydraulic permeabilities did not fall markedly but the membranes became strongly discolored. It was concluded that the hydraulic permeabilities did not change because none of the solutes were of high enough molecular weight to clog the mem-branes in a gross manner; rather, the inner pore surfaces of the membrane became fouled. Because the pores were so wide, the flow of liquid through them was but sligth-ly impeded. In other words, ultrafiltration processes involving wide pores and particles of large sizes can be accomplished with membranes which are of fouling character, provided that the strong adsorption of soluble materials to the pore walls does not decrease the pore diameter sufficiently to impede transport of solvent, and where suspended matter which is removed by ultrafiltration is very much larger than average pore diameters.

This condition actually obtains in several practical processes involving ultrafiltration of macroscopic particles, and here membrane fouling is of itself not a serious problem.

One can but speculate as to the reasons for the non-fouling character of sulfonic acid membranes. First, there is the electrostatic repulsion due to the fixed sulfonic acid groups and the negative colloidal matter present. Second, there is evidence that virtually no metallic or other cation is strongly adsorbed or forms a strong ion-pair with this fixed-charge group, with the sole exception of aromatic cations such as the cetyl-pyridinium cation or ones of similar character. In the case of metallic cations, while there is some degree of ion-pair formation and a lowered electrical conductivity with sulfonic acid membranes, nevertheless the ions are not truly fixed but are relatively mobile and these membranes retain their highly hydrophilic nature. Even with aromatic cations the adsorption is not so strong that it cannot be reversed simply by washing the membrane with water.

Another reason for the non-fouling character of sulfonic acid membranes probably lies in the unique nature of the sulfonic acid radical in its 'structure breaking' nature as regards water structure. In this respect the sulfonic acid appears to be unique; this factor alone may make for the strongly non-fouling character of the membranes containing this group in a predominant amount.

Figure 1 shows the cell used for characterizing ultrafiltration (UF) membranes. It is one that A. Michaels first made available. We also used the cell shown in Figure 2,

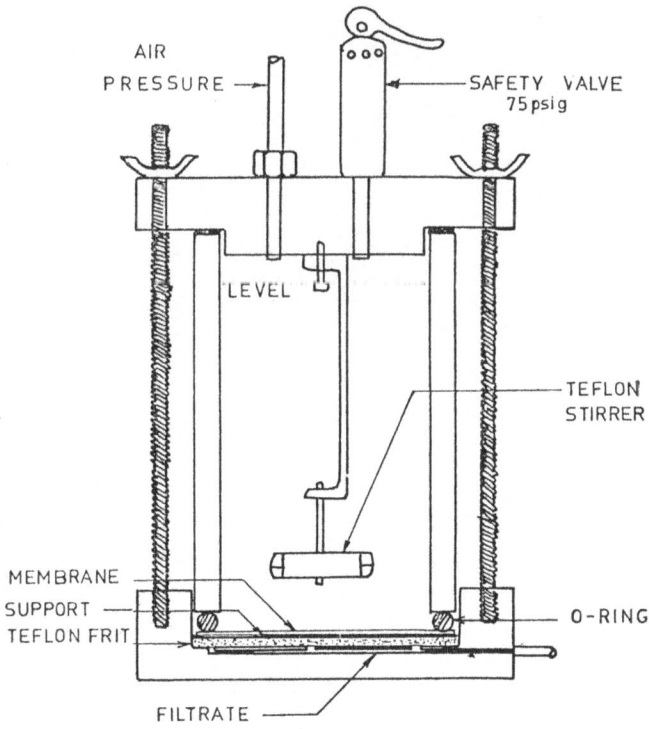

Fig. 1.

HARRY P. GREGOR

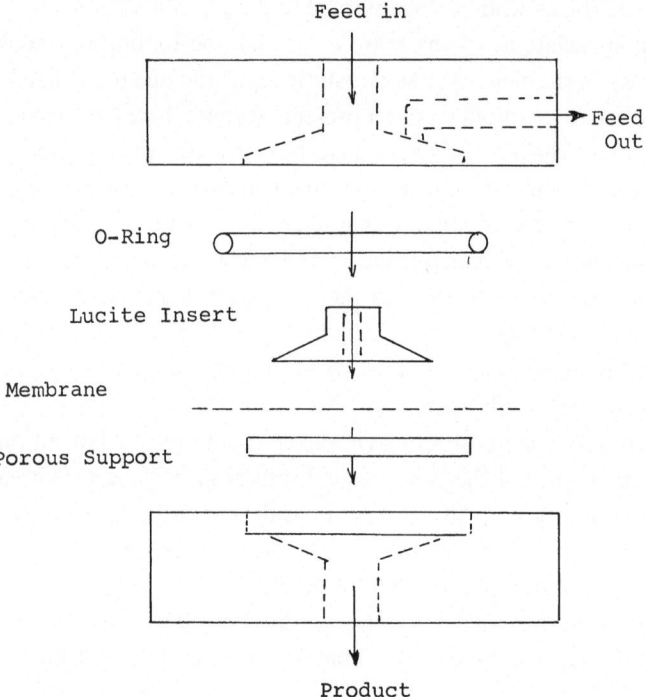

Fig. 2.

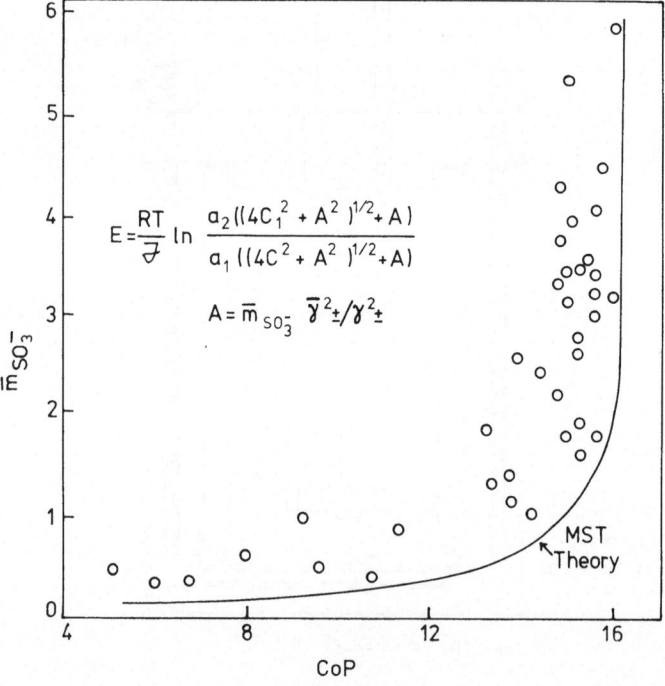

Fig. 3.

a Gelman cell which was modified with a plastic spacer to increase the flow rate, where the influent entered through the top, passed across the membrane and exited out of the side. In these small cells, about 5 cm in diameter, the rate of solution movement across the face of the membrane was rather slow, about 2 cm s^{-1}. As was evident later, a higher flow was required for optimum results.

An extensive series of membranes were prepared; the details of these procedures are in a patent [1]. These preparative procedures were usually complex ones 6, using a sulfonic acid polymer, matrix polymers and cross-linking reactions. Because of their fixed-charge nature, these are Donnan systems. Figure 3 gives a plot of the molality of fixed charge groups in the membrane as a function of the measured Concentration Potential (COP) across 0.1/0.2 M KCl. The relationship between COP and permselectivity is given below, where $\bar{t}_i$ is the transport number of an ion in the membrane and t_i its value in free solution where the ambient 1-1 salt activity is **a**. For the COP,

$$E = (\bar{t}_+ - \bar{t}_-)(RT/F)\ln a''/a'$$

For the Permselectivity, $P_i = (\bar{t}_i - t_i)/(1 - t_i)$.

One can apply the Meyer-Sievers-Teorell theory to these systems with an appropriate correction for the activity coefficient of electrolyte in the membrane phase, where $\bar{\gamma}_\pm = 0.45$ [2]. The data, as is evident, do not agree with theory because these membranes have a high water permeability, and a correction for water transport must be applied to these COP values. While these osmotic corrections were well known to Michaelis and to Donnan, Staverman is primarily responsible for the development of the theory of these irreversible processes in quantitative form. Table I summarizes

TABLE I

Membrane potentials as a function of water transport number
(0.1/0.2 N KCl)

$\bar{t}_+$ \ mV	Em $\bar{t}_w = 0$	Em $\bar{t}_w = 5.8$	Em $\bar{t}_w = 10.$	Em $\bar{t}_w = 20.$
1	16.1	15.8	15.3	14.4
0.98	15.5	15.0	14.6	13.8
0.75	8.0	7.6	7.3	6.4

these corrections and shows that if a membrane is ideally ion-selective in the absence of osmotic gradients, it shows a lowered COP when these gradients are present. With zero water transport, membranes which show 16.1 mV correspond to an ideal semipermeable or ion-selective membrane. If the transport number of water is 5 moles of water per mole of ion, the EMF falls off, etc. A typical UF membrane had a water transport of about 20 moles of water per Faraday, so its measured 14 mV corresponded to one of nearly ideal permselectivity.

Figure 4 lists some of the properties of a series of different membranes cast from similar polymers at different composition ratios. The membranes at the right are very

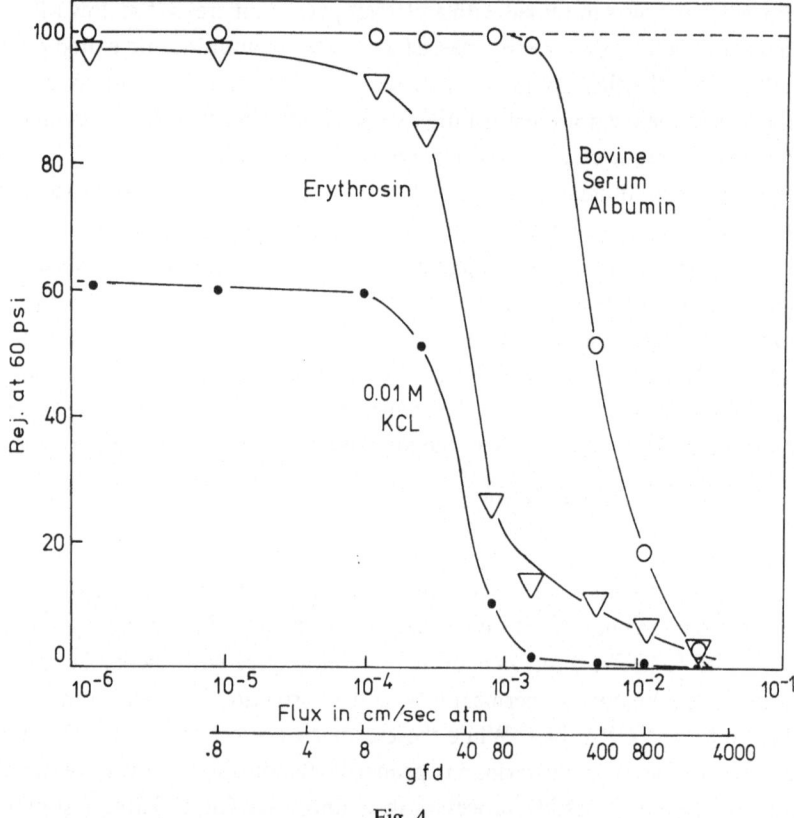

Fig. 4.

'loose' and those at the left are 'tight' membranes with the solute rejections $SR = (C_{In} - C_{out})/C_{In}$ measured at 4 atm. or 60 psi. All membranes rejected protein (bovine serum albumin) completely except for the very 'loose' ones having fluxes of the order of 400 gallons per square foot per day (gfd) at 60 psi wherein we observed substantial leakage of albumin. The rejection to dyes such as erythrosin (MW about 800 with a single sulfonic acid group) was 100% for the tight membranes with a loss in rejection for the more porous membranes. Salt rejection was about 60% in the 'tight' region and decreased with increasing porosity. All the membranes employed in our later, larger scale studies had fluxes of about 10 to 40 gfd and rejected erythrosin rather well. In general, they rejected solutes at MW of 700–1000 and greater.

Table II is a summary of a series of membranes having thicknesses of from 15 to 30 μ, fluxes of from 9 to 150 gfd at 60 psi, and corresponding rejections. Figure 5 shows the rejection – MW spectrum of membrane of intermediate porosity, measured at 60 psi and plotted as a function of the molecular weight of the solute. These solutes were present at relatively low concentrations (about 0.1% or 0.1 M) and each solute was the only one present. Figure 5 shows that the low molecular weight, uncharged solutes such as urea and glucose passed through the membrane as readily as did the solvent; in the Staverman terminology, they have reflection coefficients of zero. The

TABLE II

Summary of membrane properties

No.	L (μ)	J (gfd–4a)	KCl rej (%)	Eryth rej (%)
1	15	147	9	80
2	30	98	50	90
3	15	38	53	99
4	25	25	60	95
5	30	19	59	99
6	28	6	63	99

relationship between rejection and molecular weight was a fairly linear one for neutral solutes of low molecular weight. With ionic solutes, the rejection for sodium chloride was approximately 60% while that for sodium sulfate was 80% because the membrane is one of a fixed sulfonic acid character and selectively rejects doubly charged anions over nonvalent ones. The rejection of the membrane to erythrosin (MW 800) was good while that to vitamin B-12 (MW 1300) not as high because the latter molecule does not possess a charge. Higher molecular weight polyelectrolytes showed, at first glance, a surprisingly low rejection. For example, polyacrylic acid of MW = 50000 passed through the membrane in appreciable amounts; it must be surmised that the hydraulic conditions which obtained allowed this linear molecule to pass through in snake-like fashion. The high molecular weight, relatively spherical protein bovine serum albumin did not pass at all, as expected.

Since these membranes had a high concentration of fixed charges and are thus

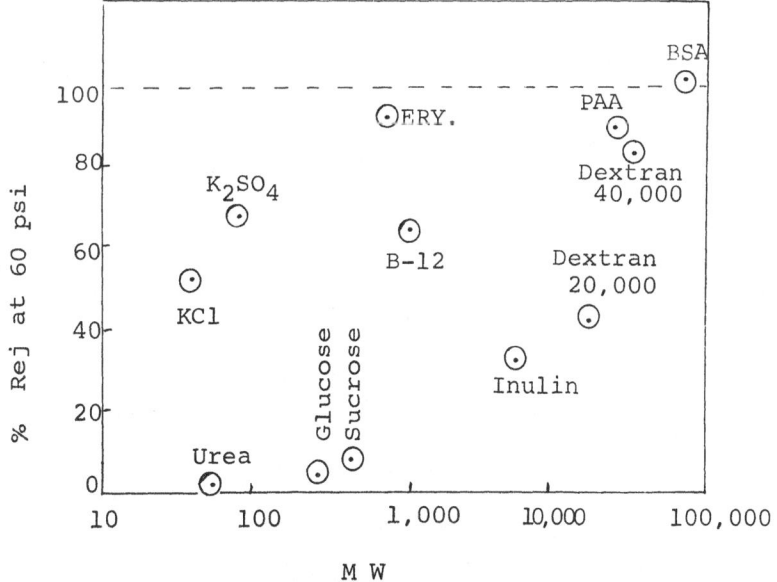

Fig. 5.

Donnan systems, their rejection to ionic solutes was a strong function of the ionic strength of the solution. Figure 6 shows the rejection by a given membrane to KCl and to the negatively charged dye erythrosin as a function of the concentration of each. In 1 M salt solution, salt rejection was virtually zero, with rejection to salt rising in a typical Donnan-like curve until it approached unity (or 100%) for 0.001 M solutions. On the other hand, when the dye was present in the absence of salt at the 1500 ppm level, its rejection was virtually complete, but when the same dye was present at the 1500 ppm level in 1 M salt, an appreciably lower dye rejection was found. Figure 6 shows the rejection for dye at different concentration levels at the stated salt concentration. Accordingly, these membranes rejected solutes on the basis of charge, size and

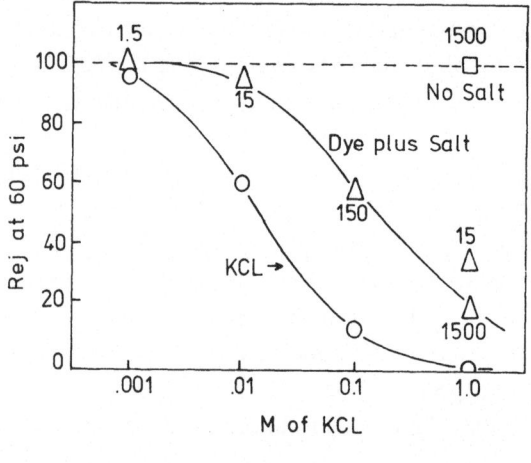

Fig. 6.

the nature of the charge itself. Negatively charged dyes were rejected more strongly than are positively charged ones, high salt concentrations lowered the rejection of all charged species while low ionic strength favored the rejection of charged species.

   This Donnan effect can be used to good advantage to ascertain the nature of an unknown species in a given solution. The rejection of a specie is measured in the presence of different concentrations of salt or at different ionic strength levels. If its rejection decreases markedly with increasing ionic strength, it is obviously charged and the nature of the charge can be ascertained by independent means. On the other hand, if the rejection of a given solute is relatively independent of ionic strength, we may conclude that the solute is not charged. However, since these are indeed Donnan gels and there is some swelling and shrinking of the membrane going to different ionic strengths, it must be expected the membranes will be somewhat more dense at high ionic strength levels and more expanded at more dilute salt solutions, so minor changes in effective pore diameters must be anticipated.

   Many membrane parameters have been measured for these systems, including their fixed charge molality, water content, hydraulic permeability and rejection to a number of different solutes. For a more functional characterization, it is often useful to have some indication of the average pore size of these membranes. Here, one makes use of the relative permeability of the membranes to ions or molecules of different sizes and uses the familiar equations relating geometrical and hydrodynamic factors from the classical studies of Ferry and of Faxen and Oseen. This treatment follows from the earlier study of Gregor and Kagawa [3] and we measure the ohmic resistance of the membrane in the potassium state and also in the tetrabutylammonium state and employ the expression,

$$\bar{\lambda}/\lambda = \overline{D}/D = h\varphi_p F'F$$
$$F'F = f(r, R)$$

where D and λ are diffusion coefficients and ionic equivalent conductances, respectively, the superscript bar refers to the membrane phase, φ_p is the pore (water) volume, **h** is the tortuosity factor or the effective pore length divided by the thickness of the membrane, and the Ferry (F) and Faxen (F') equations relate the excluded volume and drag by a function of the radius of the ion **r** and the effective radius of the pore **R**, usually expressed as a power series. While one cannot determine the pore radius from a single measurement, by having available two relative conductivities or diffusion coefficients of simple ions of spherical shape and the same charge, one can then make several simplifying assumptions and compute effective pore radii. For most of the membranes described in this report, the effective pore radius is about 8–10 AU. Of course, this measurement of effective pore radius is a highly subjective one; it has the virtue of simplicity while giving values which have practical significance.

The non-fouling characteristics of these membranes are shown best in the data of Figure 7. This information was collected early in our studies after we had made a number of different membranes, placed them into the small Gelman cells and then treated a series of different solutions with a flow rate across the membrane of about 2 cm s^{-1}. Originally, it was our intention to place a fresh set of membranes in the cell, test them with a given solution until failure and then continue with fresh membranes and a different test solutions.

We had accumulated a number of different test solutions which were known to be highly fouling, effluents from the pulp and paper industry, from the sugar industry, sewage, whey and similar solutions. Our usual practice was to condition the membrane for several days or weeks with relatively pure solutions to ascertain their characteristics and allow for a period of acclimatization, and then pass through the fouling solutions. We knew from our earlier studies that the membranes usually decreased somewhat in hydraulic permeability during the initial period and that only with the pure solutions of simple, low MW solutes could one expect a complete absence of flux loss due to fouling.

Our experience was that the membranes retained their hydraulic permeability so well that we kept the same set of membranes in the cell, in this case for about 18 months, while passing through a series of highly fouling solutions in sequence. Results with a single membrane are shown in Figure 7. The data points refer to times when

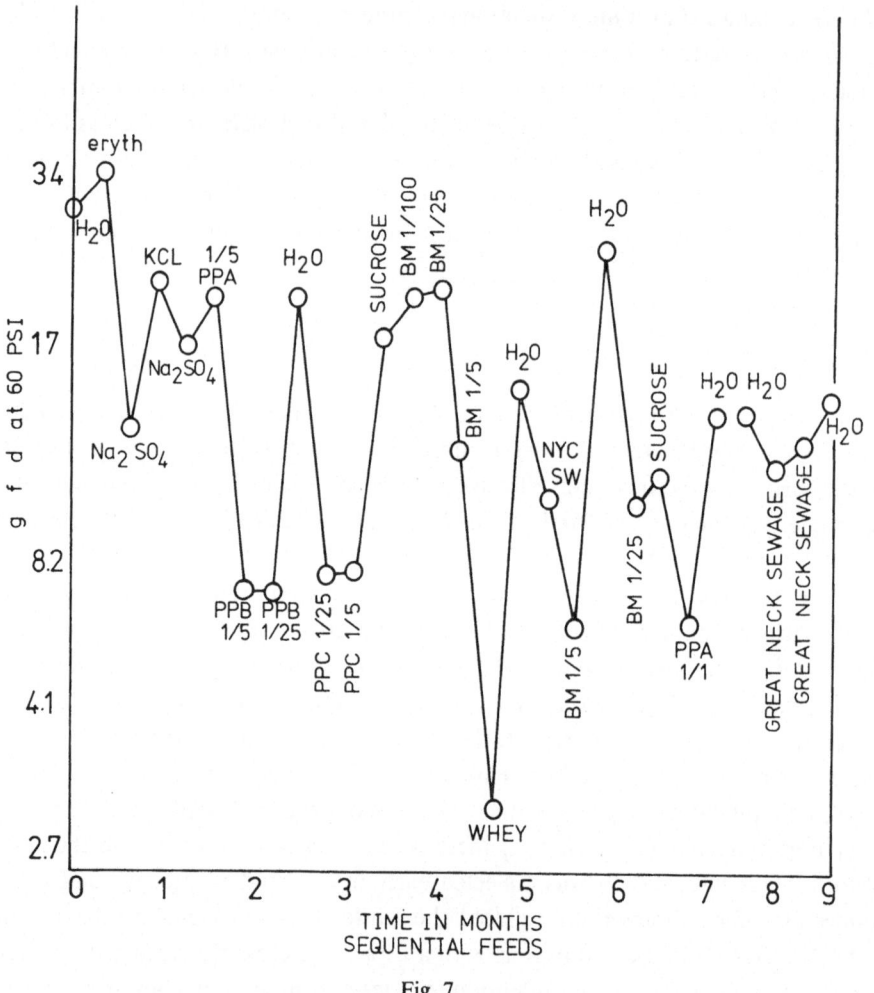

Fig. 7.

data on the effluent composition and the flux was accumulated. Figure 7 shows that even highly concentrated pulp and paper wastes, blackstrap molasses, whey and sewage primary effluent even for a period of over nine months did not produce a markedly reduced flux. The major differences in fluxes for the different test solutions were largely the result of the different counter-osmotic pressures which resulted. As was shown earlier, these membranes do reject salt appreciably. Therefore, when dealing with concentrated solutions, particularly ones which contain high concentrations of osmotically active colloids such as those encountered with pulp and paper and sugar wastes, these reverse osmotic pressure contributions are substantial. The driving pressure used was 60 psi, so relatively low concentration gradients (about 0.1 M in 1-1 electrolyte) across the membrane could give rise to zero flux by the reverse osmotic pressure effect.

Part of our research was devoted to the ultrafiltration of primary sewage effluents.

In conventional sanitary waste treatment practices, raw sewage which is approximately 200 ppm in BOD and in suspended solids is passed into a primary settling tank to remove particulate matter which settles readily. The primary effluent is then passed into an activated sludge tank containing bacteria where, under the presence of added oxygen a biochemical attack takes place, with the bacteria digesting most of the sewage, with the organic matter being converted to carbon dioxide and water, and a substantial part being converted into bacterial cell growth. These bacterial cells are then settled out in a secondary settling tank and the secondary effluent which now contains about 20 ppm of BOD and of suspended solids is then subjected to chlorination. At this point the water can either be passed into receiving water or, where necessary, it can receive tertiary treatment.

One of the principal problems of secondary waste treatment is that of solids handling. The initial primary sludge is about 5% solids. Since bacterial bodies are much more difficult to settle out, the settled secondary sludge contains only 1.5% solids. This necessitates the use of extensive (and expensive) solids handling equipment, and even so the final disposal of the sludge presents a substantial economic and environmental problem.

In our ultrafiltration studies, we first placed a number of membranes into a large Gelman rack and circulated samples of primary effluent. Some typical results a regiven in Table III where with the effluent from New York City the membranes removed on the average about 75–85% of the BOD, while from Great Neck which is a more purely residential community, the BOD removal was approximately 85–98%. These differences undoubtedly reflect differences resulting from major industrial effluent components.

It should be pointed out that in these screening experiments we were recirculating

TABLE III

Ultrafiltration of primary effluents

Memb	Flux (gfd–60 psi)	BOD ppm	Rej. (%)
New York City			
Feed	–	78	–
1	8	12	85
2	16	17	78
3	20	16	80
Feed	–	139	–
1	17	34	75
2	16	18	87
3	17	19	86
Great Neck			
Feed	–	120	–
1	12	4	96
2	7	18	85
3	12	2	98

a large volume of primary effluent and ultrafiltering a relatively small amount of product, so the composition of the solution being treated was unchanged during the process. At 60 psi flow rates, fluxes of from 8 to 20 gfd were obtained for a series of membranes. At the 2 cm s^{-1} flow rate rejection was not as favorable as under the optimum, higher flow rates used in the pilot plant.

The ultrafiltration of primary effluent was then studied in a small pilot plant wherein the sewage was charged into a 10 l tank which was then pressurized with nitrogen to 100 psi. The sewage was then recirculated across a flow-through ultrafiltration cell as

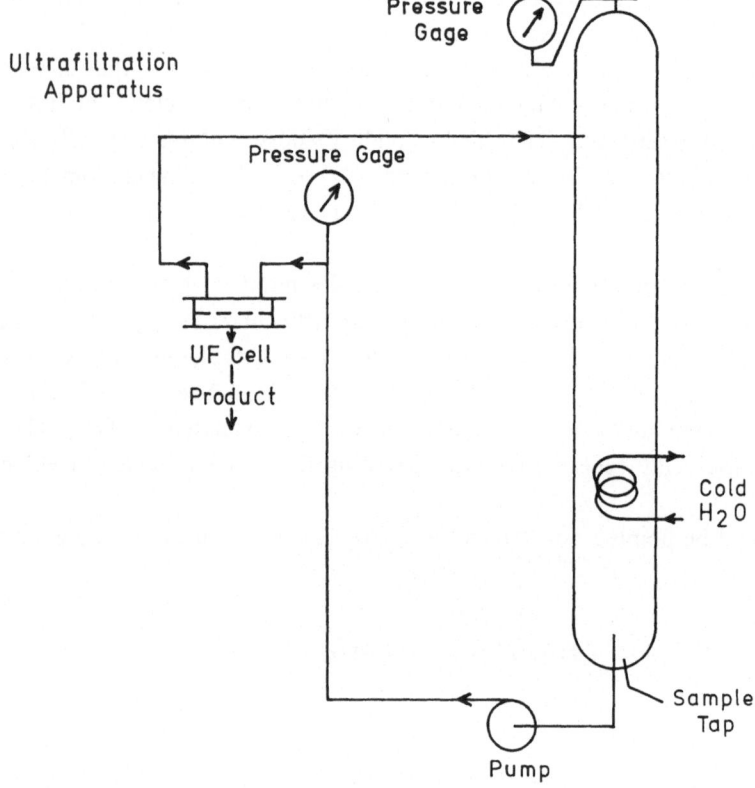

Fig. 8.

shown in Figure 8 wherein netted polypropylene plastic (Vexar) acted as a turbulence producing spacer. A small laboratory pump was adequate for recirculation because the entire system was pressurized. We observed that at recirculation velocities above approximately 40–60 cm s^{-1} an increase in the quality of the effluent and a further reduction in fouling was no longer obtained. In this pilot plant device the intake pipe is not at the very bottom of the holding tank but about three or four inches off the bottom so there was appreciable sedimentation and settling into that space during the operation of the unit.

A summary of one experiment is given in Figure 9. Approximately 98% of the water in a given charge of sewage was removed. The primary effluent contained approximately 300 ppm of total dissolved solids which rose throughout the ultrafiltration process. The total volatile solids were initially about 100 ppm and the effluent level was very much less because it is this component which was being removed. It is of particular interest to note that the removal of BOD was poor initially, but as the process progressed the rejection increased. For example, at the beginning the rejection of BOD was about 60%; after 50% treatment it was about 90%; after 98% ultrafiltration at the end of the run, the rejection was about 97%. This change in rejection with the degree of ultrafiltration is readily explainable in terms of the heterogeneous composition of the feed. Initially, the low molecular weight, non-ionic materials such as the urea and the alcohols of low molecular weight passed through the membrane and

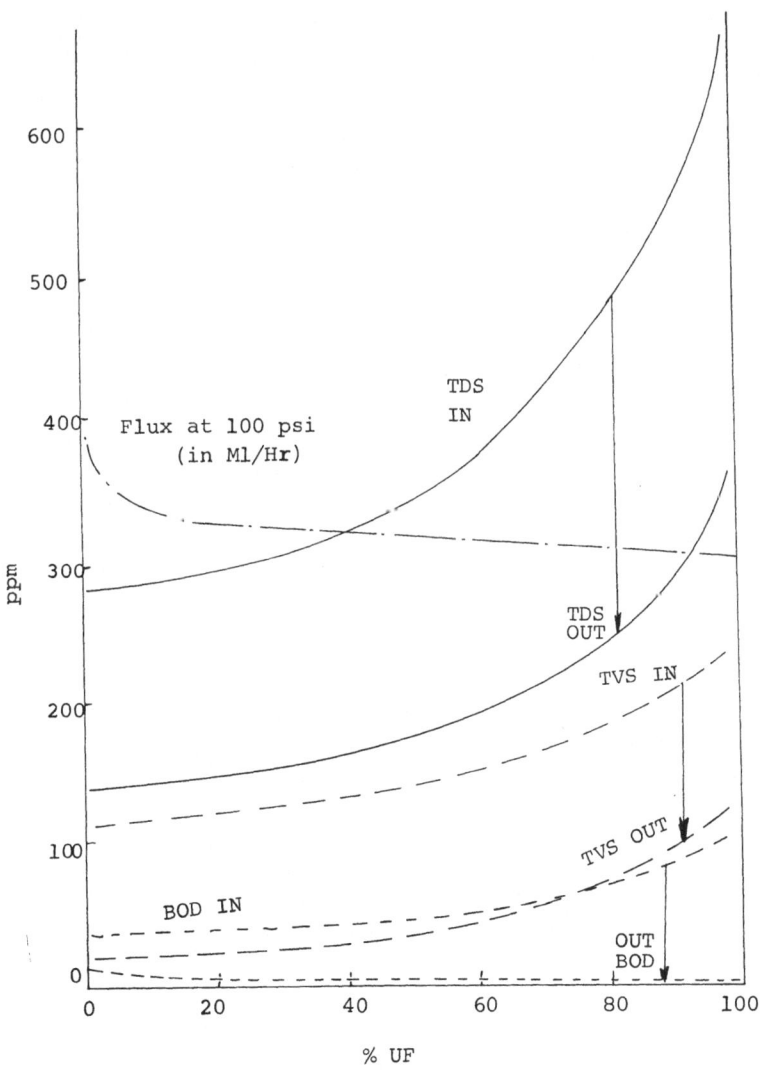

Fig. 9a.

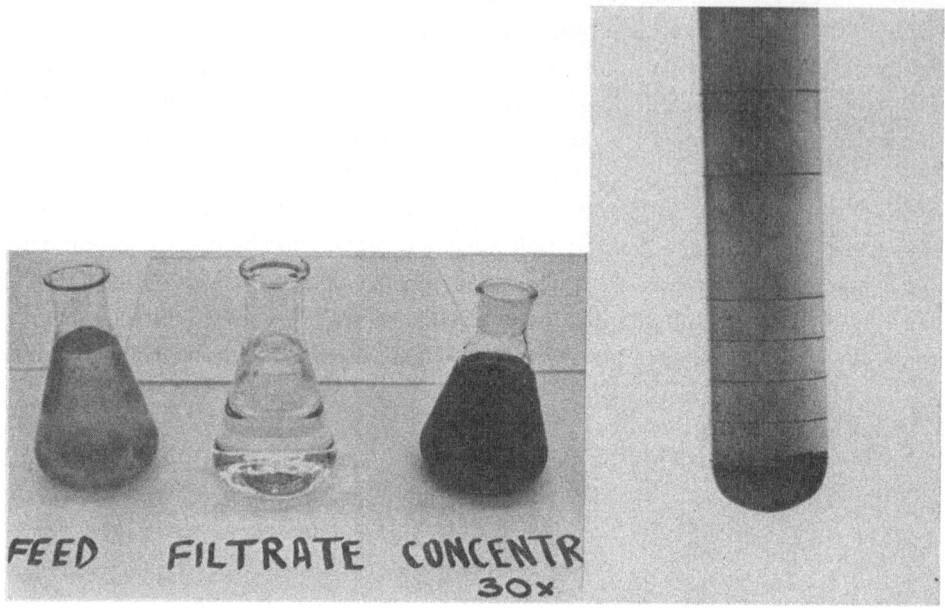

ORIGINAL PRIMARY EFFLUENT FEED,
SAMPLE OF LAST ULTRAFILTRATE TAKEN
AND FINAL 30-FOLD CONCENTRATE.

ENLARGED VIEW OF
SETTLED, 30-FOLD
CONCENTRATE

Fig. 9b.

were removed in the ultrafiltrate. Once these lower molecular weight materials had been removed, only the higher molecular weight materials which are more strongly rejected remained and these constituted most of the BOD. Accordingly, the average BOD of the ultrafiltrate was less than 8 ppm, which constitutes a high quality effluent.

Figure 9a shows an apparent discrepancy in material balance because the BOD level in the concentrate did not rise as sharply as is indicated by the ultrafiltrate composition. This was due to the fact that we allowed for solids settling at the bottom of the pressurized holding tank. At the end of the experiment this tank was drained and a reasonably good material balance in BOD was obtained.

It is interesting to note also from Figure 9 that there was an initial decrease in the rate of ultrafiltration and then a slow and virtually linear drop in ultrafiltration rate throughout the experiment. This initial decrease was due to the formation of a small amount of slime on the surface of the membrane, and even with our best membranes and using extremely high rates of stirring we were never able to eliminate this. However, this slime deposit did not increase during the experiment and the observed decrease in flow rate can be directly related to the increasing reverse osmotic pressure, because the total salt gradient across the membrane was increasing throughout the experiment due to rejection of salt.

Because these membranes had pores of molecular dimensions, the effluents were quite clear and sparkling and also quite sterile because the membrane pores excluded pathogens and bacteria. Of particular interest in this work was the fact that the concentrate was highly granular and settled rapidly. This is of particular practical importance. This effect is probably the result of the mutual coagulation of the negatively charged polysaccharide materials present in the effluent with the proteins present. No definitive experiments have been performed as yet, but it was observed in all of our experiments that the nature of the concentrate was granular, and that it settled rapidly and readily.

The photographs of Figure 9b serve also to illustrate the results obtained with the ultrafiltration of primary sewage effluent. In this particular run, 9 l of a Great Neck primary effluent (designated 'feed') was subjected to ultrafiltration to produce 8.7 l of ultrafiltrate and 300 ml of a 30-fold concentrate. The sample of the ultrafiltrate shown ('filtrate') was taken near the end of the run, and was typical of all of the product in that it was entirely clear and sparkling and also sterile because pathogens cannot pass these membranes. The concentrate sample shown in the flask had been agitated prior to taking the photograph. The test tube at the right shows the same concentrate after a few minutes of settling. The settled material had a solids content of about 20% solids. A simple filtration through a conventional filter increased the solids content to about 50%, a level at which this material can not only support its own combustion, but also can produce appreciable amounts of energy.

The quality of the effluent from this process is superior to that from conventional secondary waste treatment and indeed the equivalent of most tertiary treatment. A schematic diagram of a waste treatment process which would employ ultrafiltration entirely is given in Figure 10. A process of this kind is much more simple and has many advantages over conventional waste treatment. Since membrane modules are relatively

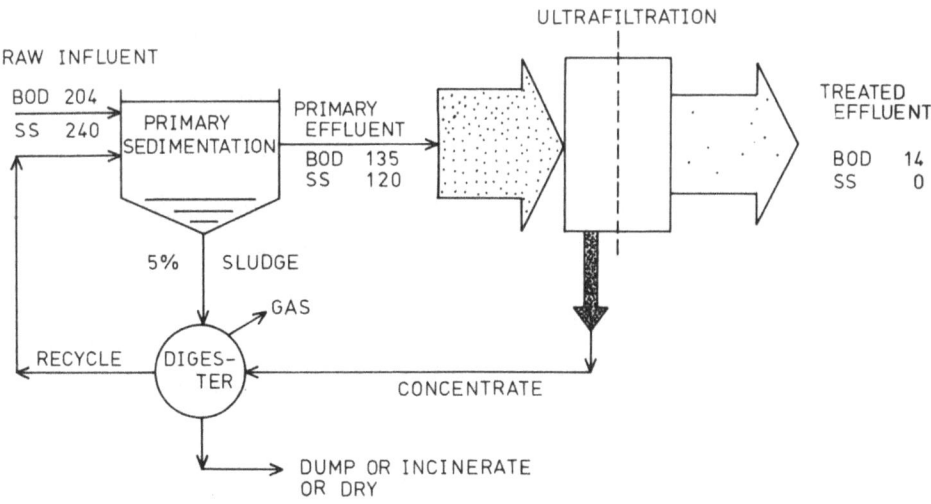

PRIMARY TREATMENT – ULTRAFILTRATION PLANT

Fig. 10.

small, this means that the cost of waste treatment of a relatively small community is not substantially greater than for the large communities. It makes unnecessary the usual requirement that small communities lay extensive pipe lines to large district sewage treatment plants because small plants of conventional nature are highly uneconomical. Conventional plants attain their quoted costs only when they are of large size.

Other industrial effluents were also studied. Whey is produced from cheese manufacture, and substantial amounts of whey are produced in the United States during the production of cottage cheese. Cottage cheese whey is potentially valuable in that it contains the valuable protein lactabumin and certain enzymes. Whey also contains about 3 or 4% lactose, smaller amounts of amino acids and other organic materials and about 1% salts. Customary practice has been to discard whey but many problems intervene because of its high BOD. While whey can be concentrated and spray dried, its economic value is largely dissapated when it is used in this form, largely as animal feed.

Samples of whey were passed through ultrafiltration membranes and here, as expected, there was complete removal of the protein in all of the membranes except the most porous ones. Following protein removal, by the use of 'tighter' membranes we could make separations of the lactose fraction from the salts which could then be passed to waste. Figure 11 shows two gel permeation chromatograms, the first of whey and the second of the ultrafiltrate. The earlier peaks for whey were due to the proteins and polypeptides present; the second, major symmetrical peak was due to lactose. For the ultrafiltrate, it was observed that all of the earlier peaks were absent except the one for lactose, with the sole exception of a small peak immediately following the void volume of the column. This had to be due to high molecular weight material, and subsequent experiments showed that it was due to the growth of bacteria because these particular flasks were not sterilized prior to the receipt of the ultrafiltered whey. When sterilized flasks were used, this initial peak disappeared and the effluent contained substantially only lactose and salts.

Extensive experiments were also carried out on effluents from pulp and paper mills, including the acid, neutral and alkaline pulps. In the usual pulp mill technology, a digestion liquor is used to treat pulp, the pulp is collected and the concentrated digestion liquor passed first into an evaporator to concentrate it and then into a furnace where the organic matter present acts to sustain combustion. The ash products thus obtained are recycled into the process. The distillate from the evaporators (which ordinarily contains a substantial amount of carry-over) is used to wash the pulp initially. The pulp is finally washed with relatively pure water and then the paper making process starts. Water pollution from pulp mills is largely due to either the pulp wash water or the overhead from the evaporators.

Since most of the contaminated solutions of this industry are essentially diluted digestion liquors, digestion liquors were obtained from pulp mills and were diluted at different ratios before being subjected to ultrafiltration. These are black odorous liquors of high solids contents (up to 15%), high in ligninsulfonic acid and a variety of other sugars and biocolloids.

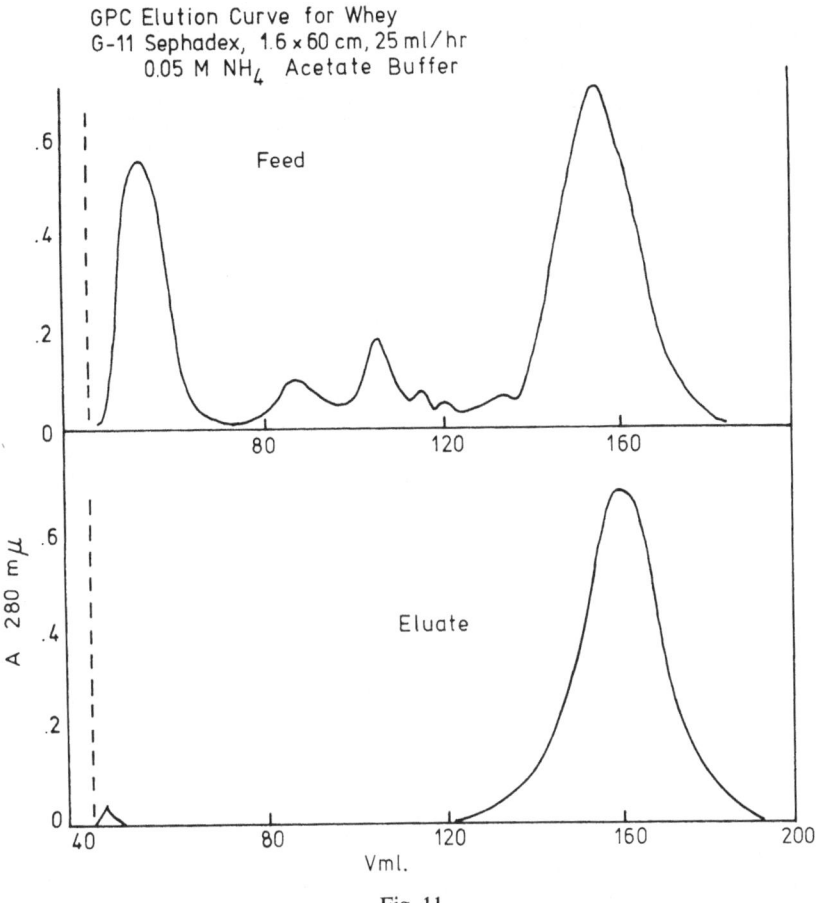

GPC Elution Curve for Whey
G-11 Sephadex, 1.6 × 60 cm, 25 ml/hr
0.05 M NH₄ Acetate Buffer

Fig. 11.

Figure 12 shows the results of a typical ultrafiltration experiment performed in the pilot plant cell on a caustic Kraft mill digestion liquor diluted 1/25. We observed that most of the color present was due to ionic materials because on dilution from 1/5 to 1/50 to 1/500, dilutions which would be encountered in industry, the rejection of color increased from 80 to 94 to 100%. Further, with the 1:5 diluted liquor the flux was 5 gfd; at 1:50 dilution the flux doubled and on further dilution to 1:500 (which approximates a fairly concentrated pulp wash water) the flux rose to 20 gfd at 60 psi, when the color rejection was essentially 100%. Color rejection was quite striking because the digestion liquor influent, even when diluted 1:500 was a black liquid, which when treated was quite clear with a slightly yellowish tinge.

3. Summary

In our experience, ultrafiltration membranes containing fixed sulfonic acid groups have many applications to problems of environmental pollution. In terms of modern sewage treatment, it is our feeling that ultrafiltration can have a substantial impact on

UF of Pulp Mill Effluent (1/25 Dilution)
at 90 psi

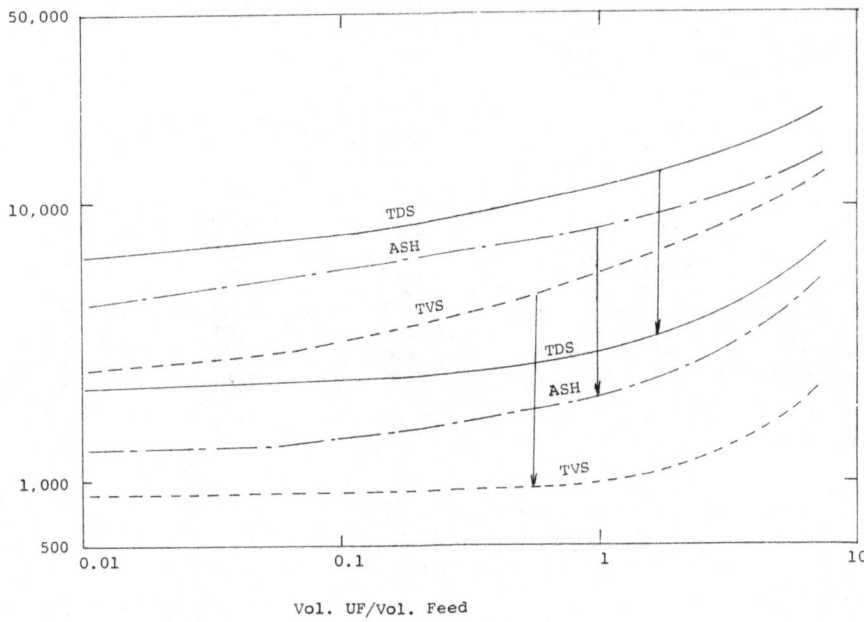

Vol. UF/Vol. Feed

Fig. 12.

water quality in the United States. It should be emphasized that ultrafiltration can handle the solids in sewage treatment in a manner much more efficient than that of conventional waste treatment. There are many other advantages to sewage treatment by ultrafiltration. Conventional sewage treatment converts the protein-like materials present to nitrates, and the high nitrates in water are undesirable for many reasons. On the other hand, if one ultrafilters out the protein and does not allow it to go through the process of nitrification, the level of nitrates are substantially reduced. This advantage alone is a major reason for considering ultrafiltration, because in many areas in the United States environmental groups have started to request tertiary treatment to reduce undesirable nitrate levels in ground waters.

There are many other products present in conventional secondary effluents which are undesirable. For example, heavy metals contaminate most municipal effluents, with much larger amounts in the effluents of industrial cities. These metals present are undesirable because they poison the bacteria of the activated sludge processes and reduce their efficiency. Second, these heavy metals return to the environment via sewage. In our experience, ultrafiltration removes a substantial amount of the heavy metals present because these are bound to the various biocolloids and proteins present and are concentrated thereby.

Ultrafiltration and filtration (as it is usually carried out) are quite different processes. In a conventional filtration process there is little lateral flow of liquid and solute, but it is almost all normal to the face of the membrane. With conventional filtration, the larger size particles or colloids form a filtration layer or 'schmutzdecke' on top of the filter media and this acts to remove more finely divided material. As the thickness of this layer builds up, the hydraulic permeability of the system decreases accordingly. In the ultrafiltration system, in contrast, the flow of fluid *parallel* to the surface of the membrane is high and always maintained to avoid forming a surface layer. Under these circumstances the rate of lateral flow is many times that of flow through the membrane so solutes do not concentrate at the membrane surface.

In the case of low molecular weight solutes whose motion away from the face of the membrane must occur via diffusion, there is a practical limit of the rate of ultrafiltration. This is defined by the extent of concentration polarization which is tolerable, and here the phenomena has been investigated in detail by many investigators, particularly A. Michaels. With relatively low molecular weight solutes which do not pass the membrane, the upper limits of ultrafiltration, except for extremely dilute solutions, is about 50 gfd. Ultrafiltration rates higher than this give rise to an increasing counter osmotic pressure such that at higher driving pressures the flux through the membrane remains virtually constant.

A somewhat different situation obtains in the ultrafiltration of particulate matter because here the pressure gradient next to the membrane acts to force particulate matter away from the face of the membrane. Under these circumstances considerably higher rates of ultrafiltration are obtainable, but again the specific, maximum rate of ultrafiltration is determined by several parameters. In our experience, any solution which could be readily pumped could be ultrafiltered, and this applied to a host of solutions and suspensions.

Ultrafiltration affords an excellent means of determining the intrinsic solubility of a crystalline material or the critical micelle concentration of a colloidal electrolyte. Solutions of various detergents were ultrafiltered and here it was found that the concentration of detergent in the ultrafiltrate closely approximated the critical micelle concentration for that material.

Another series of experiments, suspensions of finely divided precipitates were ultrafiltered. In one series, a suspension of particles of cadmium hydroxide-carbonate at pH 11 was subjected to ultrafiltration. This is an extremely fine, colloidal precipitate, such that even when filtered through hard filter paper an effluent concentration of 0.2 ppm was encountered. The same concentration is also encountered when polyelectrolytes were added to the same solution and it was clarified by flocculation.

Under ultrafiltration, the concentration of Cd in the ultrafiltrate was 0.01 ppm, closely approximating the intrinsic solubility of these salts. However, when high concentrations of ligands which form soluble complexes with cadmium (such as chloride) was present, the concentration of ultrafiltered Cd rose as high as 0.1 ppm, undoubtedly due to the presence of soluble complexes. Under these circumstances, the presence of a small concentration of sulfide acted to precipitate the highly insoluble cadmium sulfide, and an effluent concentration after ultrafiltration of 0.001 ppm was observed.

Accordingly, ultrafiltration is highly useful for determining the intrinsic solubility of even low molecular weight colloidal aggregates.

Acknowledgments

The studies described in this report were carried out with a number of different collaborators. Dr Carl Gryte collaborated on the preparation of several of the membrane systems, Mr J. Hsu and S. Mizrahi performed many of the ultrafiltration studies and Professor Gerald Palevsky of the Department of Civil and Sanitary Engineering of the City University of New York collaborated in the studies relating to sewage treatment. The National Science Foundation had, in the past, supported some fundamental studies of ours relating to fixed-charge electrodialysis membranes and helped lay the scientific basis for our ultrafiltration work. Much of the work described herein was supported by the New York State Science and Technology Foundation, and we are grateful for their support.

References

1. Gregor, H. P.: U.S. Patent 3, 808, 305, April 30, 1974.
2. Leszko, M. and Gregor, H. P.: Symposium *Thirty Years of Ion-Exchanges*, Leipzig, June, 1968, Akad. Verlag, Berlin, 1970, p. 681.
3. Kawabe, H., Jacobson, H., Miller, I. F., and Gregor, H. P.: *J. Coll. Interface Sci.* **21**, 79 (1966).

OSMOSIS AND ION TRANSPORT IN CHARGED POROUS MEMBRANES: A MACROSCOPIC, MECHANISTIC MODEL

AIN A. SONIN

Dept. of Mechanical Engineering, Massachusetts Institute of Technology Cambridge, Mass. 02139, U.S.A.

Abstract. Based on a simple mechanistic model where mass exchange takes place by continuum flow in charged, macroscopic pores, explicit equations are derived for water and ion fluxes through a charged membrane. The equations (which contain nonlinearities) show the cross coupling between the fluxes, and the coefficients are expressed explicitly in terms of the physical structure and fixed charge content of the membrane, as well as the properties of the permeant fluid. The model is applied to the hyperfiltration or reverse osmosis mode where salt is rejected as the solution is forced through the membrane under pressure.

1. Introduction

This paper discusses a mechanistic theory of osmotic flow through a membrane which is selective to a salt. An osmotic membrane can be characterized by certain phenomenological coefficients suggested by irreversible thermodynamics [1–3]. However, the approach of irreversible thermodynamics has its problems. First of all, it is an approach which treats the membrane as a 'black box', and does not enquire into how the physico-chemical structure of the membrane might determine the values or functional forms of the phenomenological coefficients which appear in it. Secondly, if the phenomenological equations are to be applied to membranes across which there are large concentration changes and through which the flow rate is high, several rather arbitrary assumptions have to be made if one is to derive relations between the fluxes and the forces applied across the membrane. The weaknesses of the resulting equations have been pointed out before [4].

Thus, mechanistic theories of osmosis and related phenomena hold some interest, to engineers who are concerned with tailoring synthetic membranes to specific needs as well as to plant and animal physiologists who want to understand the membranes they are dealing with. Several workers have addressed themselves to this problem [5–14], but the problem is far from completely resolved, partly because there are many physical mechanisms which may be involved, depending on the nature of the membrane and the permeant fluid, and partly because the underlying equations are usually too complex to allow explicit solutions for the general case. In membranes which are selective to neutral molecules, the forces acting to repel the solute are of molecular range, and it is necessary to adopt a kinetic model to describe the process [12, 14]. However, in many cases of interest the solute is in ionic form, and the interaction with the membrane may then occur primarily through long range electrostatic forces. With ionic solutes, therefore, one can conceive of a wide range of circumstances where the membrane can exert a repulsive force on the solute while there is a continuum flow of fluid through the macroscopically porous interstices of the membrane [5, 10, 11, 13]. This is the case we shall be concerned with below.

Eric Sélégny (ed.), Charged Gels and Membranes I, 255–265. All rights reserved.
Copyright © 1976 by D. Reidel Publishing Company, Dordrecht-Holland.

In what follows, we outline a mechanistic theory for osmotic flow. The membrane is pictured as being traversed by numerous macroscopic pores, through which solvent and solute can pass by continuum flow. A fixed charge distribution accounts for the selectivity. The theory is based entirely on the equations which govern continuum flow through macroscopic passages, and no extraneous conditions or boundary conditions are invoked. Our present analysis has much in common with the earlier work of Schlögl [5, 10]. However, for the case considered, we derive explicit and relatively simple analytic expressions for the water and ion fluxes in terms of the applied forces and the properties of the membrane and the permeant solution.

2. Model, Assumptions and Equations

The membrane is modelled as an electrically non-conducting solid traversed by numerous pores of constant radius a and length l, with $l \gg a$ (Figure 1). The values of l and a can be related to measurable bulk properties of the membrane [13]. Fixed charges are assumed to be distributed uniformly either within the solid parts of the membrane or on the walls of the pores*. The fluid which passes through the membrane is assumed to consist of an unionized solvent (our approximation to water) and a dissolved salt which is dissociated fully into a counter-ion whose charge is opposite

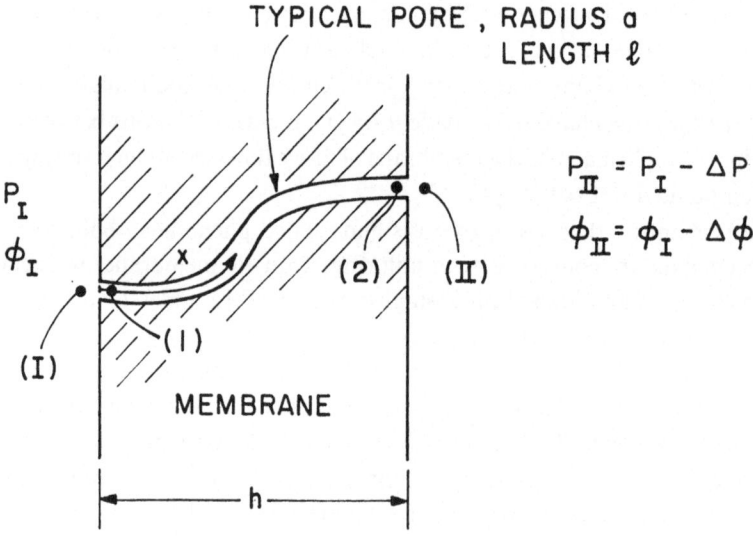

Fig. 1.   Pore model.

---

* We note that the assumption about the fixed charge distribution has a bearing on how the membrane performance depends on the exterior salt concentration. The uniform charge distribution is an obviously important special case. Another case of interest is the "constant wall potential" model used by Gross and Osterle [11] and Jacazio et al. [13], where the fixed charge distribution is approximated by a function which leaves the potential difference between the solid wall and the bounding solution the same on the upstream and downstream faces of the membrane and independent of concentration. These two models represent limiting cases which most likely bracket the behavior of real membranes with fixed charges.

to the fixed charge of the membrane and a co-ion having the same sign as the membrane. The laws for a dilute solution are assumed to apply, and only a steady state situation is considered.

The flow of the fluid as a whole in the pores is governed by the Navier-Stokes equation. In most porous membranes of interest the Reynolds number for the flow in the pores is very small, and the Navier-Stokes equation therefore takes the form

$$0 = -\nabla p + \mu \nabla^2 v + FZ_m(c_a - c_b)\nabla\phi \tag{1}$$

where p is the fluid pressure, μ the viscosity (to a good approximation equal to that of the solvent), v the *mass-centered* flow velocity, F Faraday's constant, ϕ the electric potential, and c_a and c_b are the counter-ion and co-ion concentrations, respectively, in equivalents per unit volume. Z_m represents the sign of the membrane's fixed charge, being defined to be $+1$ when the fixed charge is positive and -1 when it is negative. The three terms in Equation (1) represent, respectively, the pressure force, the viscous force, and the electrostatic body force acting on the fluid per unit volume.

In addition to Equation (1), which governs the flow of the fluid as a whole, we need equations for the fluxes of two of the three constituent species. It is convenient to use the well-established flux equations for the ion species:

$$\Gamma_a = c_a v - D_a \nabla c_a + Z_m Z_a D_a c_a \nabla\psi \tag{2}$$
$$\Gamma_b = c_b v - D_b \nabla c_b - Z_m Z_b D_b c_b \nabla\psi \tag{3}$$

Here, Γ_a and Γ_b are the counter-ion and co-ion flux densities, respectively, in equivalents per unit area per unit time, D_a and D_b are the diffusion coefficients of the counter-ions and co-ions, and

$$\psi = \frac{F\phi}{RT} \tag{4}$$

is the usual dimensionless potential, R being the universal gas constant and T the absolute temperature. In Equations (2) and (3), the terms on the right represent the convective, diffusive, and migration fluxes, respectively.

Equations (1)–(3), together with the overall mass conservation and the species conservation laws, as well as Poisson's equation between the potential and the ion concentrations, completely specify the problem at hand in the most general case. The general solution is, however, mathematically formidable and we shall make certain geometrical and operational assumptions which greatly simplify the analysis, but yet leave it appropriate for the cases of most interest in the present context. The assumptions are the following:

$$a < \lambda_D \tag{5}$$
$$l \gg \lambda_D \tag{6}$$
$$c_m \gg c_I, c_{II} \tag{7}$$
$$\frac{\bar{u}\lambda_D}{D_{a,b}} \ll 1 \tag{8}$$

Here,

$$\lambda_D = \left(\frac{\varepsilon RT}{F^2 c_I}\right)^{1/2} \tag{9}$$

is the Debye length (or Chapman-Gouy double layer thickness), ε being the permittivity of the solution, c_m is the magnitude of the membranes's fixed charge concentration, expressed in equivalents per unit *pore* volume, c_I and c_{II} are the concentrations of the bounding solutions (Figure 1) in equivalents per unit volume, $\bar{u}$ is the average flow speed in a pore, and $D_{a,b}$ represents either D_a or D_b. The assumption of low Reynolds number flow, already implicit in Equation (1), reads

$$\frac{\varrho \bar{u} a}{\mu} \ll 1 \tag{10}$$

where ϱ is the fluid density. This should be a very good approximation in most membranes of the osmotic kind.

Equation (5) is a very important simplifying assumption. It means that we restrict our attention to cases where the effect of the wall charge extends over the entire cross section of a pore [11, 13]. Only if this is so is it possible for the membrane to have high selectivity and good osmotic properties, which is the case of interest here. With Equation (5) applying, the electrostatic potential and the ionic concentrations are approximately uniform over any pore cross section, and using also the assumptions Equations (6) and (10), we can integrate Equations (1)–(3) over the pore cross section and obtain

$$\frac{8\mu}{a^2}\bar{u} = -\frac{dp}{dx} + Z_m RT (c_a - c_b)\frac{d\psi}{dx} \tag{11}$$

$$\bar{\Gamma}_a = c_a \bar{u} - D_a \frac{dc_a}{dx} + Z_m Z_a D_a c_a \frac{d\psi}{dx} \tag{12}$$

$$\bar{\Gamma}_b = c_b \bar{u} - D_b \frac{dc_b}{dx} - Z_m Z_b D_b c_b \frac{d\psi}{dx} \tag{13}$$

Here, $\bar{u}$ is the average mass-centered flow speed in a pore (the velocity profile is parabolic, as in ordinary Poiseuille flow), and $\bar{\Gamma}_a$ and $\bar{\Gamma}_b$ are the average counter-ion and co-ion flux densities, respectively. With assumptions Equations (5) and (6), furthermore, Poisson's equation for the electric potential can to a good approximation be replaced by a statement of local quasi-neutrality within the membrane as a whole,

$$c_a - c_b = c_m \tag{14}$$

Equations (11)–(14), together with the conservation laws for the fluid as a whole ($\bar{u} \simeq$ constant) and the species ($\bar{\Gamma}_a$, $\bar{\Gamma}_b$ constant), completely describe the flow *inside* each pore, that is, between points (1) and (2) in Figure 1. To complete the problem, we need also to establish how the properties change as one goes from a point just inside a pore (e.g. point 1 in Figure 1) to a point in the bounding solution just outside a pore (point I). The relationships we seek follow directly from our basic flow equations, without the need to draw on extraneous assumptions. As we go from point 1 to point I, the potential ϕ – and hence the ion concentrations – change in a distance of the order of the Debye length λ_D, which by Equation (6) is assumed to be very small compared to the characteristic length l for changes within a pore. It follows that in Equations (2) and (3), the diffusion and migration terms must be overwhelmingly large compared with the convection term and the flux term on the left, which are of the same order in the transition region 1-I as they are inside the pore. Hence, if the assumption Equation (8) applies, it follows that in the region between 1 and I (2 and II),

$$0 \simeq - \nabla c_a + Z_a Z_a c_a \nabla \psi \qquad (15)$$

$$0 \simeq - \nabla c_b - Z_m Z_b c_b \nabla \psi \qquad (16)$$

Each of these equations can readily be integrated to yield the Donnan relations between c_a and ψ and c_b and ψ, between points 1 and I and 2 and II. If we use the quasi-neutrality conditions at point (1) $(c_a + c_b = c_m)$ and point I $(c_a = c_b = c_1)$ and make use of the simplifying assumption of *high fixed charge*, Equation (7), the end result is

$$\frac{(c_b)_1}{c_m} \simeq \left(\frac{c_I}{c_m}\right)^{1+(Z_b/Z_a)} \left[1 - \frac{Z_b}{Z_a}\left(\frac{c_I}{c_m}\right)^{1+(Z_b/Z_a)}\right] \qquad (17)$$

$$Z_m Z_a (\psi_1 - \psi_I) = \ln \frac{c_m}{c_I} + \left(\frac{c_I}{c_m}\right)^{1+(Z_b/Z_a)} \qquad (18)$$

The value of c_a follows from Equation (14),

$$(c_a)_1 = c_m - (c_b)_1 \qquad (19)$$

Similar results are obtained for the relationship between points 2 and II at the other mouth of the pore.

We note that the high fixed charge assumption is quite appropriate if our primary concern is with membranes whose osmotic properties are significant. We shall see that only if the pore radius is smaller than the Debye length (Equation (5)) and the fixed charge is high (Equation (7)) can a charged membrane have a reasonably high selectivity to the solute.

Turning now to the equation of motion, Equation (1), we observe that in the transition region 1-I, the electric field term takes on a very large value because the potential drop of Equation (18) occurs over the very small double layer distance λ_D. The viscous term cannot change much in magnitude since by mass conservation the flow velocity is of the same order in the transition region as inside the pore, hence between 1 and I we must have to a good approximation

$$0 \simeq - \nabla p + F Z_m RT (c_a - c_b) \nabla \psi \qquad (20)$$

Now, if we substitute from Equations (15) and (16) into Equation (20), we obtain for the transition region 1-I

$$\nabla(p - \pi) = 0 \qquad (21)$$

that is, the difference between the pressure and the osmotic pressure

$$\pi \equiv \frac{c_a}{(Z_a} + \frac{c_b}{Z_b)} RT \tag{22}$$

is constant in the transition region. Using Equations (17) and (19), we thus get

$$p_1 - p_\mathrm{I} = \frac{c_m RT}{Z_a} - \left(\frac{1}{Z_a} + \frac{1}{Z_b}\right)\left[1 - \left(\frac{c_\mathrm{I}}{c_m}\right)^{Z_b/Z_a}\right] c_\mathrm{I} RT \tag{23}$$

and a completely analogous expression follows for $p_2 - p_\mathrm{II}$. Since $c_m \gg c_\mathrm{I}$, we see that the *pressure increases* as we go from a point I outside the membrane to a point 1 inside a pore in the membrane. The pressure gradient in the transition region is supported by an electrostatic body force, as expressed in Equation (20).

Note that the pressure increase from a point outside the membrane to a point inside a pore mouth will be larger on the side where the concentration is higher than on the side where the concentration is lower. Thus, a concentration difference creates a pressure gradient in a pore which tends to drive the fluid from the high concentration side to the low concentration side (*osmosis*). Another driving force for osmosis arises because a concentration difference also causes unequal potential jumps on the two sides (see Equation (18)). This sets up an electric field in a pore and causes flow via the body force term in Equation (11).

3. Solution

From Equations (11)–(14) for the flow inside the pore, and Equations (17), (18) and (23), which give the boundary conditions for the interior flow in terms of the properties of the bounding solutions, we can solve for the fluxes in terms of the applied pressure, the applied potential and the concentrations of the bounding solutions. It is convenient from an experimental viewpoint to obtain the solution in terms of the total mass flux, the coion flux, and the average current density in the pore

$$j = - FZ_m(\bar{\Gamma}_a - \bar{\Gamma}_b). \tag{24}$$

Without going into the details of the mathematics we shall quote here the final result for the relationship between the applied forces and the fluxes, subject to the assumption that the fixed charge density is high, as expressed in Equation (7). The following equations are correct to the leading terms in c_I/c_m and c_II/c_m:

$$\Delta p - \sigma \Delta \pi = \frac{8\mu l}{a^2}\left[(1 + \alpha)\,\bar{u} - \alpha\,\frac{Z_m j}{Fc_m}\right] \tag{25}$$

$$\Delta \psi - \lambda \frac{Z_m}{Z_a} \ln \frac{c_\mathrm{I}}{c_\mathrm{II}} = \left(\frac{1 - \gamma}{Z_a}\right)\left(\frac{jl}{Fc_m D_a} + \frac{Z_m \bar{u} l}{D_a}\right) \tag{26}$$

Here, Δ represents the difference between the value at point I and the value at point II in Figure 1. The dimensionless quantity

$$\sigma = 1 - \frac{(1 + Z_b D_b / Z_a D_a)}{(1 + Z_b / Z_a)} \frac{\left[(c_I/c_m)^{1+(Z_b/Z_a)} - (c_{II}/c_m)^{1+(Z_b/Z_a)}\right]}{(c_I/c_m - c_{II}/c_m)} \quad (27)$$

indicates how much ion leakage (as measured by the nonzero values of c_I/c_m and c_{II}/c_m) reduces the static equilibrium pressure, and the quantity

$$\lambda = 1 - \frac{D_b/D_a\left[(c_I/c_m)^{1+(Z_b/Z_a)} - (c_{II}/c_m)^{1+(Z_b/Z_a)}\right]}{\ln(c_I/c_{II})} \quad (28)$$

indicates how leakage reduces the equilibrium potential which can be developed across the membrane. The parameter

$$\alpha = \frac{a^2 c_m R T}{8\mu Z_a D_a (1 + \gamma)} \quad (29)$$

measures the coupling between mass flux and current (electro-osmosis) and indicates how much the presence of fixed charge reduces the hydraulic permeability of the membrane. (Fixed charge clearly always reduces the permeability. This effect has been observed experimentally in porous clays by Michaels and Lin [15].) The remaining undefined parameter is

$$\gamma = \left(1 + \frac{Z_b D_b}{Z_a D_a}\right)\left\{\left(\frac{c_I}{c_m}\right)^{1+(Z_b/Z_a)} - \left[\frac{1}{\xi} - \frac{1}{e^\xi - 1}\right]\left[\left(\frac{c_I}{c_m}\right)^{1+(Z_b/Z_a)} - \left(\frac{c_{II}}{c_m}\right)^{1+(Z_b/Z_a)}\right]\right\} \quad (30)$$

where

$$\xi = \left(1 + \frac{Z_b D_b}{Z_a D_a}\right)\frac{D_a}{D_b}\frac{\bar{u}l}{D_a} + \frac{Z_m Z_b}{Z_a}\frac{jl}{F c_m D_a}. \quad (31)$$

It is clear that for our present case where $c_m \gg c_I, c_{II}$, γ is *always small compared with unity*. However, to the extent that its value is important, it indicates a *nonlinear effect* between the driving forces (Δp, $\Delta \psi$) and the fluxes (j, $\bar{u}$), given the concentrations c_I and c_{II}.

Equations (25)–(26) are derived analytically as asymptotic solutions for small c_I/c_m and c_{II}/c_m, and are correct to the leading terms in those quantities. To the same accuracy, the average *coion flux* density $\bar{\Gamma}_b$ through the membrane is given by

$$\frac{\bar{\Gamma}_b l}{c_m D_b} = \frac{\xi}{e^\xi - 1}\left[\left(\frac{c_I}{c_m}\right)^{1+(Z_b/Z_a)} e^\xi - \left(\frac{c_{II}}{c_m}\right)^{1+(Z_b/Z_a)}\right] \quad (32)$$

Clearly, the co-ion flux is *nonlinear* in the driving forces. We note, however, that in the two limits $\xi \ll 1$ and $\xi \gg 1$, linear relations do obtain but with different coefficients.

Equations (25)–(32) relate the fluxes (here taken as $\bar{u}$, j and $\bar{\Gamma}_b$) to the applied pressure and applied voltage across the membrane, and the concentrations of the bounding solutions. These equations contain the various flow and coupling phenomena exhibited by charged, porous membranes. Note that we are dealing here with a physical model which is about as simple as it can be for our present purposes. The membrane is characterized by only four quantities, two physicochemical (Z_m and c_m)

and two which describe its architecture (a and l). Effective values for a and l can in principle be determined from measurements of bulk properties of the membrane [13], and j and $\bar{u}$ can similarly be related to the apparent or superficial current density and flow velocity through the membrane. The permeant fluid needs to be specified merely by its viscosity and the diffusion coefficients and charge numbers of the co-ion and counter-ion in the solvent.

As we noted, consistent with our assumption that $c_m \gg c_I, c_{II}$, the dimensionless quantities σ and λ are slightly smaller than unity, but approach unity from below as c_I/c_m and c_{II}/c_m approach zero (that is, the membrane becomes perfectly selective). Similarly, the quantity γ is very small compared with unity, and approaches zero in the limit of infinitely high fixed charge. The coupling parameter α defined in Equation (29) may in principle have an arbitrary value. However, it is a straightforward matter to demonstrate that as long as the relation between the pore radius and the fixed charge density is such that the Debye length based on the latter is larger than the former, as we have implicitly assumed, α will generally have a value small compared with unity.

Finally, it is worth reiterating that in this macroscopic model osmotic bulk mass transfer takes place by means of a bulk flow in the pores, caused in part by a pressure gradient inside the pores and in part by an electric field which acts on the space charge within the pores and thereby exerts a body force on the fluid (see Equation (11)). In the absence of an applied pressure drop across the membrane, a difference in osmolarity between the bounding solutions still induces, via a balance between electrostatic and pressure forces over the transition regions at the pore mouths, a pressure gradient inside the pores (see Equation (23) and following), and this gives rise to a bulk mass flow, that is, to osmosis.

4. Application to Hyperfiltration

Several experimental studies have been published on the characteristics of salt and water flow through membranes with fixed charges [13, 16, 17]. Most of the available data on fairly finely porous membranes, for which $\lambda_D > a$ as assumed in the present theory, is on salt rejection by the membrane in the reverse osmosis or hyperfiltration mode, where all the fluid on side II of the membrane originates from side I and no current is allowed to flow ($j = 0$). Under these conditions one has, for a steady state situation,

$$c_{II} = \frac{\bar{\Gamma}_b}{\bar{u}} \tag{34}$$

and c_{II}/c_I, Δp, and $\Delta \psi$ all become functions of $\bar{u}$ (see Equations (32), (25) and (26), respectively). The functional relationships for the salt rejection R for the case $Z_a = Z_b$ is obtained as

$$R = 1 - \frac{c_{II}}{c_I} = 1 - \frac{(e^{u^*} - 1)\, c_m/c_I}{2(1 + D_b/D_a)} \left\{ \left[1 + \frac{4(1 + D_b/D_a)^2\, (c_I/c_m)^2\, e^{u^*}}{(e^{u^*} - 1)^2} \right]^{1/2} - 1 \right\} \tag{35}$$

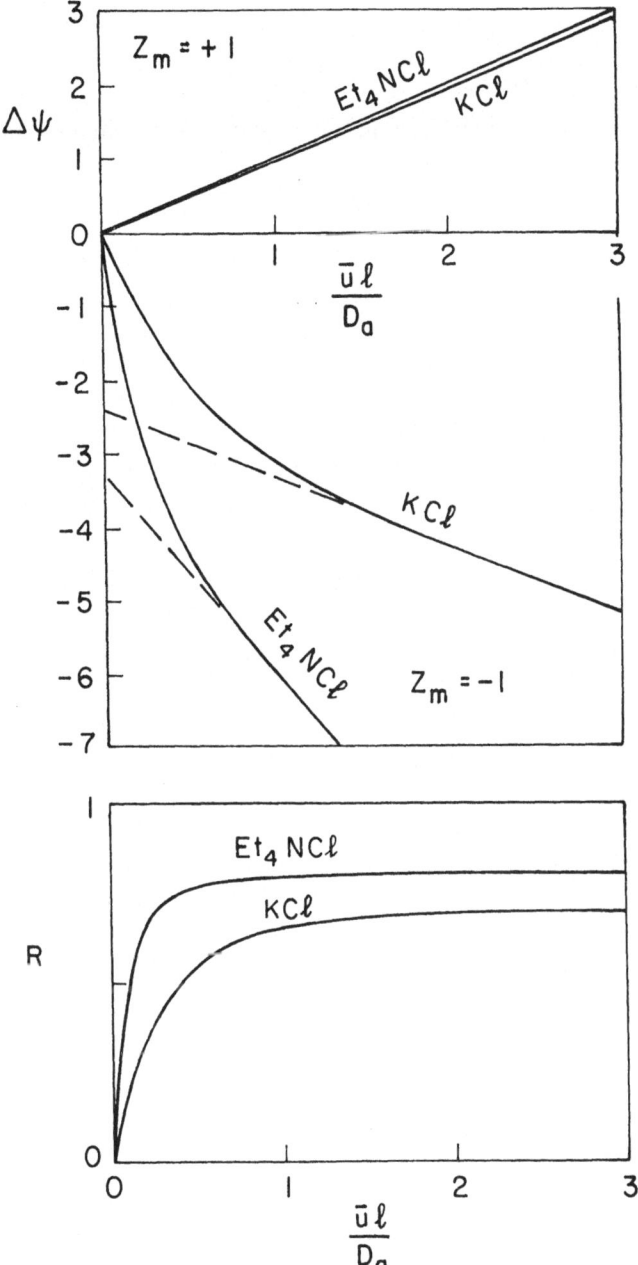

Fig. 2.   Theoretical membrane potential, as indicated by Ag-AgCl electrodes, (above) and salt rejection (below), for hyperfiltration of KCl and Et$_4$NCl solutions through a membrane with positive and negative fixed charge (cf. [17]). We have taken $c_I/c_m = 0.15$ and assumed that the diffusion coefficient of Et$_4$N$^+$ is one third that of K$^+$ and Cl$^-$ in water.

where

$$u^* = \left(1 + \frac{D_a}{D_b}\right) \frac{\bar{u}l}{D_a} \qquad (36)$$

is a dimensionless flow speed (Péclet number). The corresponding streaming potential is obtained from Equation (26) as

$$\Delta\psi = Z_m \frac{\bar{u}l}{D_a} + \lambda \frac{Z_m}{Z_a} \ln \frac{c_I}{c_{II}} \qquad (37)$$

where c_{II}/c_I is given as a function of $\bar{u}$ from (35).

Tanny et al. [17] have made extensive measurements of both streaming potential and salt rejection in charged, porous, collodion-type membranes, using 5×10^{-3} M and 5×10^{-2} M KCl and Et$_4$NCl (tetraethylammonium chloride) solutions. In their membranes the pore radius was of the order 10–100 Å, so that at the lower concentration, our theoretical assumption $\lambda_D > a$ should be a good approximation. Insufficient data is available for a complete quantitative comparison between their data and the present theory. However, in Figure 2 we show some theoretical curves for the salt rejection and streaming potential obtained from Equations (35) and (37) as functions of mass flow rate. The curve for potential has been adjusted to show the value which would be indicated by Ag-Ag Cl electrodes, as in [17]. We have taken $c_I/c_m = 0.15$ and assumed that the diffusion coefficient of Et$_4$N$^+$ is one third that of K$^+$a nd Cl$^-$ in water. Figure 2 is to be compared with the curves for the 5×10^{-3} M solutions in Figures 4 and 5 of [17]. The agreement is certainly promising. We note in particular that the theory explains the trend of the behavior with respect to the co-ion and counterion species and also the qualitative trend with respect to feed concentration c_I (the data in [17] for 5×10^{-2} M solutions cannot be compared directly with our theory, since both the assumptions $\lambda_D > a$ and $c_m \gg c_I$, c_{II} were most likely violated). However, more careful comparisons are obviously required before one can accept the quantitative accuracy of a theory based on as simple a model as the present one.

Acknowledgement

This research was sponsored by the National Science Foundation under Grant GK-35798X of the Fluid Mechanics Program.

References

1. Kedem, O. and Katchalsky, A.: Biochim. Biophys. Acta 27, 229 (1958).
2. Katchalsky, A. and Curran, P. F.: Nonequilibrium Thermodynamics in Biophysics, Harvard Univ. Press, Cambridge, Mass., 1967.
3. Spiegler, K. S. and Kedem, O.: Desalination 1, 311 (1966).
4. Bresler, E. H. and Wendt, R. P.: Science 163, 944 (1969).
5. Schlögl, R.: Z. Physik. Chem. Neue Folge 3, 73 (1955).
6. Mauro, A.: Science 126, 252 (1957); 149, 867 (1965).
7. Garby, L.: Acta Physiol. Scand. 40, Suppl. 137 (1957).

8. Ray, P. M.: *Plant Physiology* **35**, 783 (1960).
9. Scatchard, G.: *J. Phys. Chem.* **68**, 1059 (1964).
10. Schlögl, R.: *Stofftransport durch Membranen*, Steinkopf, Darmstadt, 1964.
11. Gross, R. J. and Osterle, J. F.: *J. Chem. Phys.* **49**, 228 (1968).
12. Manning, G. S.: *J. Chem. Phys.* **49**, 2668 (1968).
13. Jacazio, G., Probstein, R. F., Sonin, A. A., and Yung, D.: *J. Phys. Chem.* **76**, 4015 (1972).
14. Läuger, P.: *Biochem. Biophys. Acta* **311**, 423 (1973).
15. Michaels, A. S. and Lin, C. S.: *Ind. Eng. Chem.* **47**, 1249 (1955).
16. McKelvey, J. G., Jr., Spiegler, K. S., and Wyllie, M. R. J.: *J. Phys. Chem.* **61**, 174 (1957).
17. Tanny, G., Hoffer, E., and Kedem, O.: G. Milazzo *et al.* (eds.), in *Biological Aspects of Electro-chemistry*, Birkhauser Verlag, Basel, 1971, pp. 619–630.

MEMBRANE POTENTIALS OF
ASYMMETRIC CELLULOSE ACETATE MEMBRANES

W. PUSCH

Max-Planck-Institut für Biophysik, 6 Frankfurt am Main 70, Kennedy Allee 70, Germany

Abstract. The membrane potentials for homogeneous and asymmetric cellulose acetate membranes were measured using different electrolyte solutions. The experimental results manifest a strong difference between the membrane potential of a homogeneous membrane on the one hand and the membrane potentials of asymmetric membranes on the other hand. Furthermore, the membrane potentials of asymmetric cellulose acetate membranes depend on the boundary conditions and thus on the direction of volume flow. The results can be theoretically treated taking into account the concentration profiles within the porous sublayer of the asymmetric membranes.

1. Introduction

The membrane potential of synthetic membranes such as collodion, cellophane, and stronger ion exchange membranes has been widely discussed by several authors [1, 2, 3, 4] on the basis of the theory of Teorell [5], Meyer and Sievers [6]. From measurements of membrane potentials one readily obtains information on membrane properties such as selectivity, fixed charge concentration, and the ratio of ion mobilities within the membrane. The membrane potential, which is connected with the transport of ions across the membrane, consists of Donnan potentials at the membrane/solution phase boundaries, a diffusion potential within the membrane, and a streaming potential. If the mechanical permeability of the membrane is zero, no streaming potential exists. Thus the membrane potential can be calculated by integrating the corresponding Nernst-Planck equation. As Schlögl [7] has shown, the following relation results if the fixed charges are homogeneously distributed over the membrane

$$\Delta\Psi = \frac{RT}{\omega z \cdot F}\left\{\ln\frac{c_s'}{c_s''}\cdot\frac{\sqrt{1+4y''^2}+1}{\sqrt{1+4y'^2}+1}+\omega U\ln\frac{\sqrt{1+4y''^2}-\omega U}{\sqrt{1+4y'^2}-\omega U}\right\} \tag{1}$$

where

ω = −1 or +1 for cation or anion exchangers, respectively
X = fixed charge concentration
y = zc_s/X; c_s = electrolyte concentration; z = valency
U = $(D_+ - D_-)/(D_+ + D_-)$
D_+ = diffusion coefficient of cation within the membrane
D_- = diffusion coefficient of anion within the membrane
F = Faraday number

The first term of Equation (1) corresponds to the difference of the two Donnan potentials at the membrane/solution phase boundaries whereas the second term corresponds to the diffusion potential within the membrane. The nature of the membrane potential is schematically shown in Figure 1. Using calibration curves, it is possible to get ap-

Eric Sélégny (ed.), Charged Gels and Membranes I, 267–276. All rights reserved.

proximate values of the fixed charge concentration and the ratio of ion mobilities within a membrane by measuring $\Delta\Psi$ as a function of the salt concentration on one side of the membrane maintaining the salt concentration on the other side of the membrane constant. The membrane potential is usually measured using calomel electrodes which dip in the two solutions on both sides of the membrane.

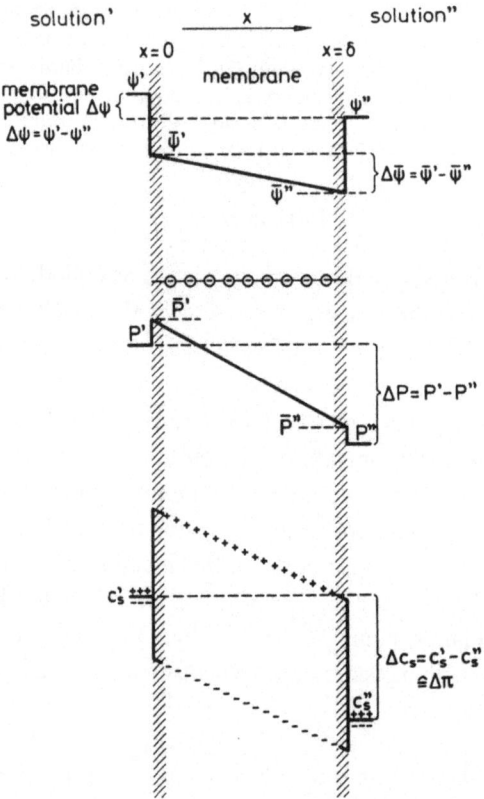

Fig. 1.   Schematic representation of the nature of membrane potential for a cation exchange membrane.

Only limited measurements [8] of membrane potentials of asymmetric cellulose acetate membranes have been reported in the literature, although Spiegler *et al.* [9] have estimated membrane potentials from the overall potential measured across an asymmetric cellulose acetate membrane in hyperfiltration experiments. Therefore, the goal of the work which will be reported was to get some information on the membrane potential of asymmetric cellulose acetate membranes.

2. Description of Equipment

Figure 2 represents a dialysis cell which was used to measure the membrane potential. The cell has been described in detail elsewhere [10]. The conductivity electrodes indicated in the figure were replaced by calomel electrodes. The potential between these electrodes was measured using a conventional potentiometer produced by Radiometer Copenhagen or Orion. The solutions on both sides of the membrane were well stirred. Since no pressure was applied, the membrane was not supported.

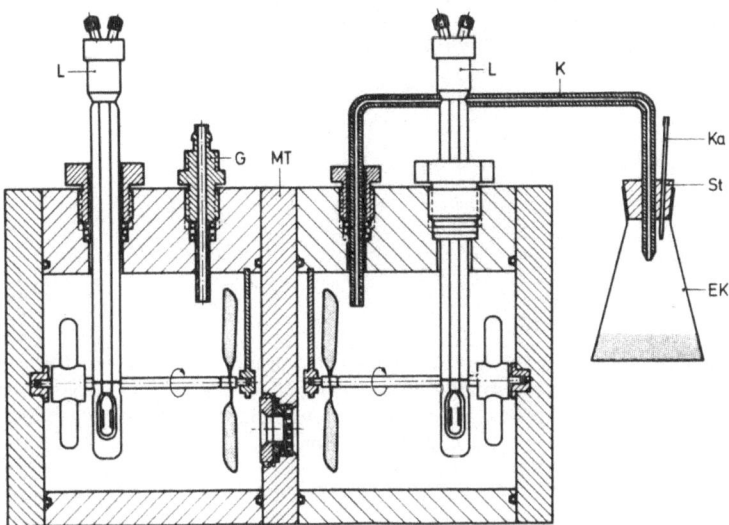

Fig. 2.   Cross-sectional view of dialysis cell.

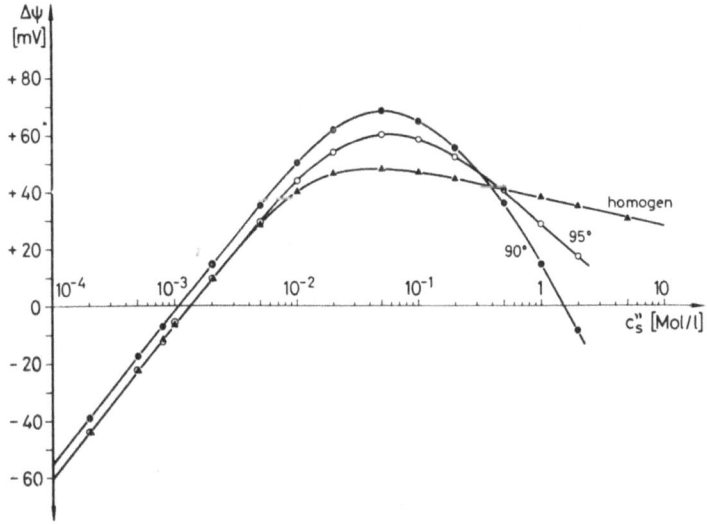

Fig. 3.   Membrane potential $\Delta\Psi$ as a function of salt concentration c_s'' using different annealed asymmetric cellulose acetate membranes and a homogeneous cellulose acetate membrane ($T = 298$ K; $c_s' = 0.001$ m NaCl; active layer adjacent to salt solution c_s').

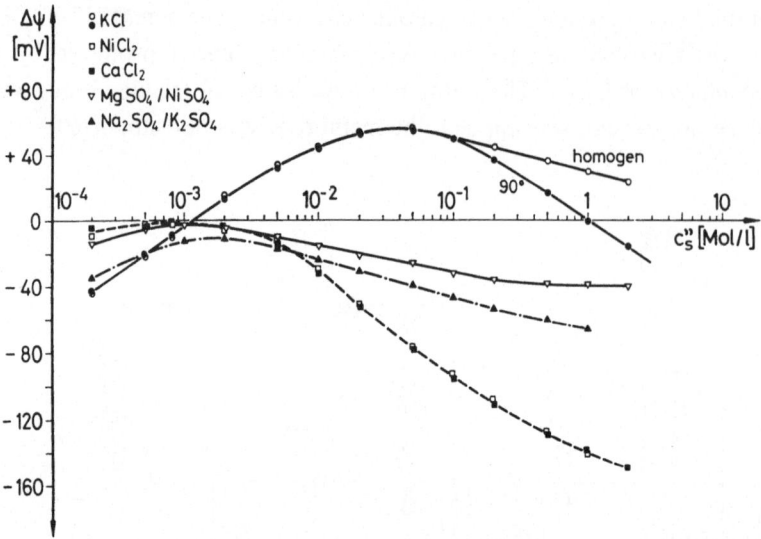

Fig. 4.   Membrane potential $\Delta\Psi$ as a function of salt concentration c_s'' for an asymmetric membrane annealed at 90 °C using different salt solutions ($T = 298$ K; $c_s' = 0.001$ m NaCl). For comparison the membrane potential as a function of concentration c_s'' of a homogeneous membrane is also shown using KCl solutions.

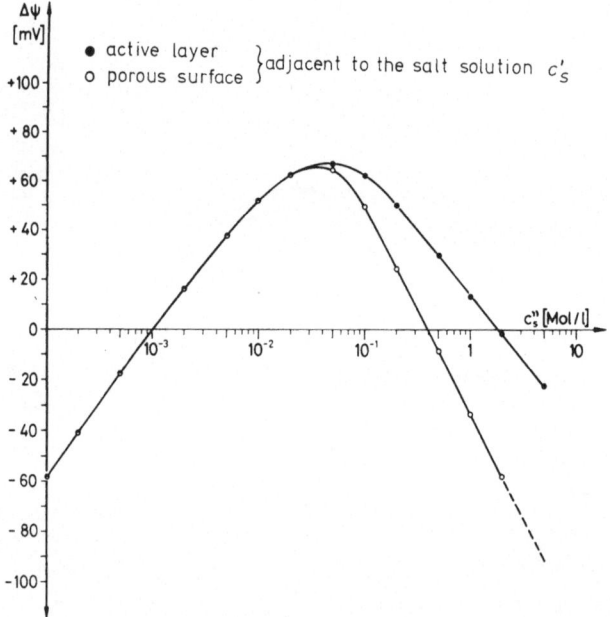

Fig. 5.   Membrane potential $\Delta\Psi$ as a function of salt concentration c_s'' using an asymmetric cellulose acetate membrane annealed at 85 °C. The data represented by the solid circles were taken with the active layer adjacent to the salt solution c_s' and the open circles, with the porous surface adjacent to the salt solution c_s' ($T = 298$ K; $c_s' = 0.001$ m NaCl).

3. Measurements and Results

Using NaCl, KCl, $NiCl_2$, $CaCl_2$, $MgSO_4$, $NiSO_4$, and K_2SO_4 solutions, the membrane potentials as a function of salt concentration were measured with homogeneous and asymmetric cellulose acetate membranes. The homogeneous cellulose acetate membranes were prepared from solutions of Bayer cellulose acetate K 700 in chloroform and the asymmetric cellulose acetate membranes were prepared in the normal way described elsewhere [11]. With sulfate solutions the calomel electrodes have been replaced by electrodes with a K_2SO_4-bridge. In Figure 3 results are graphically represented which were obtained with a homogeneous membrane and with asymmetric membranes annealed at 90 °C and 95 °C using NaCl solutions. Figure 4 represents similar results for different salts using an asymmetric cellulose acetate membrane which was annealed at 90 °C.

Furthermore, the membrane potential for an asymmetric cellulose acetate membrane annealed at 85 °C was measured using NaCl solutions. Two different boundary conditions were used. First, the active layer was adjacent to the solution with the constant salt concentration, which was 0.001 m NaCl. Second, the membrane was turned over and the porous surface was adjacent to the constant salt solution. The results obtained are shown graphically in Figure 5.

4. Discussion and Short Theoretical Analysis

As one can see from Figure 3 the three curves of $\Delta\Psi$ as a function of c_s'' have quite different shapes. With increasing annealing temperature the potential curves move toward the curve of the homogeneous membrane. The potential maximum decreases going from 90 °C annealing temperature to the homogeneous membrane.

Using calibration curves, one can determine the ratio D_+/D_- for the three membranes. Here, D_+ is the diffusion coefficient of the sodium ion within the membrane and D_- the diffusion coefficient of the chloride ion. With the homogeneous membrane one finds with NaCl $D_+/D_- \simeq 0.5$. Using the same analysis with the asymmetric cellulose acetate membranes, one finds a ratio that depends on the annealing temperature of the membrane. For a membrane annealed at 90 °C, one obtains a value of approximately 0.02. However, if one assumes that the active layer of the asymmetric membrane is similar to a homogeneous membrane, nearly the same ratio D_+/D_- would be expected for the homogeneous and the asymmetric membranes. As is discussed below, this discrepancy is not due to a real difference between the active layer and the homogeneous membrane, but rather is the result of a concentration gradient within the porous sublayer of the asymmetric membrane. Figure 4 shows that the same difference in the membrane potentials of a homogeneous and an asymmetric membrane exists for KCl solutions.

Furthermore, a difference in membrane potential exists for an asymmetric membrane depending on the boundary conditions with regard to the salt solutions. If the membrane is turned over, so that the porous surface of the asymmetric membrane is juxtaposed with the solution of low salt concentration (0.001 m NaCl), the experiment-ally determined membrane potential deviates from the membrane potential determined with the active layer adjacent to the solution of low salt concentration. The difference is pronounced at high salt concentrations. The fact that the deviation always starts at larger salt concentrations indicates that a volume flow across the asymmetric membrane may cause the difference in membrane potentials at different orientations of the membranes.

To explain these experimental results on the basis of this idea, it is necessary to look at the differences between a homogeneous and an asymmetric cellulose acetate membrane. In addition to the different structures, the most important difference is the volume flow across the asymmetric membranes which is caused by the osmotic difference at larger salt concentrations c_s''. For a better understanding of the influence of volume flow over transport properties of asymmetric membranes, the mechanical permeability l_p of an asymmetric membrane has to be considered as a function of salt concentration using different boundary conditions. Using the following linear relation between volume flux q (cm s^{-1}) and pressure difference ΔP (at) and osmotic difference $\Delta \Pi$ (at) across the membrane:

$$q = l_p(\Delta P - \sigma \Delta \Pi) \tag{2}$$

where l_p = mechanical permeability (cm s^{-1} at) and σ = reflection coefficient of the membrane, the mechanical permeability can be determined by measuring q as a function of ΔP maintaining $\Delta \Pi$ constant. The mechanical permeability was obtained with the following three different boundary conditions:
(1) The salt/concentration c_s' was varied between 0 and 0.2 m NaCl keeping $c_s''=0$ (pure water in the corresponding cell compartment). The active layer of the membrane

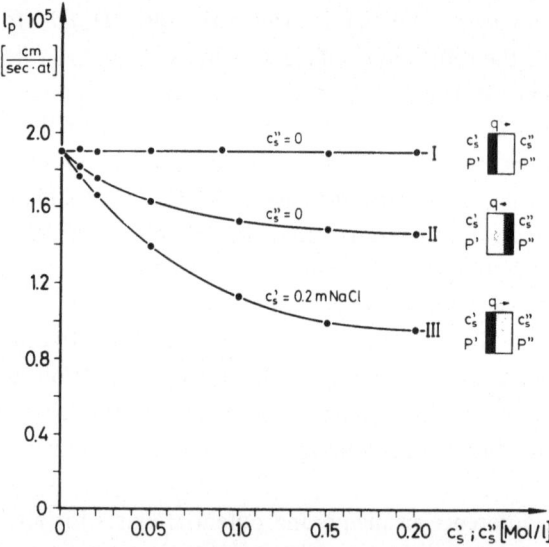

Fig. 6.   Volume flux, q, as a function of concentration c_s' or c_s'', respectively. Different boundary conditions were choosen with curves I, II, and III.

was juxtaposed with the salt solution. The volume flow was directed from phase (')
to phase (").

(2) The same arrangement as in case (1) but the membrane was turned over so that
the active layer was juxtaposed with pure water. The volume flow, therefore, was
directed from phase (") to phase (').

(3) The salt concentration c_s'' was varied between 0 and 0.2 m NaCl keeping
$c_s' = 0.2$ m NaCl. The active layer was again justaposed with phase ('), the solution
of constant salt concentration. The volume flow was directed from phase (') to
phase (").

The experimental results of these measurements are graphically represented in
Figure 6. As can be seen from this figure, l_p depends strongly on the concentration
of that solution which is adjacent to the porous surface of the asymmetric cellulose
acetate membrane. From these experimental results it was recently concluded [12]
that there exists a concentration gradient within the porous sublayer of the asymmetric
membrane. This is shown schematically in Figure 7. This concentration gradient
leads to a smaller salt concentration c_s'' at the interface active layer/porous sublayer
and an effective osmotic difference $\Delta \Pi_{eff}$ across the active layer which determines the
volume flux, q, across the active layer and thus, the volume flux through the entire
membrane. With this argument it is assumed that the reflection coefficient of the
porous sublayer is zero or nearly zero. As a result of this concentration gradient
within the porous sublayer the volume flux decreases. Thus, the reduced mechanical
permeability is a consequence of the smaller volume flux. Taking into account the

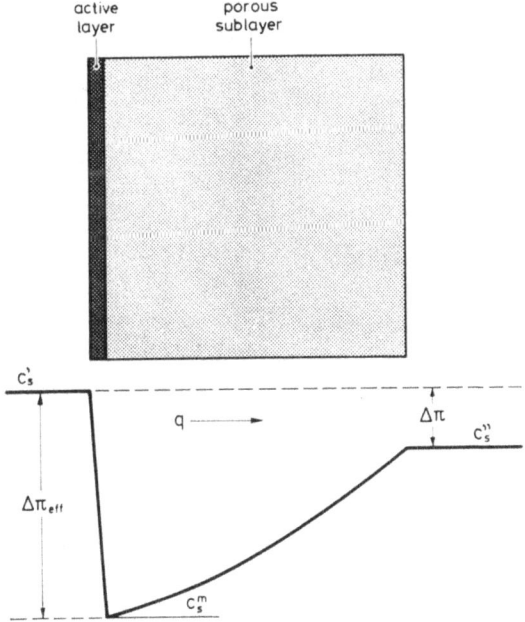

Fig. 7.   Schematic representation of the concentration profile within the porous sublayer of an asymmetric cellulose acetate membrane caused by a volume flux, q.

concentration gradient within the porous sublayer and the effect of this gradient on the volume flux, it is possible to explain the experimental results with regard to the membrane potential of asymmetric cellulose acetate membranes.

The experimentally confirmed concentration gradient within the porous sublayer of an asymmetric membrane is connected with a diffusion potential. Thus, the membrane potential of an asymmetric membrane $\Delta\Psi_a$ is composed of the membrane potential of the active layer and the diffusion potential within the porous sublayer. If it is assumed that the membrane potential of the active layer is nearly the same as that of a homogeneous membrane $\Delta\Psi_h$, the difference in membrane potential between a homogeneous and an asymmetric membrane should behave similar to a diffusion potential. In Figure 8 the difference $\Delta(\Delta\Psi)=\Delta\Psi_h-\Delta\Psi_a$ is, therefore, plotted as a function of $\ln\Pi''$. As is seen from this figure, there exists a linear relation between $\Delta(\Delta\Psi)$ and $\ln\Pi''$ at larger salt concentrations c_s''. From the slope of the corresponding straight line the ratio D_+/D_- within the porous sublayer can be determined. It results $D_+/D_-\simeq0.5$ for KCl. This is the same value as the value D_+/D_- for the homogeneous membrane. With this treatment the differences in membrane potentials between homogeneous and asymmetric membranes are attributed to the concentration gradient within the porous sublayer of the asymmetric membrane caused by the volume flow.

The difference in membrane potential of an asymmetric membrane which is created by using different boundary conditions is also caused by concentration profiles within the porous sublayer. In case 1 of Figure 5, the active layer is adjacent to phase ('), which is the salt solution of constant concentration $c_s'=0.001$ NaCl. Here, the volume

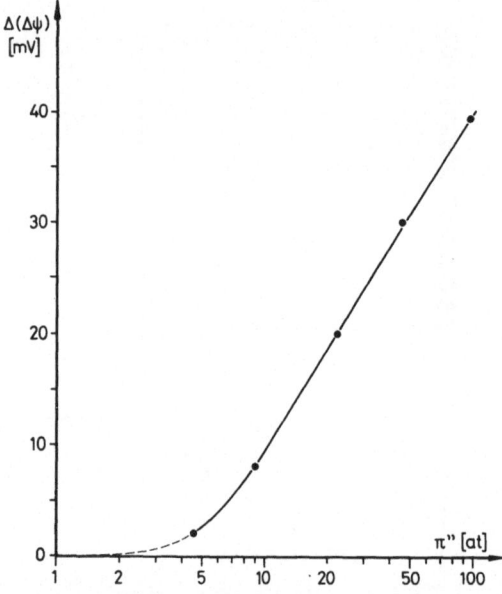

Fig. 8. Difference of membrane potentials $\Delta(\Delta\Psi)$ between a homogeneous and an asymmetric cellulose acetate membrane annealed at 90 °C using KCl ($T=298$ °C; $\Delta(\Delta\Psi)=\Delta\Psi_h-\Delta\Psi_a$; $c_s'=0.001$ m KCl).

flow across the membrane produces a concentration gradient which leads to a smaller salt concentration $c_{s_1}^m$ at the interface active layer/porous sublayer. The volume flux, q_1, in this case, is determined by the effective osmotic difference $\Delta\Pi_{\text{eff}}$ which is smaller than $\Delta\Pi = \Pi' - \Pi''$. The corresponding ratio $c_s''/c_{s_1}^m$, which determines the magnitude of the diffusion potential within the porous sublayer, can be estimated using a simplified Nernst-Planck equation. As was reported recently [12] the result is as follows:

$$c_s''/c_{s_1}^m = \exp\left(q_1\delta/D_s\right) \tag{3}$$

where δ = thickness of porous sublayer (cm); and D_s = diffusion coefficient of salt within porous sublayer $(\text{cm}^2\ \text{s}^{-1})$.

In case 2 of Figure (5), where the membrane is turned over, the active layer is adjacent to phase $(")$, in which the salt concentration c_s'' is varied. There the volume flux q_2 also produces a concentration gradient within the porous sublayer. This leads also to a salt concentration $c_{s_2}^m$ at the interface porous sublayer/active layer which is larger than c_s'. Since c_s' itself is small compared to c_s'' at larger values of c_s'', $c_{s_2}^m$ is also small compared to c_s''. Therefore, the osmotic difference across the active layer is nearly equal to the osmotic difference, $\Delta\Pi$, across the entire membrane. Because $\Delta\Pi > \Delta\Pi_{\text{eff}}$ of case 1, the volume flux $q_2 > q_1$ and thus, $c_{s_2}^m/c_s' > c_s''/c_{s_1}^m$. As a consequence of this, there results a larger diffusion potential within the porous sublayer in case 2. The ratio $c_{s_2}^m/c_s'$ is related to q_2 by the already mentioned relation:

$$c_{s_2}^m/c_s' = \exp\left(q_2\delta/D_s\right) \tag{4}$$

In deriving this relation, it was assumed that the salt rejection r of the membrane is approximately 1. If the following common relation for the diffusion potential in the porous sublayer is applied

$$\Delta\Psi_D = (URT/F)\ln\left(c_s^a/c_s^b\right) \tag{5}$$

the two diffusion potentials within the porous sublayer can be estimated. Thus, the following relations are obtained:

$$\Delta\Psi_{D_1} = (URT/F)\ln\left(c_s''/c_{s_1}^m\right) \tag{6}$$

$$\Delta\Psi_{D_2} = (URT/F)\ln\left(c_{s_2}^m/c_s'\right) \tag{7}$$

With relations (3) and (4) the following relation is obtained for the difference of the diffusion potentials between case 2 and case 1 of Figure 5 if differences in the membrane potentials of the active layer are neglected:

$$\Delta(\Delta\Psi) \simeq \Delta\Psi_{D_2} - \Delta\Psi_{D_1} = (URT/F)(q_2 - q_1)(\delta/D_s) \tag{8}$$

Thus, it is shown that the difference of the membrane potentials of an asymmetric membrane under different boundary conditions is due to the different magnitude of the corresponding volume fluxes, q_1 and q_2. The experimental findings and the semi-quantitative theoretical treatment show that it is not possible to determine parameters like D_+/D_- for asymmetric membranes from membrane potential measurements as can be done for homogeneous membranes. In this regard, the statement of Minning [9] is pertinent. He has discussed our earlier membrane potential measurements with asymmetric cellulose acetate membranes annealed at 60 °C, 80 °C, and 90 °C, and he stated that "the nonlinearities in the curves $\Delta\Psi$ as a function of c_s''/c_s' and the unusually large values of $\bar{\tau}_c^a$ (transport number of the cation) remained unexplained". On the other hand, Minning noted that the transport numbers of Na^+ ions for large salt concentrations approach the value of the Hittorf transport number of the cation t_c in free NaCl solution. This is in agreement with our own observations [13], and is consistent with the rationale proposed here, i.e. that the diffusion potential in the porous sublayer is the dominant factor in membrane potential measurements with asymmetric membranes at high salt concentrations. Furthermore, the double layer model of the asymmetric cellulose acetate membrane is confirmed by these experimental findings as well as the large concentration profiles within the porous sublayer, which strongly effect the experimentally observed transport coefficients.

The author is indebted to Professor R. Schlögl for his interest in this work as well as to Dr H. Lonsdale for reading the proofs. Furthermore, the author is very much obliged to Professor K. S. Spiegler for the invitation to the ASI meeting and also for many stimulating discussions. – The work was supported by the 'Bundesminister für Forschung und Technologie', Bonn, Germany.

References

1. Spiegler, K. S.: *J. Electrochem. Soc.* **100**, 303 C (1953).
2. Schlögl, R.: *Z. Elektrochem.* **56**, 644 (1952).
3. Schmid, G.: *Z. Elektrochem.* **54**, 424 (1950); **55**, 229 (1951).
4. Nagasawa, M. and Kagawa, I.: *Discussions of the Faraday Soc.* **21**, 52 (1956).
5. Teorell, T.: *Proc. Soc. Exp. Biol. Med.* **33**, 282 (1935); *Trans. Faraday Soc.* **33**, 1053 and 1086 (1937).
6. Meyer, K. H. and Sievers, J. F.: *Helv. Chim. Acta* **19**, 649 (1936).
7. Schlögl, R.: *Stofftransport durch Membranen*, Dr.-Dietrich-Steinkopff-Verlag, Darmstadt 1964, pp. 88–91.
8. Kepinski, J. and Chlubek, N.: *Proc. 3rd Intl. Symposium on Fresh Water from the Sea*, Vol. 2, p. 487 (1970).
9. Spiegler, K. S. and Minning, Ch. P.: *Streaming Potentials in Hyperfiltration of Saline Water*, Ph. D. Thesis of Ph. P. Minning at UCLA, Berkely, Sea Water Conversion Laboratory, 1973.
10. Pusch, W.: *Chemie-Ing.-Technik* **20**, 1216 (1973).
11. Gröpl, R. and Pusch, W.: *Desalination* **8**, 277 (1970).
12. Pusch, W.: *Proc. 4th Intl. Symposium on Fresh Water from the Sea*, Vol. 4, p. 321 (1973).
13. Drioli, E. and Pusch, W.: Unpublished results, publication in preparation.

POLARIZATION AT MEMBRANE-SOLUTION INTERFACES

IN REVERSE OSMOSIS (HYPERFILTRATION)

C. P. MINNING* and K. S. SPIEGLER**

College of Engineering, University of California, Berkeley, Calif., U.S.A.

Abstract. The study of electrical potential differences between electrolyte solutions separated by desalination membranes had led to the conclusion that not only ion-exchange membranes used in electrodialysis but also modified cellulose acetate membranes, used in reverse osmosis, often contain fixed charges.

Systematic variation of the concentrations, flow rates, and pressures of sodium chloride solutions flowing past the inner surface of a cylindrical modified cellulose acetate membrane (cured at 94°C, nominal diameter $1'' = 2.54$ cm) led to estimates of the interfacial salt concentration buildup and showed that streaming potentials of 6.2 to 8.0 mV/100 psi (0.91–1.17 mV atm^{-1}) were obtained for 0.5 M (29225 ppm) and 0.1 M (5845 ppm) sodium chloride solutions respectively. The polarity indicated a fixed negative charge.

List of Symbols

B	mean ion activity in solution
a_p	streaming potential differential, volt (unit pressure)$^{-1}$, Equation (8)
c'	concentration of high-pressure solution, mole cm^{-3}
c''	concentration of low-pressure solution, mole cm^{-3}
$c_d^\infty [=(c'')^\infty]$	concentration of low-pressure solution, *extrapolated to infinite circulation velocity*, mole cm^{-3}
c_f	feed concentration, mole cm^{-3} (because the hyperfiltration membrane test section was short, $c_f \simeq c'$)
c_s^i	salt concentration at solution-membrane interface, mole cm^{-3}
D	diffusion coefficient (diffusivity), cm^2 sec^{-1}
$E_{Ag/AgCl}$	electrode potential of a silver-silver chloride electrode, volt
$E^0{}_{Ag/AgCl}$	standard potential of a silver-silver chloride electrode, volt
E_J	junction potential (difference), volt
ΔE_m	membrane potential, volt (potential difference caused by difference of concentrations of two solutions at equal pressure separated by a membrane)
ΔE_{st}	streaming potential, volt (potential difference caused by difference of pressures of two solutions of equal concentration separated by a membrane)
ΔE_{tm}	transmembrane potential (difference), volt
ΔE^∞	electric potential difference measured between silver-silver chloride electrodes in permeate (hyperfiltrate) and high-pressure (circulating) NaCl solution respectively, volt *extrapolated to infinite circulation velocity*
ΔE_{obs}	electric potential difference measured between silver-silver chloride electrodes in permeate (hyperfiltrate) and high-pressure (circulating) NaCl solution respectively volt
f_F	Fanning friction factor
$f_{F,o}$	Fanning friction factor for outer wall of annulus
$\mathscr{F}$	Faraday's constant, 0.965×10^5, coul eq^{-1}
j	Chilton-Colburn factor
J_v	volume flux through membrane, cm sec^{-1}
l	length, cm

* Present address: Hughes Aircraft Company, Space and Communications Group, 366/C681, P.O. Box 92919, Los Angeles, Calif. 90009.
** To whom correspondence should be addressed.

Eric Sélégny (ed.), Charged Gels and Membranes I, 277–298. All rights reserved.
Copyright © 1976 by D. Reidel Publishing Company, Dordrecht-Holland.

p	pressure; units are MNewton m^{-2} (1 MNewton m^{-2}=10 bar=9.87) atm unless otherwise indicated
r	radius, cm
R	Universal gas constant, 8.317 joule (K)$^{-1}$ mole^{-1}
$\mathbf{R}$	salt rejection$\equiv(c'_s-c''_s)/c'_s$
$\mathbf{R}^\infty$	salt rejection in absence of polarization ('intrinsic' rejection)
$\mathbf{Re}$	Reynolds number$\equiv d\bar{u}/\nu_k$ (d=characteristic dimension, cm; for turbulent flow in cylindrical pipes, d is the pipe diameter)
r	radius, cm
r_i	inner radius of annulus, cm
r_o	outer radius of annulus, cm
$\mathbf{Sc}$	Schmidt number$\equiv\nu_k/D_s$
T	absolute temperature, K
$\bar{u}$	(superficial) solution flow velocity, cm sec^{-1}
$\bar{V}$	partial molal volume, cm^3 mole^{-1}
$y_\pm$	mean ionic activity coefficient
δ	thickness of diffusion layer, cm
μ	viscosity, g cm^{-1} sec^{-1}
ν_k	kinematic ('dynamic') viscosity (=viscosity/density), cm^2 sec^{-1} ($\nu_k\equiv\mu/\varrho$)
ϱ	density, g cm^{-3}
τ_o	shear stress at outer wall of annulus, dyne cm^{-2}
τ_+	cation transport number in solution
$\bar{\tau}_+$	cation transport number in membrane (apparent)

Subscripts

a	annulus
d	dilute solution
f	in electrolyte-solution feed
s	salt
v	volume

Superscripts

i	at membrane-solution interface
∞	extrapolated to infinite feed circulation velocity

Sign Conventions

(1) Positive direction is from left to right. (2) Fluxes from left to right are counted positive. (3) The operator, Δ, for finite differences refers to the value on the right minus the value on the left, as does conventionally the differential operator, d.

Driving forces are of the general form $(-d\tilde{\mu}/dz)$, $\tilde{\mu}$ being a general potential. Thus positive values of the driving force, $(-d\tilde{\mu}/dz)>0$, lead to positive fluxes. For example, Ohm's law is written as

$$i=(1/\varrho')\underbrace{(-dE/dz)}_{\text{'Driving force'}}$$

and Fick's law

$$J=D\underbrace{(-dc/dz)}_{\text{'Driving force'}}$$

1. Introduction

When solutions are compressed through ionic membranes, streaming potentials across the membranes arise, and the solution emerging on the low-pressure side often has a composition different from the high-pressure feed. Some of the pertinent previous researches on these phenomena are summarized in reference [1]. In the present work, previous findings [2, 3] were confirmed, viz. that modified cellulose acetate membranes used in industrial hyperfiltration installations [4, 5] are ionic (i.e. 'charged' polymers). These findings should be compared to contrasting previous assumptions about non-modified cellulosic membranes which were considered to exclude electrolytes only by a non-ionic mechanism [6, 7]. Measurements of electrical potentials drops resulting from the pressure difference across the membrane were used to calculate the interfacial salt concentration in the high-pressure solution. All these measurements were performed in an apparatus with annular flow geometry. The meaning of symbols used and a consistent set of units are presented in the List of Symbols.

2. Theory

In the process of hyperfiltration, the salt concentration, c_s^i, at the pressurized solution/ membrane interface increases above the bulk concentration, c_s', of the pressurized solution [8, 9]. Brian [10] derived the following expression for the ratio of these concentrations (polarization ratio) from mass-transfer theory for *turbulent* flow conditions in a cylindrical tube:

$$\frac{c_s^i}{c_s'} = \frac{\exp\left[J_v \mathbf{Sc}^{2/3}/(j\bar{u})\right]}{\mathbf{R}^\infty + (1 - \mathbf{R}^\infty)\exp\left[J_v \mathbf{Sc}^{2/3}/(j\bar{u})\right]} \simeq$$

$$\simeq \frac{\exp\left[2J_v \mathbf{Sc}^{2/3}/(f_F \bar{u})\right]}{\mathbf{R}^\infty + (1 - \mathbf{R}^\infty)\exp\left[2J_v \mathbf{Sc}^{2/3}/(f_F \bar{u})\right]} \quad (1)$$

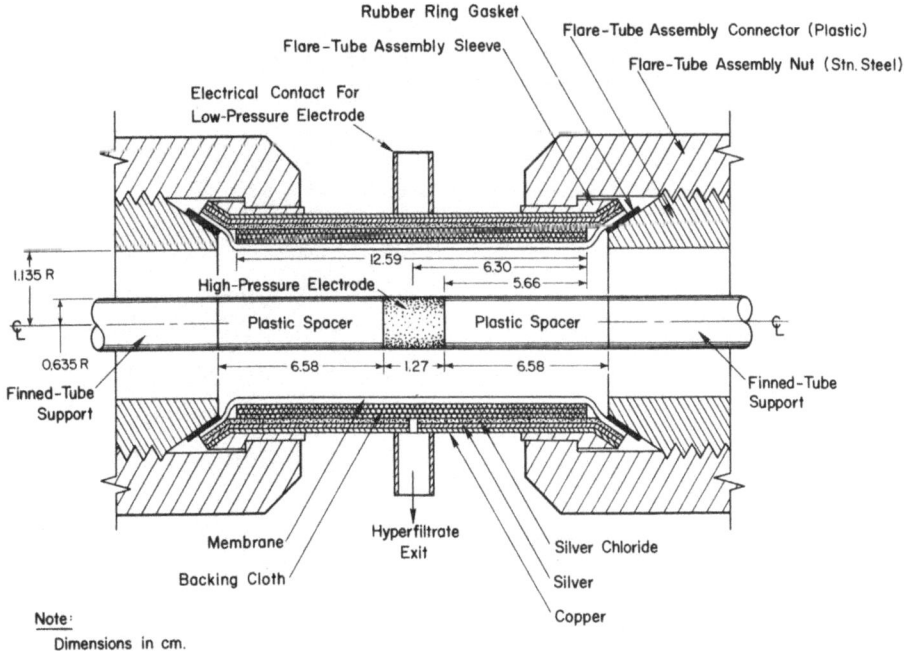

Fig. 1. Dimensions of membrane section [11].

In our hyperfiltration experiments, a cylindrical membrane of nominal diameter $1''$ was used, but since *electrical* parameters were to be measured, a cylindrical tube consisting of Ag/AgCl electrode and plastic spacers, all of diameter $0.5''$, was inserted in the axis of the cylinder. (The setup is shown in Figure 1 which shows the pertinent dimensions; further details are given in reference [11].)

The polarization equations for annuli were developed by C. P. Minning [11] and applied to this problem. After substitution of the pertinent material constant (e.g. the Schmidt number) and geometrical parameters [outer diameter of flow annulus (=inner diameter of cylindrical membrane), $2r_o = 2.27$ cm, diameter of central rod $2r_i = 1.27$ cm], the equations listed in Table I were obtained. Table II lists the solution properties.

In all calculations and graphs, the following definition of the *Reynolds number of the annulus* was used [13, p. 205].

$$\text{Re}_a \equiv 2(r_o - r_i)\bar{u}/v_k \tag{2}$$

where r_o and r_i are the outer and inner radii of the annulus respectively, $\bar{u}$ the superficial velocity, and v_k the kinematic viscosity.

The expression for the Fanning friction factor for the outer wall of the annulus in turbulent flow (in terms of the Reynolds number and the geometry of the annulus) is not simple [11, Equation III-55, p. 95], but in the range of turbulent flow rates in-

TABLE I

Polarization equations for annulus [11]

c'_s = salt concentration in high-pressure solution (mole cm^{-3})
$c_s{}^i$ = interfacial salt concentration (mole cm^{-3})
J_v = hyperfiltration rate (cm s^{-1})
$\bar{u}$ = axial flow velocity of salt solution
$f_{F,o}$ $[\equiv 2\tau_0/(\rho\bar{u}^2)]$ = Fanning friction factor. Subscript o refers to *outer* wall of annulus. Details are discussed in reference [11].
R^∞ = salt rejection of membrane in absence of polarization.

c'_s mole cm^{-3} × 10^3	Equation for $c_s{}^i/c'_s$-laminar flow[a]	Equation for $c_s{}^i/c'_s$-turbulent flow[a]
0.01	$c_s{}^i/c'_s = 1 + 2.05 \times 10^3 (J_v/\bar{u}^{1/3})$	$c_s{}^i/c'_s = \dfrac{\exp[1498 (J_v/\bar{u})(1/\sqrt{f_{F,o}})]}{R^\infty + (1-R^\infty)\exp[1498 (J_v/\bar{u}_t)(1/\sqrt{f_{F,o}})]}$
0.10	$c_s{}^i/c_s' = 1 + 2.11 \times 10^3 (J_v/\bar{u}^{1/3})$	$c_s{}^i/c'_s = \dfrac{\exp[1551 (J_v/\bar{u})(1/\sqrt{f_{F,o}})]}{R^\infty + (1-R^\infty)\exp[1551 (J_v/\bar{u})(1/\sqrt{f_{F,o}})]}$
0.50	$c_s{}^i/c'_s = 1 + 2.12 \times 10^3 (J_v/\bar{u}^{1/3})$	$c_s{}^i/c'_s = \dfrac{\exp[1577 (J_v/\bar{u})(1/\sqrt{f_{F,o}})]}{R^\infty + (1-R^\infty)\exp[1577 (J_v/\bar{u})(1/\sqrt{f_{F,o}})]}$

[a] All numerical constants refer to the following situation:
Electrolyte: Sodium chloride. Dimensions as in Figure 1. Cylindrical modified cellulose acetate membrane, cured at 94 °C. (Inner diameter 2.27 cm) In center: cylindrical silver-silver chloride electrode, diameter 1.27 cm. For 0.01 M NaCl solution, $D_{NaCl} = 1.545 \times 10^{-5}$ cm^2 s^{-1} (25 °C).

TABLE II

Properties of sodium chloride solutions used in experiments

Conc., c_s mole l^{-1}	Molality, m mole (kgH$_2$O)$^{-1}$	Density, ϱ g cm^{-3}	Viscosity, μ g cm^{-1} s^{-1}	Dynamic viscosity, ν_k cm^2 s^{-1}	Salt diffus. coeff., D_s cm^{-2} s^{-1} [a]	Schmidt number Sc	Osmotic coeff., Φ	Osmotic press., Π atm
0.50	0.5062	1.0173	9.33×10^{-3}	9.17×10^{-3}	1.474×10^{-5}	622	0.921	22.8
0.10	0.1005	1.0011	9.03×10^{-3}	9.02×10^{-3}	1.483×10^{-5}	608	0.932	4.57
0.01	0.0100	0.9974	8.95×10^{-3}	8.97×10^{-3}	1.545×10^{-5}	581	0.968	0.473

[a] From reference [12].

vestigated here, numerically quite similar values for the polarization ratio, c_s^i/c_s', are obtained when the following simple formula (similar to Blasius' formula for round tubes, [13, p. 164]) is used [14, p. 198]:

$$f_{F,o} \simeq 0.076 \, \mathbf{Re}_a^{-1/4} \tag{3}$$

Using the equations in Table I, *theoretical* curves of polarization ratio versus Reynolds number were computed [11].

The *experimental* values of the polarization ratio were determined by electrical measurements, as described in the following.

2.1. METHOD FOR THE CALCULATION OF THE POLARIZATION RATIO FROM THE MEASURED ELECTRIC POTENTIAL DIFFERENCE

Previous stages of this work [2] have shown that the modified cellulose acetate membranes used were by no means non-permselective, as had been postulated for non-modified cellulose acetate membranes [6, 7], but rather exhibited cationic selectivity for sodium ions, i.e. in sodium chloride solutions the membrane acted like a cation-exchange membrane.

In the experiments described here, the electric potential difference was measured between a silver-silver chloride electrode in pressurized sodium chloride solution (brine) circulating on the inside of the membrane tube and a silver-silver chloride electrode in the permeate (hyperfiltrate). These measurements were performed at different brine pressures and circulation rates.

As schematically shown in Figure 2, the observed potential difference between the electrodes called the 'hyperfiltration potential,' is composed of the difference of the two electrode potentials (i.e. the potential differences between electrodes and adjacent solutions), $\Delta E_{Ag/AgCl}$, a liquid junction potential, ΔE_J, between the circulating brine and the more concentrated interfacial solution, and the potential difference across the membrane or 'transmembrane potential', ΔE_{tm}. The latter, in turn, may be considered as composed of two components, viz. the 'membrane potential,' ΔE_m, i.e. the potential difference due to a concentration difference between two solutions at the same pressure separated by the membrane, and the 'streaming potential,' ΔE_{st}, viz. the potential difference between two identical solutions at different pressure separated by the membrane.

In the following, the method of evaluation of the measurement of electric potential difference for estimating the interfacial concentration is outlined.

Briefly, given the concentrations of the feed and product solutions, the electrode potentials can be readily calculated. To find the transport number of the sodium ions in the membrane from the transmembrane potentials, the total potential difference, ΔE_{obs}, was measured at different brine circulation rates, $\bar{u}$, and extrapolated on a plot of ΔE_{obs} vs. $1/\bar{u}$ to $(1/\bar{u})=0$ i.e. to infinite circulation velocity at which the polarization was assumed to be negligible. This corresponds to 'ideal stirring' in conventional measurements of the membrane potential. In this manner, the junction potential is eliminated. The streaming potential is determined by performing these measurements

at different pressures and determining that change of the measured potential which is
due to the pressure difference alone.

This reasoning is *quantitatively* expressed by summing the expressions for the
different components of the total potential as demonstrated in Figure 2. Primes and
double primes refer to the left (high-pressure) and right (atmospheric pressure) NaCl
solutions respectively in Figure 2. The superscript *i* refers to a property of the solution
at the brine-membrane interface.

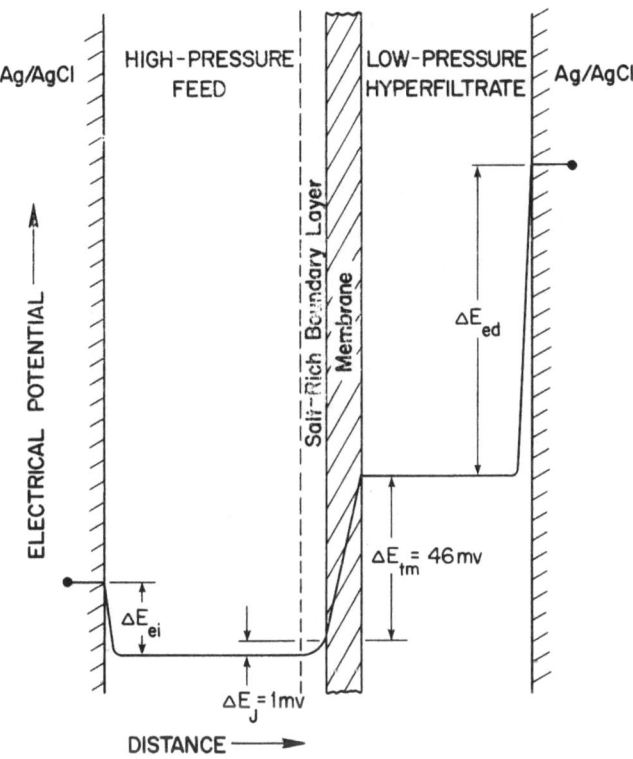

Fig. 2. Schematic of potential differences in reverse osmosis (hyperfiltration) [11]. Figure shows
potential drops in region between two Ag/AgCl electrodes immersed in solutions separated by a
moderately cation-selective membrane.

(a) For uniform pressure, the *electrode potential* is [15, 16]

$$E_{\mathrm{Ag/AgCl}} = E^0_{\mathrm{Ag/AgCl}} - (RT/\mathscr{F}) \ln a_{\mathrm{Cl}} \qquad (4)$$

where $E^0_{\mathrm{Ag/AgCl}}$ is the standard potential, a_{Cl} the mean activity of Cl^-, R the universal
gas constant, T the absolute temperature, and $\mathscr{F}$ is Faraday's constant. (The two
potential differences electrode-brine and electrode-hyperfiltrate are marked ΔE_{ei} and-
ΔE_{ed} respectively in Figure 2.) *Mean* activities rather than *individual* ion activities are
taken here, because the sum of *all* potential differences is interpretable in terms of
mean activities. For a detailed discussion of this point, see references [11] and [15].

Hence the difference between the potentials of the two electrodes is

$$\Delta E_{Ag/AgCl} = (RT/\mathscr{F})\ln(a'/a'') - \Delta p(\bar{V}_{AgCl} - \bar{V}_{Ag} - \bar{V}_{Cl^-}) \tag{5}$$

The operator Δ refers to the value of a property *on the right* minus the value of the respective property *on the left*. The pressure correction term [17] is sometimes negligible.

(b) The *junction potential* is [15, Equation (40), p. 300]

$$\Delta E_J = (1 - 2\tau_+)(RT/\mathscr{F})\ln(a^i/a') \tag{6}$$

where τ_+ is the transport number of sodium ion *in solution* ($\tau_+ \simeq 0.4$).

It is of interest that this potential difference depends only on the terminal ion activities, a' and a^i respectively, and not on the exact shape of the concentration profile in the diffusion layer.

(c) An analogous expression describes the *membrane potential*:

$$\Delta E_m = (1 - 2\bar{\tau}_+)(RT/\mathscr{F})\ln(a''/a^i) \tag{7}$$

where $\bar{\tau}_+$ is the apparent transport number of sodium ion *in the membrane* which takes water transport into account. (In these calculations, the influence of the water transport on the membrane potential is not explicitly stated [19].) For the membrane tested in the experiments described here, this transport number varied with c' from about 0.4 to unity [11, 18].

(d) The *streaming potential* is proportional to the pressure difference, Δp, maintained across the membrane:

$$\Delta E_{st} = -B_p \Delta p \tag{8}$$

The minus sign appears, because we wish to keep the parameter B_p ('streaming potential differential') positive; the streaming potential was positive when the pressure difference was negative (permeate, on right side in Figure 2, at lower pressure than brine, on left side). The streaming potential differential was found to vary only relatively little with the brine concentration.

Hence in the following development, B_p will be taken as a constant, because the (calculated) excess concentration at the interface is less than 50% of the bulk concentration of the brine.

Summing all potential differences between the two electrodes (Figure 2), we obtain

$$\Delta E_{obs} = \Delta E_{Ag/AgCl} + \Delta E_J + \Delta E_m + \Delta E_{st} =$$

$$= (RT/\mathscr{F}) \left[\ln \frac{a'}{a''} + (1 - 2\tau_+)\ln\frac{a^i}{a'} + (1 - 2\bar{\tau}_+)\ln\frac{a''}{a^i} \right] - B_p \Delta p$$

$$= (RT/\mathscr{F}) \left[2\bar{\tau}_+ \ln(a^i/a'') - 2\tau_+ \ln(a^i/a') \right] - B_p \Delta p. \tag{9}$$

Measurements of ΔE_{obs} were performed at different flow rates (at constant Δp) and extrapolated to infinite flow rate (superscript$^\infty$), where, by assumption, concentration polarization is negligible, viz.

$$(a^i)^\infty = a' \simeq a_f \tag{10}$$

Under these conditions, Equation (9) becomes

$$\Delta E^{\infty} = (RT/\mathscr{F})\{2\bar{\tau}_+ \ln[(a^i)^{\infty}/a'']\} - (B_p)^{\infty} \Delta p \tag{11}$$

Since $(a^i)^{\infty}$ is assumed to equal a' (which is known), extrapolation of the results to $\Delta p = 0$ makes it possible to calculate the unknown transport number of the cations in the membrane, $\bar{\tau}_+$. This procedure amounts to calculation of the transport number from the membrane potential under very efficient stirring of the brine. Earlier laboratory measurements [11, p. 250; 18] had shown that the value of $\bar{\tau}_+$ indicates cationic selectivity for membranes *cured at* $90\,^{\circ}C$. For average solution concentrations of 0.01, 0.1, and 1.0 N NaCl, the transport numbers were $\bar{\tau}_+ = 0.90, 0.77$, and 0.56 respectively.

To obtain an equation for the ratio a^i/a', we subtract Equation (9) from (11). In this operation, it is assumed that the 'intrinsic' rejection of the membrane, $\mathbf{R}^{\infty}$, is independent of the concentration, i.e. that the ratio of the solution concentrations in direct contact with the two membrane faces is independent of the concentration

$$(c_s^i/c_s'') \simeq (c_s^i/c_s'')^{\infty} = c_s'/(c_s'')^{\infty} \tag{12}$$

and that the variation of the activity coefficients, $y_{\pm}$, can be neglected in the small concentration range c^i to $(c^i)^{\infty}$. Hence the following relationship is also approximately valid:

$$(a^i/a'') \simeq (a^i/a'')^{\infty} \tag{13}$$

It is also assumed that the variation of the streaming potential differential, B_p, with the concentration can be neglected over the limited range of interfacial concentration changes taking place when the circulation velocity, $\bar{u}$, is varied at constant pressure difference, Δp:

$$(B_p)^{\infty} \simeq B_p \text{ (for limited concentration range)} \tag{14}$$

It is believed that these assumptions are justified since, contrary to the variation of the activities, a, with the salt concentration which is a first-order variation, the variations of $y_{\pm}, \mathbf{R}^{\infty}$, and B_p with the salt concentration are of much smaller order for homogeneous membranes.* For heterogeneous membranes, such as the modified cellulose acetate used here, these assumptions are more questionable, and there is indeed evidence that in some cases (primarily in osmosis) internal concentration polarization in the membrane may develop [18].

With these assumptions, we obtain by subtracting Equation (9) from (11):

$$\Delta E^{\infty} - \Delta E_{obs} = 2(RT/\mathscr{F})\tau_+ \ln(a^i/a') \tag{15}$$

The experimental polarization ratio is found from this equation by measuring (1) the potential difference across the membrane, ΔE_{obs}, at the desired circulation velocity, then determining its value, ΔE_{obs}^{∞}, at the same pressure difference, Δp, and at infinite

---

* This is not to say that the rejection, **R**, is independent of the *effective pressure difference*, across the membrane; the pressure influence is particularly pronounced at low applied pressures [11, Appendix D], [5, p. 140], [20].

circulation velocity (by extrapolation), substituting the transport number, τ_+, of Na^+ in solution from tables [12] and thus obtaining (a^i/a'). After substitution of the appropriate activity coefficients, $y_\pm$, (12) by an iterating procedure, we obtain the polarization ratio:

$$c^i/c' = (a^i/a')\,(y'_\pm/y^i_\pm).\tag{16}$$

The streaming potential differential, $B_p[\simeq(B_p)^\infty]$, was determined for each salt concentration from the change of ΔE^∞ with pressure, $-\Delta p$ [Equation (11)]. ΔE^∞ was obtained by extrapolating the observed potential differences for each given pressure to infinite flow velocity $(1/\bar{u}=0)$, i.e. assumed absence of polarization. In this case, the first term in Equation (11) is constant and the slope of the plot of ΔE^∞ versus $(-\Delta p)$ is equal to the streaming potential differential.

3. Experimental Methods

The general layout of the apparatus [11, 21] and the details of its construction and of the experimental procedure [11] have been published. Although modified cellulose acetate cured at 94 °C was used for hyperfiltration in this work, it is believed that the apparatus and the methods developed here are suitable for other membranes also. Briefly, pressurized saline solution was circulated through a short annulus bounded by a cylindrical modified cellulose-acetate membrane cured at 94 °C under carefully controlled flow conditions. The apparatus was designed so that the following quantities could be measured: observed potential difference, (ΔE_{obs}), at the desired circulation velocity, $\bar{u}$, and (by extrapolation), $\Delta E^\infty(\equiv\lim_{\bar{u}\to\infty}\Delta E_{obs})$; feed concentration, c'_s; hyperfiltrate concentration, c''_s; hyperfiltrate flux, J_v; feed pressure $(=-\Delta p$, the static pressure difference); and feed temperature. The values of ΔE_{obs}, ΔE^∞, $\bar{u}$, c'_s, c''_s, and J_v obtained with this apparatus were used to calculate the concentration polarization ratio, c^i/c', by two methods: (a) substitution of values of the observed potential difference into Equations (15) and (16) which yielded the experimental polarization ratio, and (b) substitution of values of the hyperfiltrate flux, circulation velocity, feed concentration, and the hyperfiltrate concentration into the equations for laminar or turbulent flow (Table I). This yielded the 'theoretical' polarization ratio.

The electrodes used to measure the electrical potential difference between low- and high-pressure solutions (Figure 2) had to be (a) electrically well-insulated from each other and from other metal parts, and (b) sturdy enough to withstand the high pressures to which the inner electrode would be subjected.

Figure 3 is a longitudinal cutaway view of the composite test section used during the experiments; the membrane section was approximately 6 in. long and contained one high-pressure electrode. The basic geometry was a concentric annulus in which pressurized saline feed flowed in the annular space between the tubes. A standard copper tube (1-in. OD, outer diameter, nominal, and 0.035-in. wall thickness) with an electropolated silver-silver chloride layer on its inner surface served as the low-pressure (outer) electrode and membrane support. The silver chloride layer extended over

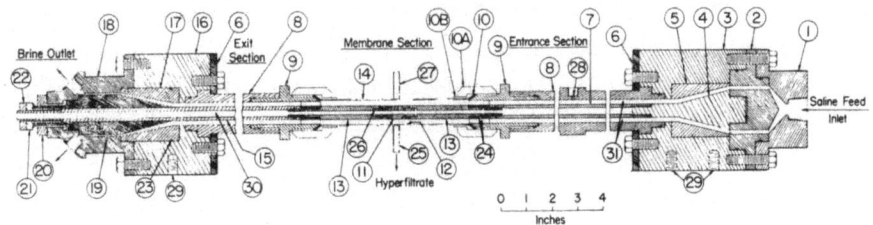

MEASUREMENT OF HYPERFILTRATION POTENTIAL. LONGITUDINAL CUTAWAY VIEW OF COMPOSITE TEST-SECTION

PART NO.	PART NAME	PART NO.	PART NAME
1	Feed inlet	16	Exit header
2	Feed distributor	17	Flow diffuser
3	Entrance header	18	Brine exit
4	Support for inner tube (entrance)	19	Electrode support (exit)
5	Flow reducer	20	Holding nut
6	Outer tube support	21	Positioning nut
7	Inner tube (stainless steel)	22	Locking nut
8	Entrance-section outer tube	23	O-ring
9	Flare-tube connector	24	Finned-tube support
10	Rubber ring gasket	25	Hyperfiltrate exit
10A	Flare-tube assembly nut	26	Electrical contact for high-pressure electrode
10B	Flare-tube assembly sleeve	27	Electrical contact for low-pressure electrode
11	High-pressure (inner) electrode	28	Pressure tap
12	Membrane and backing material	29	Holes for mounting bolts
13	Inner tube (Delrin plastic)	30	Conduit for lead to high-pressure electrode
14	Low-pressure (outer) electrode	31	Annular flow channel for saline feed
15	Inner tube (positioning)		

Fig. 3. Measurement of hyperfiltration potential. Longitudinal cutaway view of composite test-section [11].

the entire tube length. The inner tube, which served as a support for the high-pressure (inner) electrode, was fabricated from sections of 'Delrin' plastic rod (E. I. Dupont de Nemours Co., Wilmington, Delaware) and silver rod (99.9+% pure silver, Engelhard Industries, Newark, N.J.) joined end-to-end in the following manner: plastic-silver-plastic. The outer diameter of the inner tube was 0.5 in. The portion of the silver rod exposed to the pressurized saline feed was coated with a layer of silver which was then partially converted to silver chloride. A thin insulated wire was soldered to one of the non-exposed ends of the silver rod (item #26, Figure 3) and brought out through the hollow portion (open to the atmosphere) of the inner tube (Figure 3, left end). The result was a cell in which electrodes consisting of the same material were placed on both sides of the membrane. Membrane section dimensions are shown in Figure 1.

The tubular membrane configuration used in this apparatus was essentially that designed by Loeb and coworkers [4]. Saline feed flowed axially through the annulus while the hyperfiltrate (desalted water) flowed radially through the membrane and then through the porous backing material to the hyperfiltrate exit which was located at the midlength position of the outer tube. A piece of clear plastic tubing connected to the hyperfiltrate exit (a short piece of 1/4-in. OD copper tubing positioned around a 1/16-in. diameter hole drilled through the outer tube) conducted hyperfiltrate to a sample collection bottle. A short copper nipple located at the midlength position of the outer tube, but on the side opposite the hyperfiltrate exit, served as the electrical connection to the low-pressure electrode.

In the inner tube assembly, the 'Delrin' plastic spacers served to position the silver electrode along the axis of the membrane section and to prevent metallic contact between the electrode and the rest of the composite test-section. The spacers were designed so that the center of the electrode was located at the midlength position, opposite the hyperfiltrate exit, of the membrane section. Adjacent parts of the inner-tube assembly fitted together with O-ring slip joints. The plastic male ends of the inner tube assembly fitted into the female ends of the stainless steel finned-tube supports (item #24, Figure 3) in the entrance and exit sections.

4. Results and Discussion

4.1. POLARIZATION

For each saline-solution circulation rate, $\bar{u}$, and pressure difference, $-\Delta p$, across the membrane, the hyperfiltrate concentration, c_d'', hyperfiltrate flux, J_v, and the observed potential difference, ΔE_{obs}, were measured and plotted versus $1/\bar{u}$. Points correspond-

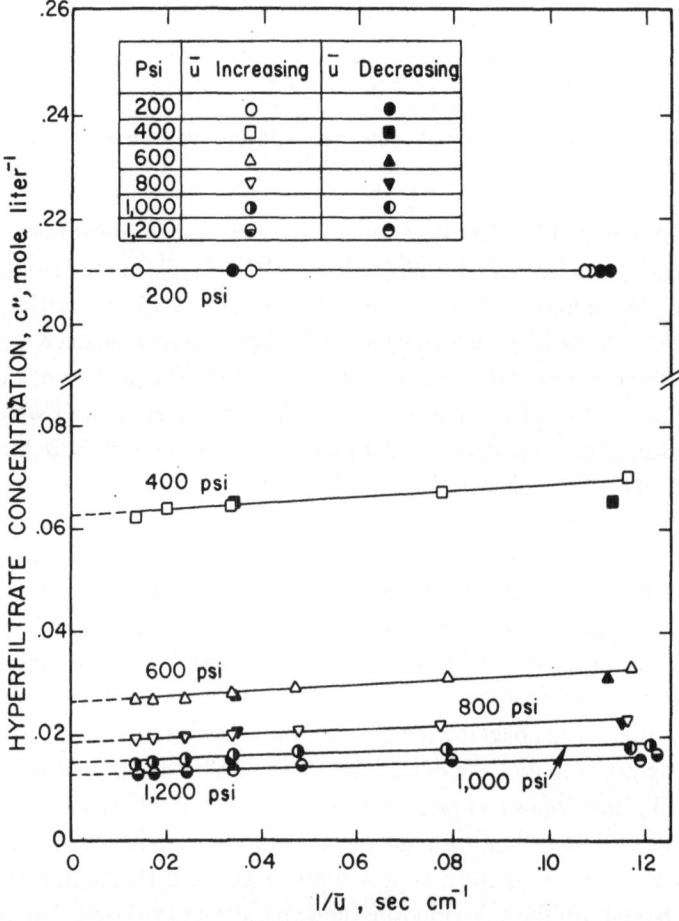

Fig. 4. Hyperfiltrate concentration versus reciprocal circulation velocity, $1/\bar{u}$, for several values of $-\Delta p$ (0.5 M NaCl).

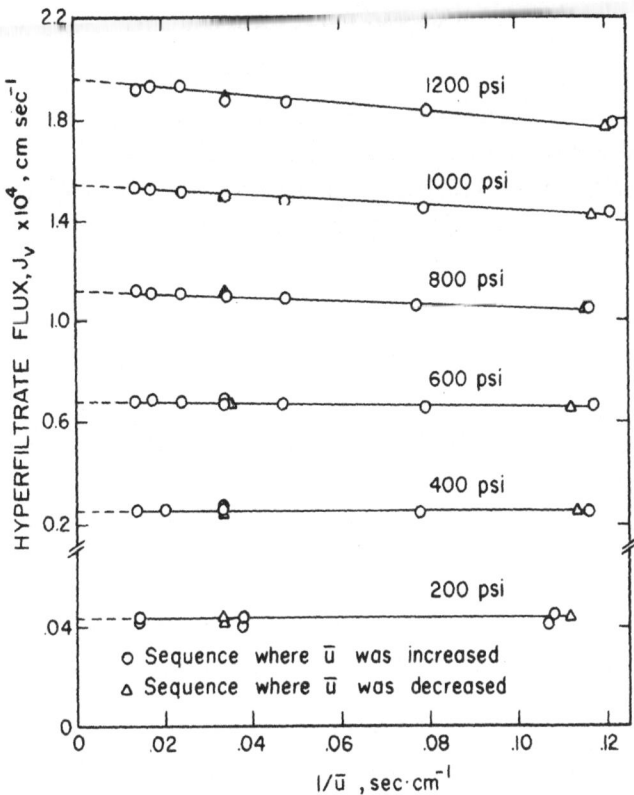

Fig. 5. Hyperfiltrate flux, J_v, versus reciprocal circulation velocity, $1/\bar{u}$, for several values of $-\Delta p$ (0.5 M NaCl).

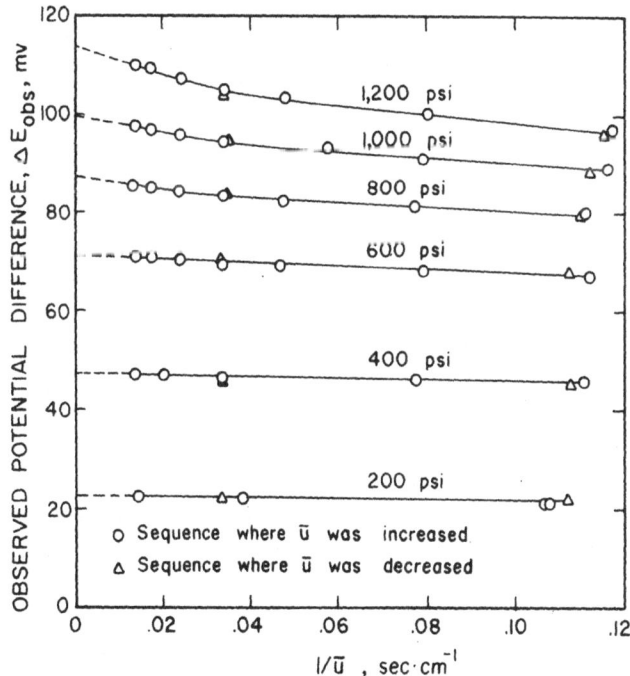

Fig. 6. Observed potential difference versus reciprocal circulation velocity, $1/\bar{u}$, for several values of $-\Delta p$ (0.5 M NaCl).

ing to the same pressure difference were connected and the resulting lines extrapolated to $1/\bar{u}=0$, i.e. to infinite circulation velocity at which concentration polarization was assumed to be absent. Figures 4, 5, and 6 illustrate the extrapolation procedure for experiments with 0.5 M NaCl (25°C). A similar procedure was used for the results obtained with 0.1 and 0.01 M NaCl solutions. The resulting extrapolated values, c_d^∞, J_v^∞, and ΔE_{obs}^∞ respectively are listed in Table III. This table also lists the salt rejection calculated from the extrapolated hyperfiltrate concentration

$$\mathbf{R}^\infty \equiv (c' - c_d^\infty)/c'$$

This extrapolated salt rejection is often called 'intrinsic' salt rejection of the membrane [22], although it depends also on parameters which are not intrinsic to the membrane (e.g. pressure).

Table IV shows a sample calculation for obtaining the polarization ratio from the measured electrical potential differences in one specific case.

The experimental and theoretical values of $c^i/c_f (\simeq c^i/c')$ are shown in Figures 7 to 12 for 400 psi g and 1000 psi g. In these figures, the uncertainties in the experimental values of c^i/c_f, due to the uncertainties in the values of $\Delta E_{obs}^\infty - \Delta E_{obs}$ for each point,

TABLE III

Hyperfiltrate concentrations, hyperfiltrate fluxes, and observed potential differences extrapolated to infinite circulation velocity of NaCl feed (Feed temperature = 25 °C).

$-\Delta p$, ψ	c', mole l^{-1}	$c_d^\infty{}^a$, mole l^{-1}	$\mathbf{R}^\infty$	$J_v^\infty \times 10^5$, cm s^{-1}	ΔE_{obs}^∞ mV
200	0.0098	1.32×10^{-4}	0.986	5.20	237.5
400	0.0098	6.80×10^{-5}	0.993	10.3	280
600	0.0098	5.50×10^{-5}	0.994	15.2	309.5
800	0.0097	4.80×10^{-5}	0.995	20.1	328
1000	0.0098	4.90×10^{-5}	0.995	24.3	342.5
1200	0.0098	5.30×10^{-5}	0.995	27.9	354
200	0.108	9.60×10^{-3}	0.911	3.81	90.0
400	0.109	5.10×10^{-3}	0.953	8.99	119.5
600	0.106	3.25×10^{-3}	0.969	13.5	142
800	0.103	2.62×10^{-3}	0.974	17.9	160
1000	0.102	2.20×10^{-3}	0.978	21.8	176
1200	0.103	1.73×10^{-3}	0.983	25.8	193
200	0.513 ^{b}	2.10×10^{-1}	0.591 ^{c}	0.430	25
400	0.506	6.25×10^{-2}	0.876	2.55	47.5
600	0.507	2.65×10^{-2}	0.948	6.81	71.5
800	0.508	1.80×10^{-2}	0.964	11.2	87.5
1000	0.509	1.50×10^{-2}	0.970	15.4	100
1200	0.512	1.25×10^{-2}	0.976	19.6	114

^{a} c_d^∞ is the hyperfiltrate (permeate) concentration in the absence of concentration polarization, found from the extrapolation procedures illustrated in Figures 4–6.
^{b} Average of five values.
^{c} Based on average value of feed concentration.

TABLE IV

Sample calculation No. 1. Determination of the concentration polarization ratio, c^i/c_f,[a] using values of the observed potential difference (Run #36, 0.5 M NaCl, 1000 psi g) [based on Equation (15)].

(A) Data: $c_f = 0.509$ M NaCl, $(y_\pm)_f = 0.680$
            $\Delta E_{obs} = 91.38$ mV
            $\Delta E^\infty_{obs} = 100$ mV (Table III)

(B) Other constants: $\tau_+ = 0.385$,
                   $RT/\mathscr{F} = 25.69$ mV

(C) Calculate c^i/c_f using Equations (15) and (16)
     (1) Starting an iteration procedure, assume $(y_\pm)_f = (y_\pm)^i$
[$= 0.680$; ref. [12], p. 490]

$$\frac{c^i}{c_f} = \exp\left[\frac{100 - 91.38}{2(0.385)\,(25.69)}\right] = \exp\left[\frac{8.62}{19.79}\right] = \exp\,(0.436)$$

$$\frac{c^i}{c_f} = 1.546$$

     (2) Calculate $c^i = 1.546 \times 0.509 = 0.787$; the activity coefficient, $(y_\pm)^i$, for this concentration is 0.662.
     (3) Recalculate c^i/c_f using $(y_\pm)^i = 0.662$ and $(y_\pm)_f = 0.680$

$$\frac{c^i}{c_f} = \frac{0.680}{0.662}\,(1.546) = (1.028)\,(1.546) = 1.589$$

     (4) Iterate: $c^i = 1.589 \times 0.509 = 0.809$; the activity coefficient, $(y_\pm)^i$, for this concentration is 0.661.
     (5) Recalculate c^i/c_f using $(y_\pm)^i = 0.661$ and $(y_\pm)_f = 0.680$

$$\frac{c^i}{c_f} = \frac{0.680}{0.661}\,(1.546) = 1.589$$

     (6) Therefore, $c^i/c_f = 1.589$

[a] Subscript f refers to properties of saline-solution feed. Since the test section was short, the saline-solution concentration changed very little, viz. $c_f \simeq c'$.

are illustrated by the size of the vertical lines representing each measurement. For each value of $\Delta E^\infty_{obs} - \Delta E_{obs}$, this uncertainty was estimated to be $\pm 0.4\,mV$. The equations used for calculation of the polarization ratio are listed in Table I.

It was found [11] that for 0.1 M and 0.5 M NaCl feed solutions, the experimental values of c^i/c_f increased with the static pressure difference across the membrane, $-\Delta p$, and decreased with the circulation velocity, $\bar{u}$, as expected. For 0.01 M NaCl feed, the experimental values of c^i/c_f increased with the feed pressure in the range 200 psi g–600 psi g; in the range 600 psi g–1200 psi g, the experimental values decreased. The reason for this behavior is not immediately obvious since c^i/c_f, and therefore $(\Delta E^\infty_{obs} - \Delta E_{obs})$, were expected to increase due to the increase in the hyperfiltrate flux with pressure. It should be noted that the hyperfiltrate concentration, c'', was quite small in these experiments (less than $10^{-4}\,M$) and that deviations from the predicted electrode potential of the low-pressure electrode are not impossible.

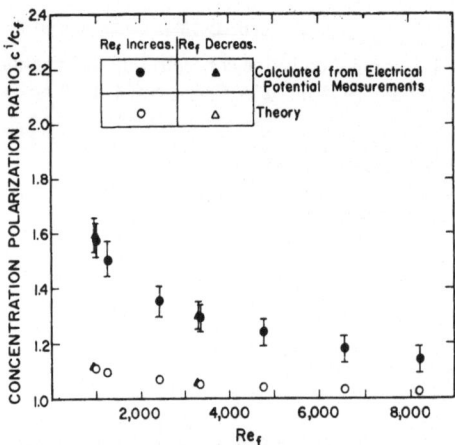

Fig. 7.   0.01 M NaCl.

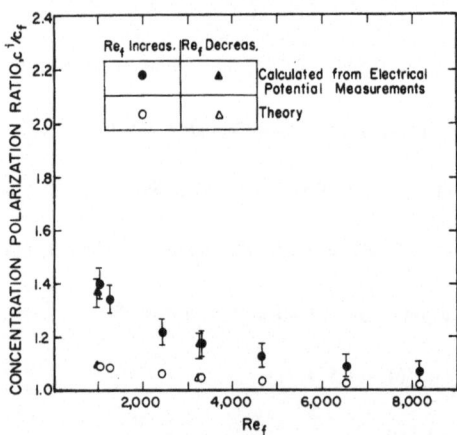

Fig. 8.   0.1 M NaCl.

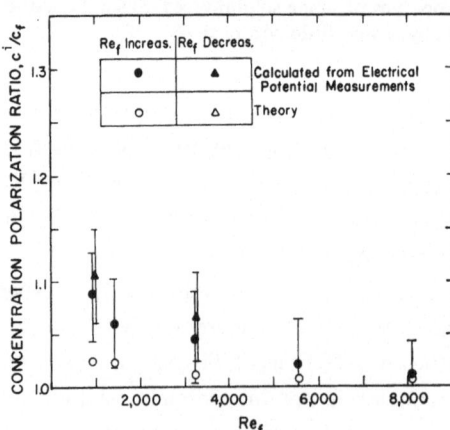

Fig. 9.   0.5 M NaCl.

Figs. 7–9. Concentration polarization ratios in reverse osmosis (hyperfiltration) vs. Reynolds number of feed. Comparison of ratios calculated from electrical potential measurements with theory (400 psi g).

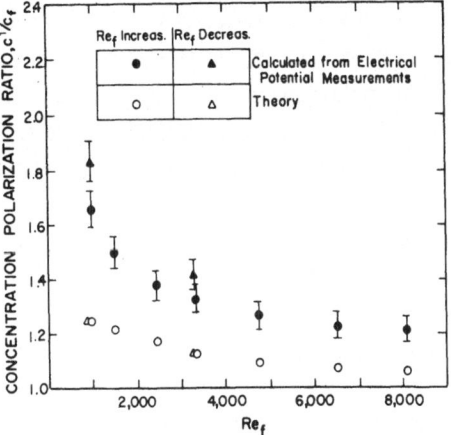

Fig. 10.   0.01 M NaCl.

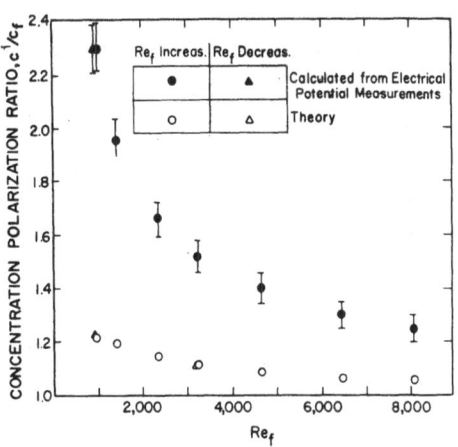

Fig. 11.   0.1 M NaCl.

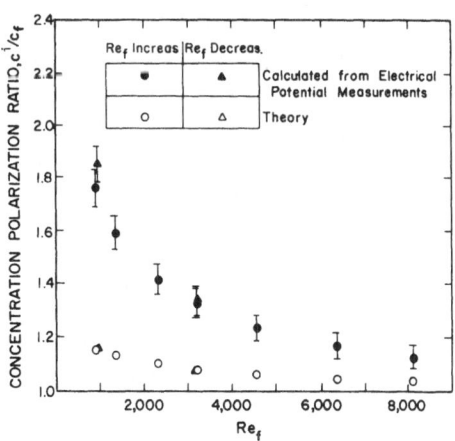

Fig. 12.   0.5 M NaCl.

Figs. 10–12.   Concentration polarization ratios in reverse osmosis (hyperfiltration) vs. Reynolds number of feed. Comparison of ratio calculated from electrical potential measurements with theory (1000 psi g).

It is important that the polarization ratios calculated from electrical measurements followed in a semi-quantitative manner the trends predicted by the theory, but the experimental values of the polarization ratio, c^i/c_f ($\simeq c^i/c'$), were larger than the theoretical values predicted by the equations listed in Table I. The agreement between theoretical and experimental values was best for 0.01 M NaCl and worst for 0.1 M NaCl at 1000 psi g.

To assess the significance of these deviations, it is important to consider that the calculation of the experimental polarization ratio from electrical measurements involves finding the difference $\Delta E_{obs}^{\infty}-\Delta E_{obs}$ [Equation (15)], which is a small number when compared with either ΔE_{obs} or ΔE_{obs}^{∞}. A relatively small error in the determination of ΔE_{obs}^{∞} obtained by extrapolation (Figure 6) can lead to a larger error in the calculation of the polarization ratio. The deviations are not random, however, but quite systematically the measured polarization seems to be higher than the predicted one.

It should also be kept in mind that in the derivation of Equation (15) from which the 'experimental' polarization was calculated, concentration-independence of streaming potential coefficient B_p, of the cation transport number in the membrane $\bar{\tau}_+$, and of the activity ratio, a''/a', was assumed *over the limited range of variation of the interfacial concentration which is caused by changes of the circulation velocity, $\bar{u}$*. The validity of these assumptions and their influence on the calculated polarization ratios is discussed in detail in reference [11]. It is well known that counterion transport numbers of permselective membranes do decrease somewhat with increasing concentration of the solution in contact with them [15, 23, 24] and so does the streaming potential coefficient, B_p. Moreover, the non-homogeneous structure of the modified cellulose acetate membranes in which a thin dense skin faces the high-pressure solution, while the rest of the membrane is much more permeable and apparently serves primarily as mechanical support [4, 5] causes the actual concentration profile in the system to be more complex than assumed in Figure 2 which served as a basis of the calculations. It should also be noted that while the 'intrinsic' rejection, $\mathbf{R}^{\infty}$, as listed in Table III, is reasonably constant with respect to the hydrodynamic pressure, $-\Delta p$, for 0.01 M NaCl solutions, considerable decreases of $\mathbf{R}^{\infty}$ with decreasing pressure were noted for 0.5 M NaCl solutions. The results for 0.1 M NaCl are intermediate. This variation with pressure can influence data interpretation. It is well known that the rejection in general decreases with decreasing pressure [5, p. 140], but the relevant parameter determining this decrease is the hyperfiltration (permeation) rate rather than the pressure [20]. Since in past work, decreasing pressure usually caused a decrease in the huperfiltration rate, the two parameters influence the rejection similarly. In this work, however, the circulation rate was varied at *constant pressure* and yet the hyperfiltration rate changed in accordance with the polarization. Therefore, even at constant pressure, $-\Delta p$, the rejection may be expected to vary. A discussion of this effect [11] leads to the conclusion that the assumption of substantially constant 'intrinsic' rejection, $\mathbf{R}^{\infty}$, is valid for all the experiments described in this paper, except for those with 0.5 M NaCl at pressure above 600 psi (Figure 12), where some deviations from this assumption are expected.

In spite of these qualifications, it is believed that the general method of determining concentration polarization in hyperfiltration systems by measurements of electric potential differences can be refined by taking all these factors into account.

4.2. STREAMING POTENTIALS

By evaluating measurements of the electric potential difference as a function of pressure for circulating NaCl solutions of different concentrations (approximately 0.01, 0.1, and 0.5 M), the streaming potential differential, B_p, could be determined. According to Equations (11) and (13), a plot against pressure – Δp, of the electric potential difference (between two silver-silver chloride electrodes separated by the membrane), extrapolated to infinite circulation velocity should be a straight line whose ordinate intercept is related to the cation transport number in the membrane, $\bar{t}_+$, and whose slope is the streaming potential differential, B_p (also called the 'streaming potential coefficient'). Since the concentrations, $c_s (= a/y_\pm)$ of the high-pressure and the low-pressure solutions, c'_s and c''_s respectively were determined by chemical

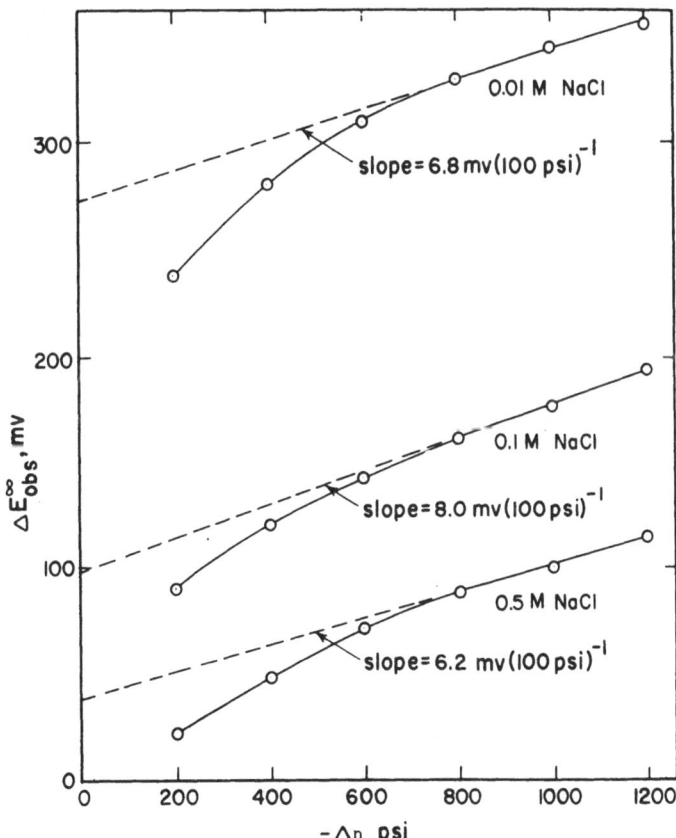

Fig. 13.   Extrapolated values of the observed potential difference vs. $-\Delta p$. Points represent potential differences extrapolated to infinite circulation velocity, where concentration polarization is presumably negligible.

analysis, and since the corresponding mean activity coefficients, $y_\pm$, can be found by interpolation of literature data [12], it is possible to calculate the cation transport number from such plots. Note that at high flow velocities the activity $(a^i)^\infty$ appearing in Equation (11) is assumed to be equal to the known mean ion activity, a', in the circulating brine [Equation (10)].

Figure 13 shows the plots for three concentrations. The lack of linearity at low pressures may be due to variation of the rejection (and hence of the ratio a'/a'') and/or the lack of compaction of the membrane. Reasonably straight lines are obtained when only the experiments in the high-pressure region are considered. Evaluation of the straight lines obtained in this manner by Equation (11) yields the streaming potential coefficients and the cation transport numbers, $\bar{\tau}_+$, presented in Table V. The fact that in one case the computation yielded a value about two percent *higher* than unity is believed to be due to the uncertainty inherent in the extrapolation procedure, as values of $\bar{\tau}_+$ larger than unity have no simple physical meaning.

TABLE V

Evaluation of extrapolated electric potential differences for the calculation of the streaming potentia coefficient, B_p, and the cation transport number, $\bar{\tau}_+$, in the membrane.[a]

$c_t, (=c')$ mole l^{-1}	$\mathbf{R}^{\infty \, b}$	$c^\infty_d, (=c'')$ [b] mole l^{-1}	$(y_\pm)_t$	$(y_\pm)^\infty_d$	B_p, mV$(100 \, \Psi)^{-1}$	$\bar{\tau}_+$
0.0098	0.995	5.30×10^{-5}	0.904	0.992	6.8	1.02
0.103	0.983	1.73×10^{-3}	0.776	0.955	8.0	0.49
0.512	0.976	1.25×10^{-2}	0.680	0.894	6.2	0.38

[a] Membrane cured at 94 °C.
[b] Evaluated at $\Delta p = -1200$ psi g.

It is seen that as the solution concentration on the skin side of the membrane increases by almost two orders of magnitude, the cation transport number decreases from approximately unity to approximately its value in free solution (0.38). This is characteristic of membranes of moderate fixed-charge density, and is also reflected in the variation of the streaming potential with the solution concentration. For comparison, consider corresponding experiments with membranes of high charge density (commercial ion-exchange membranes). In the latter, the streaming-potential coefficient is more than one order of magnitude lower and depends little on the concentration [1]. The counterion transport number also depends little on the concentration. The lower the fixed-charge concentration (in terms of molality), the higher the streaming-potential coefficient and the more pronounced are the decrease of both streaming-potential coefficient and counterion transport number with the concentration [25, 26]. Since the streaming-potential coefficient in the modified cellulose acetate membranes is about one order of magnitude higher than in the commercial ion-exchange membranes investigated in the past, one concludes from Saxén's law that the electro-osmotic transfer (per Faraday) is also higher by about one order of magnitude [1] and so is the water/fixed charge ratio in the membrane.* This means that the fixed-charge molality (which is the reciprocal of this ratio) is almost one order of magnitude less in the relevant portion of the modified cellulose acetate membrane than in typical commercial cation-exchange membranes.

Acknowledgments

The authors express their thanks to professors D. N. Bennion and J. W. McCutcheon, and their group at the University of California, Los Angeles, for supply of the modified cellulose acetate membranes. The authors thank the Office of Saline Water (now part of the Office of Water Research and Technology), U.S. Department of the Interior, and the Sea Water Conversion Project, Water Resources Center, University of California for support of this work. Thanks are due to Vera L. Dean for her conscientious help in typing and editing this paper.

References

1. McKelvey, J. G., Spiegler, K. S., and Wyllie, M. R. J., *Chem. Eng. Progr. Ser.* **55**, 199 (1959).
2. Minning, C. P., 'Electrochemical Studies on Cellulose Acetate Hyperfiltration Membranes', M. S. Report, Department of Mechan. Eng., University of California, Berkeley, 1966.
3. Bennion, D. N. and Rhee, B. W., *Ind. Eng. Chem. Fundamentals* **8**, 36 (1969).
4. Loeb, S., *Desalination* **1**, 35 (1966).
5. Merten, U. (ed.), *Desalination by Reverse Osmosis*, M.I.T. Press, Cambridge, Mass., 1966.
6. 'Third Annual Report of the Secretary of the Interior on Saline Water Conversion', U.S. Government Printing Office, Washington, D.C., Jan., 1955, p. 95.
7. Reid, C. E. and Spencer, H. G., *J. Polym. Sci.* **4**, 354 (1960).
8. Sherwood, T. K., Brian, P. L. T., Fisher, R. E., and Dresner, L., *Ind. Eng. Chem. Fundamentals* **4**, 113 (1965).
9. Meares, P. (ed.), *Membrane Separation Processes*, Chapter 4 'Reverse Osmosis (Hyperfiltration) in Water Desalination', by F. L. Harris, G. B. Humphreys, and K. S. Spiegler, Elsevier, Amsterdam, 1976; in press.
10. Brian, P. L. T., Chapter 5 in *Desalination by Reverse Osmosis*, U. Merten (ed.), M.I.T. Press, Cambridge, Mass. (1966).
11. Minning, C. P., 'Streaming Potentials in Hyperfiltration (Reverse Osmosis) of Saline Waters', Ph.D. thesis, College of Engineering, University of California, Berkeley (1973).
12. Robinson, R. A. and Stokes, R. H., *Electrolyte Solutions*, 2nd ed. (revised), Butterworth's, London, 1959.
13. Bennett, C. O. and Myers, J. E., *Momentum, Heat and Mass Transfer*, 2nd ed., McGraw-Hill, New York, 1974, (a) p. 205, (b) p. 164.
14. Knudsen, J. G. and Katz, D. L., *Fluid Dynamics and Heat Transfer*, McGraw-Hill, New York, 1958, p. 198.
15. Spiegler, K. S. and Wyllie, M. R. J., 'Electrical Potential Differences', Chapter 7 in *Physical Techniques in Biological Research*, 2nd ed., Vol. II, part A, Dan H. Moore (ed.), Acad. Press, New York, 1968.
16. MacInnes, D. A., *The Principles of Electrochemistry*, Reinhold, New York, 1939.
17. Spiegler, K. S., *Desalination* **15**, 135 (1974).

---

* The amount of water carried by each counterion in electro-osmosis is not equal to the ratio: total water in membrane/counterions in membrane (as determined by chemical analysis), but there is a strong correlation. About one half of the total water in cation-exchange membranes seems to be carried by counterions, when the latter are alkali metal ions [25, 27, 28].

18. Pusch, W., Private communication; also his article in *Permeability of Plastic Films and Coatings to Gases, Vapors and Liquids*, H. B. Hopfenberg (ed.), *Polymer Sci. and Technol.*, Vol. 6, Plenum Press, New York, N.Y., 1974.
19. Scatchard, G. J., *Amer. Chem. Soc.* **75**, 2883 (1953).
20. Spiegler, K. S. and Kedem, O., *Desalination* **1**, 311 (1966).
21. Research and Development Report No. 613, Office of Saline Water, U.S. Department of the Interior, Superintendent of Documents, U.S. Government Printing Office, Washington, D.C. 20402, 1971.
22. Shor, A. J., Kraus, K. A., Johnson, J. S., and Smith, W. T., *Ind. Eng. Chem.* (*Fundamentals*) **7**, 44 (1968).
23. Teorell, T., *Proc. Soc. Exptl. Biol. Med.* **33**, 282 (1935).
24. Meyer, K. H. and Sievers, J. F., *Helv. Chim. Acta* **19**, 649 (1936).
25. Spiegler, K. S., *J. Electrochem. Soc.* **100**, 303C (1953).
26. Gray, D. H., 'Coupled-Flow Phenomena in Clay-Water Systems', Ph.D. thesis, Department of Civil Engineering, University of California, Berkeley (1966).
27. Spiegler, K. S., *Trans. Farad. Soc.* **54**, 1408 (1958).
28. Paterson, R., Proc. 'Study Week on Membranes and Desalination', Pontificia Academia Scientiarum, Vatican City, 1975, in press.

INDEX